Studies in Mathematical Physics

NATO ADVANCED STUDY INSTITUTES SERIES

Proceedings of the Advanced Study Institute Programme, which aims at the dissemination of advanced knowledge and the formation of contacts among scientists from different countries

The series is published by an international board of publishers in conjunction with NATO Scientific Affairs Division

A	Life Sciences	Plenum Publishing Corporation
B	Physics	London and New York
C	Mathematical and Physical Sciences	D. Reidel Publishing Company Dordrecht and Boston
D	Behavioral and Social Sciences	Sijthoff International Publishing Company Leiden
E	Applied Sciences	Noordhoff International Publishing Leiden

Series C – Mathematical and Physical Sciences

Volume 1 – Studies in Mathematical Physics

Studies in Mathematical Physics

Lectures Presented at the NATO Advanced Study Institute on Mathematical Physics held in Istanbul, August, 1970

edited by

A. O. BARUT

Institute for Theoretical Physics, University of Colorado, Boulder, Colo., U.S.A.

D. Reidel Publishing Company

Dordrecht-Holland / Boston-U.S.A.

Published in cooperation with NATO Scientific Affairs Division

First printing: December 1973

Library of Congress Catalog Card Number 73–88587

ISBN 90 277 0405 8

Published by D. Reidel Publishing Company
P.O. Box 17, Dordrecht, Holland

Sold and distributed in the U.S.A., Canada, and Mexico
by D. Reidel Publishing Company, Inc.
306 Dartmouth Street, Boston, Mass. 02116, U.S.A.

Printed in The Netherlands by D. Reidel, Dordrecht

PREFACE

Mathematical physics has become, in recent years, an independent and important branch of science. It is being increasingly recognized that a better knowledge and a more effective channeling of modern mathematics is of great value in solving the problems of pure and applied sciences, and in recognizing the general unifying principles in science. Conversely, mathematical developments are greatly influenced by new physical concepts and ideas. In the last century there were very close links between mathematics and theoretical physics. It must be taken as an encouraging sign that today, after a long communication gap, mathematicians and physicists have common interests and can talk to each other. There is an unmistakable trend of rapprochement when both groups turn towards the common source of their science—Nature. To this end the meetings and conferences addressed to mathematicians and physicists and the publication of the studies collected in this Volume are based on lectures presented at the NATO Advanced Study Institute on Mathematical Physics held in Istanbul in August 1970. They contain review papers and didactic material as well as original results. Some of the studies will be helpful for physicists to learn the language and methods of modern mathematical analysis—others for mathematicians to learn physics. All subjects are among the most interesting research areas of mathematical physics.

I should like to express my gratitude to the Scientific Affairs Division of NATO for its support of international scientific collaboration in this area and to lecturers and participants for their efforts. Finally I should like to thank Mrs. Ann Cofer for her excellent typing of the manuscript.

Boulder, May 1972 A. O. Barut

CONTENTS

BASIC NOTIONS IN TOPOLOGICAL VECTOR SPACES†

Daniel Sternheimer
CNRS
Paris

Foreword

It is only recently that theoretical physicists have become aware of the existence and possible relevance for physics of spaces other than Hilbert (or at best Banach) spaces. One even has sometimes the impression that not all of them are fully aware of the difficulties that may arise from the fact that not all Hilbert spaces are finite-dimensional. It is true that Dirac introduced and worked with δ and its derivatives long before mathematicians defined the notion of distribution, and used symbols such as the ket $|p\rangle$ before the notion of generalized eigenvector was made rigorous. But not everyone has the same physical intuition as Dirac had to protect him from deriving wrong conclusions by heuristic arguments, especially in the present state of physics. Of course, one cannot expect from every physicist to develop the mathematical tools he needs (though this might benefit mathematics). But one can expect from them (especially in the absence of very precise physical motivations, which seems to be the case nowadays) to know what are the existing tools, and to try and use these to make rigorous the heuristic arguments they may use as a first step.

These lectures aim at making people more familiar with the quite abstract theory of locally convex spaces. Those who have already heard about the so-called "rigged Hilbert spaces" may consider it as some kind of generalization which, we hope, will make more clear and more precise some aspects in the theory of these "rigged spaces." We did not give many proofs; in fact, some proofs will be presented in the beginning, but later we shall be content with giving sketches of proof, or just the results, referring for more details to the existing literature. When getting deeper in the theory, the proofs often get more involved and complicated but are more or less of the same kind and we do not think that seeing all of them

†Presented at the NATO Summer School in Mathematical Physics, Istanbul, 1970.

would be of great help for the non-specialist; and besides, these lecture notes are already long enough this way.

We begin by recalling definitions and elementary results of topology and linear spaces. Then we introduce topological vector spaces and concentrate on the study of locally convex spaces. We show, also with examples, their structure, their relations with normed spaces, and define some important classes of such spaces. We then give, and partly prove, the basic theorems for these spaces and linear operators on them. Then we concentrate on the important notion of duality, define and study topologies on tensor products, in relation with the very important class of nuclear spaces for which we give the major results. We end by some other examples and some applications, for which the knowledge of a great deal of the notions and results we presented is quite necessary.

LECTURE I
DEFINITIONS AND FIRST BASIC NOTIONS

A. Preliminaries

1) Zorn's lemma. If an ordered set is inductively ordered (i.e., each totally ordered subset has an upper bound), then it has a maximal element.

Filter on a set X: it is a family $\mathfrak{F}$ of subsets of X such that:

(F I) $G \supset F,\ F \in \mathfrak{F} \Rightarrow G \in \mathfrak{F}$.

(F II) $F, G \in \mathfrak{F} \Rightarrow F \cap G \in \mathfrak{F}$.

(F III) $\mathfrak{F}$ is non-void and $\emptyset \notin \mathfrak{F}$ (By F I, $X \in \mathfrak{F}$).

Filter base: It is a family $\mathcal{B}$ of subsets of X generating a filter, i.e., satisfying (F III) and (B I): $B_1, B_2 \in \mathcal{B} \Rightarrow \exists B_3 \in \mathcal{B}$, $B_3 \subset B_1 \cap B_2$. (Equivalently: $\forall F \in \mathfrak{F}$, $\exists B \in \mathcal{B}$ such that $B \subset F$.)

2) Topology on a set X: It is defined by a family $\underset{\sim}{O}$ of subsets (called open) invariant under finite intersection and arbitrary union. $(X, \underset{\sim}{O})$ is then a topological space. The complements of elements of $\underset{\sim}{O}$ are called closed.

Definitions. If $A \subset X$, we define

$\mathring{A}$ = interior of A = union of all open subsets of A,

$\overline{A}$ = closure of A = intersection of all closed sets containing A.

Neighbourhood of $x \in X$: Any set U such that $x \in \mathring{U}$ (i.e., containing an open set containing x).

The collection of all neighbourhoods of a point is a filter.

If $A, B \subset X$ and $A \subset \overline{B}$, we say that B is dense relative to A (in A if in addition $B \subset A$).

Definition: A topological space X is called (do not mix!):
—separable if it contains a countable dense subset;
—separated (or Hausdorff) if any two distinct points have disjoint (i.e., with $\emptyset$ intersection) neighbourhoods;
—connected if it is not the union of two disjoint open subsets (which would then be also closed).

3) Convergence: A filter base on X is said to converge at $x \in X$ if it is finer than the neighbourhood filter of x (i.e., contains all neighbourhoods of x). A map $f: X \to Y$ is said to be continuous at $x \in X$ if the inverse image $f^{-1}(V)$ of any neighbourhood V of $y=f(x)$ is a neighbourhood of x, and continuous if it is continuous at every point (i.e., if f^{-1}(open) = open).

Homeomorphism: It is a bijective (one-to-one) map, continuous together with its inverse.

If A is a directed set (every finite subset has an upper bound), one can define a (generalized or not, according to the cardinality of A) sequence x_α, $\alpha \in A$, of points in X. Such a sequence is said to converge to x if its section filter base $\{\{x_\alpha; \alpha \leqq \alpha_0, \alpha_0 \in A\}\}$ converges to x. A point x is a limit point of a filter $\mathfrak{F}$ if $x \in \overline{F}$ for all $F \in \mathfrak{F}$.

4) Comparison of topologies. A topology $\mathcal{C}_1$ on a set X is said to be finer than another one $\mathcal{C}_2$ (or $\mathcal{C}_2$ coarser than $\mathcal{C}_1$) if "$\mathcal{C}_1$ has more open sets," i.e., if every $\mathcal{C}_2$-open subset of X is also $\mathcal{C}_1$-open.

Projective and inductive topologies (important ways of defining topologies). If X is a set, X_α a family of topological spaces, and if we are given mappings f_α: X $\overrightarrow{\text{into}}$ X_α [resp., g_α: X_α $\overrightarrow{\text{into}}$ X], then the projective (or initial) topology [resp., the inductive, or final, topology] on X relatively to this collection of maps and spaces is the coarsest [resp., the finest] such that all maps are continuous.

Example of projective topology: The product topology on the set $X = \Pi X_\alpha$, the f_α being the projections $X \to X_\alpha$.

5) A metric (or distance) on a set X is a positive function $d: X \times X \to \mathbb{R}_+$ such that $d(x,y)=0$ iff (= if and only if) $x=y$, $d(x,y)=d(y,x)$, and (triangle inequality) $d(x,z) \leqq d(x,y)+d(y,z)$.

Metric (resp. metrizable) space: a topological space the topology of which is derived (resp. can be derived) from a metric d, a neighbourhood base of x being composed of (e.g.) the open balls $\{y; d(x,y) < r, r > 0\}$.

6) Compact space: A separated topological space X, on which every filter has a limit point (equivalently: every open cover of X contains a finite subcover). If it is not necessarily Hausdorff, we call X "quasi-compact."

Locally compact space: A separated space in which every point has a compact neighbourhood.

7) A subset $A \subset X$ is rare if its interior $\mathring{A}$ is void. It is meager (= 1st category of Baire) if it is a countable union of rare sets. If not, it is called of Baire's 2nd category.

Baire space: A topological space in which every non-void subset is of Baire's 2nd category. Every locally compact space is Baire space, and so are complete (cf. Section C) metrizable spaces (Baire's theorem).

8) Vector spaces. A (left) vector space L on a field K is an additive group with scalar (left) multiplications by elements of K, with the expected distributivity properties.

Each vector space L has (Zorn's lemma) a maximal set of linearly independent elements x_α, $\alpha \in A$ ($\sum_{\alpha \in I} \lambda_\alpha x_\alpha = 0$ for each finite subset $I \subset A \Rightarrow$ all $\lambda_\alpha = 0$), called a basis of L, all bases having the same cardinality.

When the field is (e.g.) $\mathbb{R}$ or $\mathbb{C}$, we say, if $A, B \subset L$ and $\exists \lambda_0$ such that $B \subset \lambda A$ for $|\lambda| \geqq |\lambda_0|$, that A absorbs B. A is absorbing if it absorbs every point. $A \subset L$ is called circled (or balanced) if $\lambda A \subset A$ for all $|\lambda| \leqq 1$.

9) Uniform spaces, Cauchy filters, complete topological spaces: see (e.g.) Bourbaki or Schaefer. Since we shall be concerned almost exclusively with locally convex spaces, we shall define completeness directly for these spaces, without explicit use of the notion of uniform structure.

Convex set in a linear space: It is a subset A such that $x, y \in A$ implies $\lambda x + (1-\lambda) y \in A$ for $0 < \lambda < 1$.

B. Topological Vector Spaces

1) Definition. A topological vector space (t.v.s.) in a topological field K is a vector space endowed with a topology compatible with its vector-space structure, i.e., such that $(x, y) \to x - y$ and $(\lambda, x) \to \lambda x$ are continuous mappings $L \times L \to L$ and $K \times L \to L$ (respectively).

(On $\mathbb{R}$ or $\mathbb{C}$, the topology is usually defined with the absolute value.)

2) Topological vector subspace. Notion obvious.

Quotient space. If M is a subspace of L (with the induced topology), the quotient topology on the quotient space L/M is the final topology relative to the projection $L \to L/M$ (it is easy to see that this is a vector-space topology), and is separated if and only if M is a closed subspace. [Therefore, for any t.v.s. L, $L/\overline{\{0\}}$ is separated and is called the separated space associated with L.]

Product. $L = \Pi L_\alpha$ is endowed with the product (initial) topology defined by the projections $L \to L_\alpha$.

Remarks.

a) If E is a vector space, E_α a family of t.v.s. with maps $f_\alpha: E \to E_\alpha$, then the initial topology on E is a vector space topology.

b) If

$$E = \sum_{i=1}^{n} E_i$$

is the algebraic direct sum of the vector spaces E_i, the map $(x_i) \to \sum x_i$ from ΠE_i onto E is always continuous (but not necessarily its inverse) if E is given any vector space topology, the E_i being given the subspace topology. We say that E is a topological direct sum if this map is a homeomorphism.

3)a) A linear manifold in a vector space L is the translated $F = x_0 + M (x_0 \in L)$ of a vector subspace M. A hyperplane is a maximal linear manifold, which occurs iff codim $M = \dim\{N;\ M + N = L,$ direct sum$\} = 1$.

b) A set B in a t.v.s. L is called bounded if it is absorbed by each 0-neighbourhood.

c) Theorem.

i) A separated t.v.s. L is metrizable iff it has a countable neighbourhood base at 0; and then it has a metric satisfying also $d(\lambda x, 0) \leqq d(x, 0)$ for $|\lambda| \leqq 1$ and all $x \in L$.

ii) If a t.v.s. has a bounded neighbourhood of 0, it is metrizable.

iii) L metrizable, M closed in L $\Rightarrow$ L/M metrizable.

iv) A t.v.s. is locally compact iff it is separated and finite-dimensional.

4) Properties of convex sets in t.v.s.

a) A convex, $x \in \mathring{A}$, $y \in \overline{A} \Rightarrow \{\lambda x + (1-\lambda)y;\ 0 < \lambda < 1\} \subset \mathring{A}$.

b) A, B convex sets $\Rightarrow$ so are $\mathring{A}$, $\overline{A}$, $A + B$, αA.

c) A convex not rare $\Rightarrow \overline{A} = \overset{\circ}{\overline{A}}$.

5) Semi-norms. Definition. A semi-norm on a vector space L is a (positive) function $p: L \to \mathbb{R}_+$ such that $p(x+y) \leqq p(x) + p(y)$ and $p(\alpha x) \leqq |\alpha| \, p(x)$.

It is a norm if in addition $p(x) = 0$ iff $x = 0$.

Definition. The gauge of an absorbing subset M in a vector space L is the function $x \to p_M(x) = \inf\{\lambda > 0;\ x \in \lambda M\}$.

Proposition. The gauge of an absorbing convex circled set M is a semi-norm, and conversely.

Proof. "Conversely": define $M = \{x: p(x) < 1\}$, the unit semi-ball for M. "Directly": If $x, y \in L$, $\lambda_1 > p(x)$, $\lambda_2 > p(y)$, then

$$x+y \in \lambda_1 M + \lambda_2 M = (\lambda_1 + \lambda_2)\left(\frac{\lambda_1}{\lambda_1 + \lambda_2} M + \frac{\lambda_2}{\lambda_1 + \lambda_2} M\right) \subset (\lambda_1 + \lambda_2) M.$$

Therefore $p(x+y) \leqq \lambda_1 + \lambda_2$, and thus $p(x+y) \leqq p(x) + p(y)$. Moreover, if $\lambda \neq 0$, $p(\lambda x) = \inf\{\mu > 0,\ x \in (\mu/|\lambda|)M\}$ since $\lambda x \in \mu M$ iff $|\lambda| x \in \mu M$ (M is circled). Therefore $p(\lambda x = \inf_{\mu > 0}\{|\lambda|\mu;\ x \in \mu M\} = |\lambda| p(x)$. For $\lambda = 0$, $p(0) = 0$.

6) Lemma. Let p be a semi-norm on a t.v.s. E. Then it is continuous at 0 iff its unit semi-ball M is open, or iff it is uniformly continuous.

Definition. The set of all semi-norms on a vector space E is denoted by SPEC E. The set of all continuous semi-norms on a t.v.s. E is denoted by Spec E (the spectrum of E).

Lemma. The sets $S_1 = \text{Spec } E$ and $S_2 = \text{SPEC} =$ on a t.v.s. E are irreducible sets of semi-norms; this means that they are cones (if $p \in S_i$, $\lambda > 0$, then $\lambda p \in S_i$), that if $p \in S_i$ and $q \in \text{SPEC } E$ and $q \leqq p$ (the unit semi-ball of p is contained in that of q), then $q \in S_i$, and that if $p, q \in S_i$ then $\sup(p,q) \in S_i$ ($i = 1, 2$; the last two statements are trivial for S_2).

7) Theorem. A separated t.v.s. E is normable iff it has a bounded convex 0-neighbourhood.

Proof. Necessity: The unit closed ball $M_1 = \{x;\ ||x|| \leqq 1\}$ is a convex neighbourhood, which is bounded since $\{(1/n)M_1 : n \in N\}$ is a base of 0-neighbourhoods (N is the set of natural numbers).

Sufficiency: Let V be a convex bounded 0-neighbourhood. From the continuity at $(0,0)$ of $(\lambda, x) \to \lambda x$, $\exists \alpha > 0$ and W, 0-neighbourhood, such that $\lambda W \subset V$ for $|\lambda| \leqq \alpha$. Then

$$U_0 = \bigcup_{|\lambda| \leqq \alpha} \lambda W$$

is circled ($\mu U_o \subset U_o$ for $|\mu| \leqq 1$) and contained in V, and the convex hull U of U_o (the intersection of all convex sets containing U_o) is also circled and contained in V, thus bounded. Therefore $\{\frac{1}{n} U\}$ is a base of 0-neighbourhoods and the gauge p of U, which is a norm since E is separated, generatés the topology of E.

8) Definition. A t.v.s. $\mathbb{E}$ over $\mathbb{R}$ or $\mathbb{C}$ is called locally convex (we shall write l.c.s.) if any 0-neighbourhood contains a convex 0-neighbourhood (or iff it has a base of convex 0-neighbourhoods, or a base of convex circled absorbing 0-neighbourhoods: the equivalence of these conditions is straightforward). The topology of a l.c.s. E can be defined by Spec E (take finite intersections of semi-balls).

If Γ is a set of semi-norms on a vector space E, the finite intersections of multiple of unit semi-balls are a base for a locally convex topology on E, called the locally convex topology associated with Γ. The topology is separated iff $p(x) = 0$ for all $p \in \Gamma$ implies $x=0$. In particular, SPEC E defines a topology, the finest locally convex topology on E (it is not metrizable if $\dim E = \infty$).

9) Counter-example. The metric space of functions f on [0, 1] such that

$$\int_0^1 |f(t)|^p dt = d(f, 0) < \infty$$

is not locally convex, when $0 < p < 1$. This space is quite pathological: every continuous linear functional on it is zero.

C. Elementary Properties of Locally Convex Spaces: Metrizability, Completeness

1) Because (among other reasons) of the pathological character of t.v.s., which are not l.c.s. (as shown by the last counter-example), we shall be concerned almost exclusively, from now on, with l.c.s. which we shall suppose also (unless otherwise stated) separated.

Lemma 1. A (separated) l.c.s. is metrizable iff its topology can be defined by a countable set of semi-norms, which can therefore be taken increasing (defining $p \propto q$ if $\exists \alpha > 0$ such that $p(x) \leqq \alpha q(x)$ for all $x \in E$).

Proof: follows from Th(B 3 c i), from the fact that the metric can be defined, if $\{p_n\}$ is a sequence of semi-norms defining the topology of E, by:

$$d(x, 0) = \sum_n \frac{1}{2^n} \frac{p_n(x)}{1+p_n(x)}$$

(and, by translation, $d(x, y) = d(x-y, 0)$) since

$$\frac{a+b}{1+(a+b)} \leqq \frac{a}{1+a} + \frac{b}{1+b}$$

(look at the function $t/(1+t)$ for $t \geqq 0$), and from the inclusions $\{x;\ d(x, 0) \leqq 2^{-k}\} \supset \{x;\ p_{k+1}(x) \leqq 2^{-(k+2)}\}$ which show that the defined metric indeed generates the topology of E.

Lemma 2. A (separated) metrizable l.c.s. E is normable iff it has a countable fundamental system of bounded sets.

The condition is obviously necessary. It is sufficient since, if we denote by (p_n) and (B_k) countable systems of semi-norms defining the topology and of bounded sets (respectively), by

$$\mu_{nk} = \sup_{x \in B_k} p_n(x) \quad \text{and} \quad \mu_n = \sup_{k=1, \ldots, n} \mu_{nk},$$

then the set $B = \{x;\ p_n(x) \leqq \mu_n \text{ for all } n \in \mathbb{N}\}$ is bounded and absorbs all bounded sets (because $\forall k$, $\exists \lambda_k > 0$ such that $\mu_{nk} \leqq \lambda_k \mu_n$, so that $B_k \subset \lambda_k B$); the norm is then (e.g.) the gauge of the convex circled hull $\tilde{B}$ of B, which absorbs all bounded sets and is therefore a 0-neighbourhood since E is metrizable (if V_n is a base of 0-neighbourhoods and if $\forall n\ \exists x_n \in \frac{1}{n} V_n$, $x_n \not\in \tilde{B}$, then $A = \{nx_n\}$ is bounded, but would not be absorbed by $\tilde{B}$ since $x_n \not\in (\lambda/n)B \subset B$ for $n \geqq \lambda$). [A simpler proof could be given, knowing that the completion of E is a Baire space.]

Remark. There are non-metrizable l.c.s. having a countable fundamental system of bounded sets (i.e., every bounded set is contained in some member of the system).

Lemma 3. A set $B \subset E$ (separated l.c.s.) is bounded iff

$$\sup_{b \in B} p(b) < \infty$$

for all $p \in \Gamma$, a family generating the topology.

2) Definition. A Cauchy sequence in a l.c.s. E, with a generating family Γ of semi-norms, is a sequence $\{x_n\}$ such that $p(x_n - x_m) \to 0$ for $n, m \to \infty$ and all $p \in \Gamma$.

A filter $\mathfrak{F}$ on a l.c.s. E is said a Cauchy filter if $\forall p \in \Gamma$ and $\epsilon > 0$, $\exists F \in \mathfrak{F}$ such that $p(x-y) < \epsilon$ for all $x, y \in F$. The filter of sections of a Cauchy sequence is a Cauchy filter—a base of the section filter is given, so we have seen, by the sets $\{x_n;\ n \leqq n_0,$ n_0 fixed$\}$. It is equivalent to consider Cauchy filters or generalized sequences (x_α), $\alpha \in A$, a directed set (= each finite subset has an upper bound), the section filter of which is a Cauchy filter iff the generalized sequence is a Cauchy (generalized) sequence (this is the rigorous definition of a Cauchy generalized sequence).

Definition. A l.c.s. is said complete if every Cauchy filter converges.

Remarks.

i) For general t.v.s., one has to introduce the associated uniform structure in order to define Cauchy filters, and therefore the notion of complete space.

ii) For a metrizable space, it is enough to deal with Cauchy sequences.

iii) A general t.v.s. is said semi-complete if every Cauchy sequence converges.

Definition. A t.v.s. is said quasi-complete if every bounded closed subset is complete.

A quasi-complete t.v.s. is semi-complete, but the converse is false (cf. ℓ^1 with the weak topology).

3) Completion. If L is a separated l.c.s. (or t.v.s.), one defines a space $\hat{L}$, which is unique up to isomorphism, by the property that it is separated and complete, and that L is isomorphic to a dense subspace of $\hat{L}$. $\hat{L}$ is called the completion of L.

(If L is not separated, one first constructs the separated space $L/\{\overline{0}\}$, and then its completion $\hat{L}$, the "separated completed" space of L.)

$\hat{L}$ can be constructed as the set of all Cauchy filters on L modulo the equivalence relation $\mathfrak{F} \sim \mathfrak{F}'$ iff $\exists$ a 0-neighbourhood V and sets $A \in \mathfrak{F}$, $B \in \mathfrak{F}'$ such that $A-B \subset V$ (which expresses that $\mathfrak{F}$ and $\mathfrak{F}'$ "converge to the same point in L": one adds to L formal limits of Cauchy filters). $\hat{L}$ is l.c.s. whenever L is l.c.s., and its topology can be defined by the extensions by continuity to $\hat{L}$ of the semi-norms of a defining family of semi-norms of L; $\hat{L}$ is metrizable whenever L is.

Remark. One can also define a quasi-completed space associated to L, in a similar way.

Definition. A (F)-space, or Fréchet space, is a complete l.c.s.

Definition. A set A in a Hausdorff t.v.s. E is said precompact if its closure $\bar{A}$ in the completion $\hat{E}$ of E is compact.

Remark. Fréchet (hence Banach) spaces are Baire spaces (cf. A.7).

D. Projective and Inductive Topologies

1) If E_α is a family of l.c.s. and if we are given a family of maps $f_\alpha: E \to E_\alpha$ from a vector space E into E_α, then the projective (initial) topology on E is locally convex (we recall that it is defined as the coarsest such that all f_α are continuous). The finite intersections of $f_\alpha^{-1}(U_\alpha)$, U_α neighbourhoods of $x_\alpha = f_\alpha(x)$ for some $x \in E$, give a base of neighbourhoods of x. Therefore, the projective topology on E is separated iff $\forall x \neq 0$, $\exists \alpha$ and U_α, 0-neighbourhood in E_α, such that $f_\alpha(x) \in U_\alpha$.

Universal property of projective topologies. If a set E is endowed with the projective topology relative to the topological spaces E_α and the maps $f_\alpha: E \to E_\alpha$, then a map u from any topological space F into E is continuous iff $\forall \alpha$, the composed map $f_\alpha \circ u$ is continuous from F to E_α. Indeed, if G_α is open in E_α, then $u^{-1}(f_\alpha^{-1}(G_\alpha))$ is open in F (if $f_\alpha \circ u$ is continuous), and so will be the unions of families of finite intersections of such sets, which are exactly the sets $u^{-1}(G)$, G being any open subset of E— whence the continuity of u (the converse part is trivial).

2) If E_α is a family of l.c.s., and if we are given maps $g_\alpha: E_\alpha \to E$, one can define the inductive locally convex topology on the vector space E as the finest locally convex topology such that all g_α are continuous. Since the g_α are continuous when E is given the trivial topology, in which only $\emptyset$ and E are open, the class of l.c. topologies on E for which the g_α are continuous is non-void; its upperbound is the l.c. inductive topology, a base of neighbourhoods of which is given by the convex circled absorbing sets V such that $\forall \alpha$, $g_\alpha^{-1}(V)$ is a 0-neighbourhood in E_α. The inductive l.c. topology of a family of l.c. topologies has also a universal property (a map u: $E \to F$ is continuous iff each $u \circ f_\alpha$ is continuous). However, it need not be separated, even if all E_α are so.

3) If E_α is a family of l.c.s. the algebraic direct sum $E = \bigoplus_\alpha E_\alpha$ (the subspace of ΠE_α for which only a finite number of components are $\neq 0$) can be endowed with the l.c. inductive topology defined by the injections $E_\alpha \to E$: we get the locally convex direct sum. One sees easily that it is complete iff each E_α is complete.

In particular, if each E_α is one-dimensional, one gets the finest l.c. topology on E, which can be defined by the set SPEC E of all semi-norms on E. It is complete (and not metrizable if dim E = ∞: therefore, the Hilbert direct sum of an infinite number of Hilbert spaces is not the l.c. direct sum); any linear map from E (with the finest l.c. topology) into any l.c.s. is continuous: therefore, the topological dual E' (space of all linear continuous functionals $E \to \mathbb{C}$) is equal, in this particular case, to the algebraic dual E^* (space of all linear functionals $E \to \mathbb{C}$), and every subspace is closed; a set $B \subset E$ is bounded iff it is contained and bounded in a finite-dimensional subspace.

4) Projective limit. Let E_α be a family of l.c.s., with $\alpha \in A$ a directed set (the order relation of which we shall write $\leqq$), and suppose given a family of maps $g_{\alpha\beta}$ from E_β into E_α, for $\alpha \leqq \beta$. The subspace

$$E \subset \prod_{\alpha \in A} E_\alpha$$

of the "tuples" $x = (x_\alpha)$ such that $x_\alpha = g_{\alpha\beta} x_\beta$ for $\alpha \leqq \beta$ is called the projective limit of the E_α (relatively to the maps $g_{\alpha\beta}$), and we denote $E = \varprojlim g_{\alpha\beta} E_\beta$, or $E = \varprojlim E_\alpha$ if no confusion is possible (sometimes, for typographical reasons mainly, one writes limproj instead of $\varprojlim$). The topology on $E = \varprojlim g_{\alpha\beta} E_\beta$ is (of course) the projective topology relatively to the family (E_α, f_α) where f_α is the restriction to E of the projection of

$$\prod_{\beta \in A} E_\beta$$

onto E_α. If all E_α are quasi-complete (or complete), then so is E.

Example. Suppose we have a family of l.c. spaces E_α; then the product ΠE_α can be considered as a projective limit as follows: denote by

$$E_H = \prod_{\alpha \in H} E_\alpha ,$$

where H is any finite subset of the index set A, and by g_{HK} the projection of E_K onto E_H when $H \subset K$ (the family of finite subsets of A is ordered by inclusion); then $\Pi E_\alpha = \varprojlim g_{HK} E_K$. Any projective limit is a closed subspace of the product. Other examples will be seen later (of the type $E = \bigcap E_\alpha$).

Semi-norms. Let us denote by $f_{\alpha*}$ the map SPEC $E_\alpha \to$ SPEC E defined by the $f_\alpha: E \to E_\alpha$ as follows: for all $p \in$ SPEC E_α, we define $f_{\alpha*}(p) = p \circ f_\alpha \in$ SPEC E. [This is a general procedure: to any linear map $f: E \to F$ one can similarly associate a map f_*: SPEC F $\to$ SPEC E, and when E and F are given l.c. topologies by choice of some irreducible sets of semi-norms Spec E and Spec F, respectively, then f is continuous iff $f_*(\text{Spec } F) \subset \text{Spec } E$, in which case f_* preserves the order of continuous semi-norms]. Then one verifies easily that the projective limit topology on E is defined by the irreducible hull, in SPEC E, of the set

$$\underline{\bigcup_{\alpha \in A} f_{\alpha*}(\text{Spec } E_\alpha)}$$

—which defines the topology of E in the normal manner (finite intersections of semi-balls relative to the semi-norms in this set); the hull is obtained by closing the set under the following operations: upper bounds of finite families, multiplication by positive numbers and addition of "smaller" semi-norms (cf. B.6).

5) Inductive limit. We shall define it only in the following "special" case (which is still very general and includes all interesting examples; for a slightly more general definition, cf. Schaefer):

Let E_α be a family of l.c. subspaces of a vector space E, $\alpha \in A$ directed by inclusion of the corresponding subspaces ($\alpha \leqq \beta$ iff $E_\alpha \subset E_\beta$), with $E = \bigcup_\alpha E_\alpha$. Suppose that if $E_\alpha \subset E_\beta$, the topology of E_α is finer than the topology induced on E_α by E_β. If the inductive topology on E relatively to the inclusions $g_\alpha: E_\alpha \to E$ is Hausdorff, then it is called the inductive limit topology and we write (when no confusion arises as to the maps) $E = \varinjlim E_\alpha$ (or limind E_α sometimes). A base of 0-neighbourhoods in E is given by the convex circled absorbing sets V such that $V \cap E_\alpha$ is a 0-neighbourhood in E_α for all $\alpha \in A$. If we consider the maps $g_{\alpha*}$: SPEC E $\to$ SPEC E_α, then the inductive limit topology is defined by the (irreducible) set of semi-norms

$$\bigcap_{\alpha \in A} g_{\alpha*}^{-1}(\text{Spec } E_\alpha).$$

E can also be considered (cf. Grothendieck) as a quotient of the l.c. direct sum ΣE_α, via a map $\Sigma x_\alpha \to \Sigma g_\alpha(x_\alpha)$ from ΣE_α onto E.

6) The inductive limit is said strict if the topology of E_α is equal to the topology induced on E_α by E_β, when $E_\alpha \subset E_\beta$ (and, according to most authors—like Bourbaki but not Schaefer—if the index set A is countable).

Theorem. The strict inductive limit $E = \bigcup_n E_n$ of a sequence of subspaces induces on each E_n its original topology.

Proof. Let V_n be any convex 0-neighbourhood in E_n. There exists a sequence V_{n+p} of convex 0-neighbourhoods in E_{n+p} such that $V_{n+p} \cap E_n = V_n$ for all $p \geqq 1$; indeed, by hypothesis there exists a convex 0-neighbourhood W_{n+p} in E_{n+p} such that $W_{n+p} \cap E_n \subset V_n$, and it is easily checked that the convex hull V_{n+p} of $W_{n+p} \cup V_n$ in E_{n+p} has the required property (it is composed of the points $z = \lambda x + (1-\lambda)y$, $x \in W_{n+p}$, $y \in V_n$, $0 \leqq \lambda \leqq 1$, and if $z \in E_n$, $x \in E_n \cap W_{n+p} \subset V_n$ and thus $z \in V_n$, if $\lambda \neq 0$; the conclusion holds trivially also for $\lambda = 0$). Then

$$V = \bigcup_{p \geqq 0} V_{n+p}$$

is convex and a 0-neighbourhood in E (since $V \cap E_k$ is a 0-neighbourhood in E_k for all k), and since $V \cap E_n = V_n$, the theorem is proved.

Theorem. Let $E = \bigcup E_n$ be the strict inductive limit of a sequence of l.c. spaces. Then if each E_n is complete, E is complete; and if each E_n is closed in E_{n+1}, a subset $B \subset E$ is bounded iff it is contained and bounded in some E_k.

(Hint of) Proof. Let $\mathfrak{F}$ be a Cauchy filter in E, $\mathcal{V}$ the 0-neighbourhood filter; then $\mathfrak{F} + \mathcal{V} = \{F + V;\ F \in \mathfrak{F},\ V \in \mathcal{V}\}$ is a Cauchy filter base in E. One can show (by usual l.c.s. arguments—cf. Schaefer) that for some k, $(\mathfrak{F} + \mathcal{V}) \cap E_k$ is a filter base, thus a Cauchy filter base, which therefore converges to some $x_k \in E_k$ which is also a limit point for $\mathfrak{F} + \mathcal{V}$, and therefore also for $\mathfrak{F}$. The second part of the theorem is also seen by standard arguments (the "if" is trivial and the "only if" proved ab absurdo).

Definitions. (LB)-space: strict inductive limit of a sequence of Banach (= complete normed) spaces.

(LF)-space: strict inductive limit of a sequence of Fréchet (= complete l.c. metrizable) spaces.

[These definitions are the nowadays most commonly accepted ones for these notions.]

LECTURE II
LOCALLY CONVEX AND NORMED SPACES. LINEAR OPERATORS. MAIN THEOREMS.

A. Normed Spaces Associated with Locally Convex Spaces

1) Let $\{U_\alpha;\ \alpha \in A\}$ be a base of convex circled 0-neighbourhoods in a l.c.s. E, and direct A by inclusion of the

U_α: $\alpha \leqq \beta$ if $U_\beta \subset U_\alpha$. Let p_α be the gauge of U_α (the set $\{p_\alpha;\ \alpha \in A\}$ of semi-norms in then also directed and generates the topology of E) and write $V_\alpha = \ker p_\alpha = p_\alpha^{-1}(0)$: this is a closed ($p_\alpha$ is continuous) subspace of E. We now define $F_\alpha = E/V_\alpha$, and give this vector space the topology generated by the unique norm p_α, which is obviously coarser than the topology of quotient space on F_α [in the latter we may have other, non-equivalent, continuous semi-norms]. If $\alpha \leqq \beta$, $V_\beta \subset V_\alpha$ and each equivalence class x_β modulo V_β is contained in an unique class modulo V_α, x_α, the mapping $g_{\alpha\beta}$: $x_\beta \to x_\alpha$ being a continuous map between normed spaces $F_\beta \to F_\alpha$. [If $p \in \operatorname{Spec} E$ is the gauge of a convex circled 0-neighbourhood U, we write $E_U = E/\ker p$.]

<u>Theorem</u>. <u>A complete l.c.s. is isomorphic with a projective limit of Banach spaces</u> (a sequence of them for a—"countably normed"—Fréchet space) and any (separated) l.c.s. is isomorphic with a subspace of a product of Banach spaces.

<u>Proof</u> (sketch). Set $F = \varprojlim g_{\alpha\beta} F_\beta$. The map $x \to (x_\alpha)$ from E to F is linear, and bijective (one-to-one) since E is separated, and onto if E is complete. Moreover, this map and the inverse map are easily seen to be both continuous. Now if $\hat{F}_\alpha$ is the Banach space, completion of F_α, and $\bar{g}_{\alpha\beta}$ the continuous extensions of $g_{\alpha\beta}$ to $\hat{F}_\beta$, F is obviously dense in $\varprojlim \bar{g}_{\alpha\beta}\hat{F}_\beta = \hat{F}$. But we have seen that $E \approx F$ if E is complete, and $\hat{F}$ is complete; thus $E \approx \varprojlim \hat{F}_\beta$ then. In the general case, E is isomorphic to a subspace of $\hat{F}$, thus of $\Pi \hat{F}_\beta$. The maps $g_{\alpha\beta}$ have in general no special properties. In the case of nuclear spaces (cf. Lecture IV) "enough" of them will be "of trace class"; an intermediate situation is possible in which they are compact (transform some 0-neighbourhood to a precompact set), namely the case (introduced by Grothendieck) of Schwartz spaces.

<u>Definition</u>. A l.c.s. E is said a <u>Schwartz</u> space if given any U, convex circled 0-neighbourhood in E, there exists another one V the image of which in the normed space E_U is precompact. We call (FS) space a Fréchet and Schwartz l.c.s. Products, subspaces, countable inductive limits of Schwartz spaces are Schwartz spaces. In a Schwartz space, bounded sets are relatively compact.

2) Let B be a closed convex circled bounded set in a (separated) locally convex space E. We shall denote by $E_B = \bigcup_n nB$ the normed vector space generated by B, with B as unit ball (the norm being then the gauge p_B of B; if $B = \emptyset$, we take $E_B = \{0\}$ and $p_B = 0$). This topology is finer than that induced by E (since, if $\mathring{B} = \emptyset$—which will be the case by theorem (I.B.7) iff E is not normable—B is not a 0-neighbourhood for the induced topology). If E is

quasi-complete (and, more generally, if B is complete), then E_B is a Banach space.

Let $\mathcal{B}$ be the family of all closed convex circled bounded sets, ordered by inclusion. Define on E the inductive topology relative to the family (E_B, g_B) where g_B is the inclusion $E_B \to E$. We call it the <u>bornological topology</u> associated with the original one. The obtained space, E_o (= E set-theoretically) is thus $\varinjlim E_B$.

3) <u>Definition</u>. A l.c.s. is called <u>bornological</u> if its topology is equal to the associated bornological topology, i.e., if $E = \varinjlim E_B$. It is said <u>ultrabornological</u> if it is an inductive limit of Banach spaces (e.g.: a quasi-complete bornological space).

<u>Theorem</u> (usual definition). A l.c.s. E is bornological iff every convex circled set that absorbs every bounded set (a "<u>bornivore</u>" set) is a 0-neighbourhood, i.e., iff every semi-norm $p \in$ SPEC E that is bounded on bounded sets is continuous.

<u>Proof</u> (sketch). (A continuous semi-norm is obviously bounded on bounded sets, but the converse need not be true.) The bornological topology is (by either definition) the finest l.c. topology whose family of bounded sets is identical with B.

<u>Corollary</u>. Every metrizable l.c.s. is bornological. Every separated quotient, l.c. direct sum, or inductive limit of bornological spaces is bornological.

<u>Proof</u>. If E is metrizable, choose $\{V_n; n \in N\}$ a (decreasing) countable 0-neighbourhood base, and consider any bornivore set A: $V_n \subset nA$ for some n (hence A is a 0-neighbourhood) since otherwise $\exists \{x_n\}$, $x_n \in V_n$, and hence $x_n \to 0$, such that for all n, $x_n \notin nA$, which is impossible since by the first condition $\{x_n\}$ is bounded. The second part follows from the (easy to check) fact that the inductive topology relative to a family of bornological spaces is bornological.

4) <u>Remarks</u>.

a) Closed subspaces of bornological spaces need not be bornological (there are counter-examples).

b) From a theorem attributed to Mackey and Ulam, the products $\mathbb{R}^d$ and $\mathbb{C}^d$ (and therefore E^d if E is a bornological space) are bornological if d is smaller than the smallest strongly non-accessible cardinal [a strongly non-accessible cardinal d_o is such that $d_o > \aleph_o$, $\Sigma\{d_\alpha; \alpha \in A\} < d_o$ whenever card $A < d_o$ and $d_\alpha < d_o$, and that $d < d_o$ implies $2^d < d_o$; it is not known if there exists such cardinals] and, in particular, for $d = \aleph_o$, $d = \aleph$, and $d = 2^{\aleph}$.

c) It follows from the corollary that (LB), (F) and (LF) spaces are (ultra) bornological.

d) Some properties of bornological spaces can be studied algebraically, by axiomatizing the notion of bounded sets ("bornology").

Theorem.

a) Let u be a linear map from a bornological space E into any l.c.s. F. Then u is continuous iff it is bounded on bounded sets, or iff $u(x_n) \to 0$ whenever $x_n \to 0$.

b) A l.c.s. E is bornological iff every linear map from E into any Banach space F that carries bounded sets to bounded sets is continuous.

Proof.

a) If u is bounded on bounded sets, and $B \subset E$ is any bounded set, given a convex circled neighbourhood of 0 in F, V, it absorbs u(B); hence $u^{-1}(V)$ absorbs all B, hence is a 0-neighbourhood in E (bornological), i.e., u is continuous.

If $B \subset E$ is bounded and $u(y_n) \to 0$ whenever $y_n \to 0$, given $\{x_n\}$, $x_n \in B$ (arbitrary sequence), then whenever $\lambda_n \to 0$ in the base field (scalars) $\lambda_n x_n \to 0$ in E, hence $\lambda_n u(x_n) \to 0$ in F and thus u(B) is bounded. The other implications are then obvious.

b) (Hint) The identity map from E onto E_0 (bornological space associated to E) is continuous.

5) Definitions. A barrel in a topological vector space is an absorbing convex circled closed subset.

A l.c.s. is said barreled if every barrel is a 0-neighbourhood.

A l.c.s. is said infrabarrelled if every bornivore barrel is a 0-neighbourhood.

Of course, a barreled space is infrabarreled, but the converse need not be true (only "big" barrels, that absorb bounded sets —not only finite sets—, are required to be 0-neighbourhoods).

Theorem. Let E_α be a family of barreled l.c.s., E a vector space, and give E the inductive l.c. topology relative to some given maps g_α: $E_\alpha \to E$. Then E is barreled.

Proof. Let T be a barrel in E; $\forall \alpha$, $g_\alpha^{-1}(T)$ is a barrel in E_α, hence a 0-neighbourhood, and therefore T is 0-neighbourhood.

In particular, inductive limits, separated quotients, l.c. direct sums of barreled spaces are barreled. But projective limits of barreled spaces need not be barreled, and closed subspaces of barreled spaces are in general not barreled. However, products and completions of barreled spaces can be proved to be barreled.

Theorem. Every l.c.s. which is a Baire space (e.g., Banach and Fréchet spaces) is barreled.

Proof. If T is any barrel (absorbing and circled in particular),

$$E = \bigcup_{n \in \mathbb{N}} nT;$$

since E is Baire, at least one of the (closed) nT, say $n_0 T$, has an interior point y; hence $\mathring{T} \ni \frac{1}{n_0} y$, and also (T is circled) $-\frac{1}{n_0} y$, hence also (convexity) $\mathring{T} \ni 0$ and E is barreled.

Remarks.

a) (LF) spaces (that are not F-spaces) are not Baire spaces, but still are barreled.

b) Ultrabornological spaces are barreled.

c) There are normed (hence bornological) spaces which are not barreled, and there are barreled spaces which are not bornological.

d) Every quasi-complete infrabarreled space is barreled (more generally, in any l.c.s., any barrel absorbs each bounded convex circled complete subset).

e) Every bornological space is infrabarreled.

This category of spaces will be important in the study of spaces of linear maps.

6) Examples.

a) Let $\mathcal{D}_\ell^m$ be the space of (complex-valued) m-times continuouslv differentiable functions on $\mathbb{R}^n$ having support in the ball $||x|| \leqq \ell$ ($\ell > 0$), endowed with the usual Banach space topology, that we shall here define equivalently by the semi-norms $p_D(f) = \max_x |(Df)(x)|$, where D is any partial differential operator with constant coefficients of order $\leqq m$.

b) As usual, we define $\mathcal{D}^m = \lim_\ell \text{ind}\, \mathcal{D}_\ell^m$, i.e., $\mathcal{D}^m = \bigcup_\ell \mathcal{D}_\ell^m$, and a base of 0-neighbourhoods in $\mathcal{D}^m$ is given by convex sets N for which $N \cap \mathcal{D}_\ell^m$ is a 0-neighbourhood in $\mathcal{D}_\ell^m$ for every ℓ. It is an (LB) space.

c) We can also define $\mathcal{D}_\ell = \lim_m \text{proj}\, \mathcal{D}_\ell^m = \bigcap_m \mathcal{D}_\ell^m$; a set $N \subset \mathcal{D}_\ell$ is a 0-neighbourhood if it is the intersection with $\mathcal{D}_\ell$ (= C^∞ functions with compact support inside $||x|| \leqq \ell$) of some 0-neighbourhood in some $\mathcal{D}_\ell^m$; its topology can be defined by all seminorms p_D, the order of D being unrestricted, and is metrizable in the usual way (cf. I.C.1); this is, moreover, a Fréchet space.

d) Following L. Schwartz, we now define $\mathcal{D} = \lim_\ell \text{ind}\, \mathcal{D}_\ell = \bigcup_\ell \mathcal{D}_\ell$. A convex set $N \subset \mathcal{D}$ is a 0-neighbourhood if its intersection with each $\mathcal{D}_\ell$ is a 0-neighbourhood in $\mathcal{D}_\ell$. It is an (LF) space and a Schwartz space. The topology of $\mathcal{D}$ can (cf. Schwartz) be defined by the semi-norms:

$$p[\{m_\ell\}, \{\epsilon_\ell\}](f) = \sup_\ell \Big(\sup_{\substack{||x|| \geqq \ell \\ |q| \leqq m_\ell}} |D^q f(x)/\epsilon_\ell| \Big),$$

where D^q is a derivation monomial of total order $|q|$ (= sum of the orders in each variable), $\ell = 1, 2, \ldots, \epsilon_\ell \searrow +0$, and $m_\ell \nearrow +\infty$ (the m_ℓ being integers). Notice that the topology of $\mathcal{D}^m$ can be defined similarly, via the semi-norms $p[m, \{\epsilon_\ell\}]$, in which we make all m_ℓ equal to a fixed integer m instead of $\nearrow +\infty$).

e) We can also define $\mathcal{D}_F = \lim_m \mathrm{proj}\, \mathcal{D}^m = \bigcap_m \mathcal{D}^m$; it will be topologized by all semi-norms $p[m, \{\epsilon_\ell\}]$, m being any integer. A convex set $N \subset \mathcal{D}_F$ is a 0-neighbourhood if it is the intersection with $\mathcal{D}_F$ of a 0-neighbourhood in some $\mathcal{D}^m$.

f) Set theoretically, $\mathcal{D}$ and $\mathcal{D}_F$ are equal; both consist of C^∞ functions with compact support. Moreover, a set is bounded in $\mathcal{D}$ (relatively to its topology) iff it is contained and bounded in some $\mathcal{D}_\ell$ (cf. L. Schwartz), and the same argument shows that the same is true for $\mathcal{D}_F$ (with its own topology): $\mathcal{D}$ and $\mathcal{D}_F$ have the same bounded sets and convergent sequences (their "pseudo-topologies," as one used to say in the early 50's, coincide). Both also induce on the subspace $\mathcal{D}_\ell$ its original topology. But they do not coincide as topological vector spaces: the topology of <u>$\mathcal{D}$ is finer than that of $\mathcal{D}_F$</u>. This is easily seen from the above sets of defining semi-norms, and made even more clear by the following (in a sense equivalent) "geometric" argument: A convex set N is in the usual base of 0-neighbourhood in $\mathcal{D}_F$ iff $\exists m$ and (convex) N^m, 0-neighbourhood in $\mathcal{D}^m$, such that $N = \mathcal{D} \cap N^m$ (we identify set-theoretically

$$\mathcal{D}_F = \bigcap_m \bigcup_\ell \mathcal{D}_\ell^m = \bigcup_\ell \bigcap_m \mathcal{D}_\ell^m = \mathcal{D}),$$

i.e., iff $\exists m$ and (convex) N_ℓ^m, 0-neighbourhood in $\mathcal{D}_\ell^m$, such that $\forall \ell$, $(N \cap \mathcal{D}_\ell) = (N_\ell^m \cap \mathcal{D}_\ell) \cap \mathcal{D} = N_\ell^m \cap \mathcal{D}_\ell = N_\ell^m$. It is in the usual base of 0-neighbourhoods in $\mathcal{D}$ iff $\forall \ell$, $N \cap \mathcal{D}_\ell = \tilde{N}_\ell$, (convex) 0-neighbourhood in $\mathcal{D}_\ell$, i.e., iff $\forall \ell$, $\exists m(\ell)$ (it may—and will in general—depend on ℓ) and $\tilde{N}^{m(\ell)}$, (convex) 0-neighbourhood in $\mathcal{D}_\ell^{m(\ell)}$, such that $N \cap \mathcal{D}_\ell = \tilde{N}_\ell^{m(\ell)} \cap \mathcal{D}_\ell$. Therefore any 0-neighbourhood in $\mathcal{D}_F$ is a 0-neighbourhood in $\mathcal{D}$, but not conversely—and this merely follows from the non-commutativity of the symbols $(\forall \ell)$ and $(\exists m)$, for $\mathcal{D}_\ell^m$. One could also see this more abstractly: $\mathcal{D}$, being an (LF) space, is ultrabornological (and barreled); therefore, $\mathcal{D}_F$, which has the same bounded sets, must have a coarser topology, strictly coarser since (cf. above or below) one can exhibit a semi-norm or a functional continuous on $\mathcal{D}$ but not on $\mathcal{D}_F$.

g) In summary, we have shown that

$$\mathcal{D} = \varinjlim_{\ell} \varprojlim_{m} \mathcal{D}_\ell^m \subsetneq \varprojlim_{m} \varinjlim_{\ell} \mathcal{D}_\ell^m = \mathcal{D}_F$$

the "inclusion" being the identity operator between

$$\bigcup_\ell \bigcap_m \mathcal{D}_\ell^m \quad \text{and} \quad \bigcap_m \bigcup_\ell \mathcal{D}_\ell^m,$$

and being continuous. [This phenomenon of non-commutativity of $\varprojlim$ and $\varinjlim$ is quite general because of the above-mentioned non-commutativity of the quantifiers $\exists$ and $\forall$.] This is also easily seen on the duals (spaces of continuous linear functionals from the t.v.s. to $\mathbf{C}$). The dual of $\mathcal{D}_F$ is

$$\mathcal{D}_F' = \bigcup_m \mathcal{D}'^m$$

(this equality, which is for the time being set-theoretical but is also true, as we shall see later, when the union is considered as an inductive limit, follows from the Hahn-Banach theorem which we shall present now). Each $T \in \mathcal{D}_F'$, distribution of finite order, can be written

$$T = \sum_{j=1}^{m} D_j \mu_j,$$

where the D_j are linear partial differential operators with constant coefficients and the μ_j measures (continuous functionals on the space $\mathcal{D}^0$ of continuous functions with compact support, endowed with the usual (LB) topology), the sum being finite. On the contrary, each $T \in \mathcal{D}'$ can be written as

$$T = \sum_{\iota \in I} D_\iota \mu_\iota$$

where the sum (infinite in general) is only locally finite (for any compact $L \subset \mathbb{R}^n$, $\operatorname{Supp} \mu_\iota \cap L \neq \emptyset$ for a finite subset of the index set I only). For example $(n = 1)$

$$T = \sum_{i=1}^{\infty} d^i \delta(x-i)/dx^i$$

belongs to $\mathcal{D}'$ but not to $\mathcal{D}_F'$.

B. Hahn-Banach Theorem and Related Results

1) Theorem. Let E be a topological vector space (over $\mathbb{R}$ or $\mathbb{C}$), $A \subset E$ a convex set, V a linear manifold such that $V \cap A = \emptyset$. Then there exists a closed hyperplane H containing V and disjoint from A if either

a) A is open (geometric form of the Hahn-Banach theorem); or

b) E is a l.c.s., A is (convex) compact and V closed [here, V need only be convex, not a linear manifold, and then H can be chosen such that V and A will be in different open semispaces defined by H].

Proof a). By translation, one may suppose that V is a subspace. The set M of closed real subspaces of E containing V and disjoint from A is not void (denoting by V_o the real subspace associated with V, $\overline{V}_o \in M$) and is inductively ordered (by inclusion), thus (Zorn's lemma) has a maximal element H_o [an upper bound for totally ordered subset of M is, e.g., the closure of the union of the elements of this subset]. If E_o is the real vector space associated with E, E_o/H_o is separated and $\dim (E_o/H_o) \geqq 1$ (since $A \neq \emptyset$). The image $\pi(A)$ of A in E_o/H_o is convex and $0 \notin \pi(A)$. The a) part will then follow (in the real case) from the following (proved below):

Lemma. If L is a real separated t.v.s. of dimension $\geqq 2$, $B \not\ni 0$ a convex open set, then there exists a one-dimensional subspace of L disjoint from B.

Indeed, if H_o is not an hyperplane, take $N \subset L_o/H_o$ (of $\dim \geqq 2$), with $\dim N = 1$ and $N \cap \pi(A) = \emptyset$. Then $\pi^{-1}(N) \cap A = \emptyset$ and $\pi^{-1}(N)$ is closed and strictly contains H_o, which cannot be (H_o is maximal).

In the complex case, since $V = iV$ (complex subspace), $H = H_o \cap iH_o$ is the wanted hyperplane.

Proof of the lemma. Let P be a two-dimensional subspace of L; if $P \cap B = \emptyset$, the lemma is trivial. If not, $B_1 = P \cap B$ is convex, open, $\not\ni 0$. Taking coordinates (x,y) in P, one projects B_1 onto a subset of the unit circumference by

$$f: (x,y) \to \left(\frac{x}{\sqrt{x^2+y^2}}, \frac{y}{\sqrt{x^2+y^2}}\right).$$

Now B_1 is connected (it is convex) and f is continuous on B_1; therefore $f(B_1)$ is connected, and since it is also open, $f(B_1)$ is an open arc of length $\leqq \pi$ (for its closure) since if two diametrical points were in $f(B_1)$ we would have $0 \in B_1$ since B_1 is convex, i.e., is contained in an open half-plane, which will also contain B_1.

To prove b), one shows that there is a convex open 0-neighbourhood U such that $(U+A)\cap V = \emptyset$, and since $U+A$ is also open convex, we are reduced (when V is a linear manifold) to a). (This follows mainly from purely general topological considerations.)

Corollary (of b). Every non-empty closed convex subset of a l.c.s. is the intersection of all closed semi-spaces containing it.

Corollary (of a). $E' \neq \{0\}$ iff $\exists$ open convex set A, $\emptyset \neq A \neq E$.

$\Rightarrow$: let $f \in E'$, $f \neq 0$. Then $A = \{x; |f(x)| < 1\}$ is a nontrivial open convex set.

$\Leftarrow$: let $x_0 \in A$. Then $\exists$ H closed hyperplane, with $H \ni x_0$ and $H \cap A = \emptyset$. H can be written $H = \{x; f(x) = \alpha\}$ for some $f \in E^*$ and is closed iff $f \in E'$, whence the result.

This shows the importance of l.c.s. among t.v.s.

2) Definition. Let A be a convex set in a vector space E. A point $x \in A$ is said extreme (or extremal) if it is not interior to any (open) segment contained in A and containing x.

Theorem. (Krein-Milman.) In a (separated) l.c.s. E, any convex compact set A is the closed convex hull of its extreme points (i.e., the intersection of all closed convex sets containing the extreme points).

The proof is based on the preceding theorem (cf. Bourbaki or Schaefer).

3) Theorem. (Hahn-Banach, analytic form.) Let E be a vector space, $p \in$ SPEC E; M a subspace of E and $f \in M^*$ such that $|f(x)| \leqq p(x)$ $\forall x \in M$. Then $\exists f_1 \in E^*$ with $f_1(x) = f(x)$ for $x \in M$ and $|f_1(x)| \leqq p(x)$ $\forall x \in E$.

Proof. The case $f=0$ is trivial. If $f \neq 0$, $H = \{x \in M; f(x) = 1\}$ is an hyperplane in M and a linear manifold in E. The $V_n = \{x \in E;\ p(x) < \frac{1}{n}\}$ are a base of 0-neighbourhoods for some t.v.s. topology on E. $A = V_1$ is open, $A \cap H = \emptyset$, and H is closed. Therefore, there exists an hyperplane $H_1 \supset H$ in E such that $H_1 \cap A = \emptyset$. Since $0 \in A$, $0 \notin H_1$ and $H_1 \cap M \neq M$; moreover, $H (\subset H_1)$ and $H_1 \cap M$ are both hyperplanes in M and thus $H = H_1 \cap M$. Since H_1 is an hyperplane, it can be written $H_1 = \{x \in E;\ f_1(x) = 1\}$ for some $f_1 \in E^*$, and for $x \in M$, we shall have $f(x) = f_1(x)$ since $H = H_1 \cap M$. Since $H_1 \cap A = \emptyset$, we shall also have $|f_1(x)| \leqq p(x)$ for all $x \in E$.

Corollary 1. Let E be a l.c.s., M a subspace of E, $f \in M'$. Then $\exists f_1 \in E'$ that extends f. [Note: f_1 is in general not unique!]

Proof. $V = \{x \in M : |f(x)| \leqq 1\}$ is a circled convex neighbourhood in M; therefore, $\exists$ U, circled convex 0-neighbourhood in

E such that $U \cap M = V$. The gauge p of U is in Spec E and $|f(x)| \leqq p(x)$ for $x \in M$. The above obtained extension $f_1 \in E^*$ is continuous since $\forall\, \epsilon > 0$, $x-y \in \epsilon U$ implies

$$|f_1(x) - f_1(y)| \leqq \epsilon\, p\left(\frac{x-y}{\epsilon}\right) \leqq \epsilon$$

(since $\frac{x-y}{\epsilon} \in U$ and $p(z) \leqq 1$ for $z \in U$).

Corollary 2. Let E be a vector space, $p \in$ SPEC E, $x_0 \in E$. Then $\exists\, f \in E^*$ such that $f(x_0) = p(x_0)$ and $|f(x)| \leqq p(x)$.

Proof. Apply the theorem for $M = \mathbf{C}x_0$ and $f: \lambda x_0 \to \lambda p(x_0)$.

Example of application. $E = \mathcal{C}(\mathbf{C}^n)$, $M = \mathcal{H}(\mathbf{C}^n)$ (entire functions), both with the topology of uniform convergence on compact sets K (semi-norms

$$p_K(f) = \sup_{x \in K} |f(x)|):$$

they are (FS) spaces, M is closed in E, and $\forall T \in \mathcal{H}'$ (analytic functional), $\exists\, \mu \in \mathcal{C}'$ (measure with compact support) such that $T(f) = \int f d\mu$ for all $f \in \mathcal{H}(\mathbf{C}^n)$.

C. Linear Operators: Definitions; the Closed Graph Theorem

1) Continuity. If E and F are two t.v.s., we denote by L (E,F) the space of linear continuous operators from E into F, and by L (E) the space L (E,E).

Theorem. If E and F are l.c.s., then a linear map $u \in L(E,F)$ iff $\forall p \in$ Spec F, $\exists\, q \in$ Spec E such that $p(u(x)) \leqq q(x)$ for all $x \in E$.

Indeed, this just means that $u \in L(E,F)$ iff $\forall$ convex circled absorbing 0-neighbourhood V in F (the gauge of which is p) $\exists$ U convex circled absorbing 0-neighbourhood in E (the unit semi-ball for q) such that $u^{-1}(V) \subset U$.

Equivalently, if Γ is a set of semi-norms generating the topology of E (the irreducible hull of which, in the sense of (I.B.6) and (I.D.4), is Spec E), then $u \in L(E,F)$ iff $\forall p \in$ Spec F, $\exists q_i \in \Gamma$ $(i = 1, \ldots, n < \infty)$ and a number $c > 0$ such that

$$p(u(x)) \leqq c \sup_i q_i(x), \quad \forall x \in E.$$

Remark. When F is the base field ($\mathbf{C}$, e.g.) then L(E,F) is usually written E', the topological dual of E.

2) Factorization. If u is any linear operator between two vector spaces E and F, it can be factorized in the following way:

$$E \xrightarrow{\pi} (E/\ker u) \xrightarrow{\bar{u}} \operatorname{Im} u \xrightarrow{i} F$$

where π is the canonical projection on the quotient space ($\ker u = u^{-1}(0)$), i the natural injection of $\operatorname{Im} u = u(E)$ into F, and $\bar{u}$ is one-to-one (bijective). If E and F are t.v.s., then π and i are (by definition) continuous, and so is $\bar{u}$ iff $u \in L(E,F)$. Similarly, we check easily that $\bar{u}$ is open (i.e., $\bar{u}(U)$ is open in F for every U open in E) iff u is open.

<u>Definition</u>. $u \in L(E,F)$ is said a topological <u>homomorphism</u> if $\bar{u}$ is a (t.v.s.) isomorphism.

Equivalently, $u \in L(E,F)$ is an homomorphism iff it is open.

<u>Lemma</u>. If u is a linear map from a t.v.s. E into a t.v.s. F and $\operatorname{Im} u$ is separated and finite-dimensional (in F), then u is an homomorphism iff $u \in L(E,F)$, or iff $\ker u$ is closed in E.

<u>Proof</u>. $u \in L(E,F) \Rightarrow \ker u$ is closed (since $\{0\}$ is closed in the separated space $\operatorname{Im} u$) $\Rightarrow E/\ker u$ is a separated t.v.s. of finite-dimension, and therefore the bijection $\bar{u}$ is an isomorphism. In particular, if E is any t.v.s., any $u \in E'$ is an homomorphism.

3) <u>Theorem</u>.

i) If E and F are complete metrizable t.v.s., then $u \in L(E,F)$ is an homomorphism iff $\operatorname{Im} u$ is <u>closed</u> in F; and, if not, $\operatorname{Im} u$ is meager in its closure in F, $\overline{\operatorname{Im} u}$.

ii) If E and F are (locally convex) (LF) spaces and $u \in L(E,F)$, then $\operatorname{Im} u = F$ (u is onto, or surjective) implies that u is an homomorphism.

[Part i) is known as the <u>Banach homomorphism theorem</u>, and Part ii) as the <u>Dieudonné-Schwartz theorem</u>; for Part i), cf., e.g., Schaefer; we shall sketch here the proof of Part ii) as a corollary of i).]

Proof of ii). Write $E = \varinjlim E_m$, $F = \varinjlim F_n$, where E_m and F_n are (F) spaces (hence Baire spaces). The space $G_{mn} = E_m \cap u^{-1}(F_n)$ is closed in E_m, hence complete and metrizable; now

$$F_n = \bigcup_m u(G_{mn})$$

since $u(E) = F$, and thus $u(G_{mn})$ is non-meager in F_n for some $m = \mu$; therefore, from i), $u(G_{\mu n}) = F_n$. Now $\forall n$, for any 0-neighbourhood U in E, $U \cap G_{\mu n}$ is a 0-neighbourhood in $G_{\mu n}$ and (from i)) so is its image by u in F, which is contained in $u(U) \cap F_n$—and therefore $u(U)$ is a 0-neighbourhood in F.

4) Closed graph theorem. We shall give several formulations of this theorem. The first one is a corollary of the Banach homomorphism theorem.

Theorem. Let E and F be complete metrizable t.v.s. or (LF) l.c.s. Then a linear map $u: E \to F$ is continuous iff its graph $\Gamma(u) = \{(x, u(x)); x \in E\}$ is a closed subspace of $E \times F$.

Proof. It is trivial that if $u \in L(E,F)$, then $\Gamma(u)$ is closed. Conversely, $\Gamma(u)$, being closed, is metrizable and complete; the projection $(x, u(x)) \to x$ is a bijective continuous map from $\Gamma(u)$ onto E, and therefore open from the last theorem, and the reciprocal map (hence also u) is continuous.

Generalizations. a) If E is ultrabornological and F is obtained by countable inductive and/or projective limits from Banach spaces, then a linear map $u: E \to F$ is continuous iff its graph is closed. [This includes the case of E and F being (LF) spaces.]

5) Lemma. If every closed map from E into every Banach space F is continuous, then E is barreled.

Proof (due to Mahowald) of the lemma. Let T be any barrel in E and π the projection $E \xrightarrow[\pi]{} \hat{E}_T$. Since T is closed,

$$T = \{x \in E; \|\pi(x)\| \leqq 1\}$$

and therefore will be a 0-neighbourhood (in view of the hypotheses) if the graph Γ of π in $E \times \hat{E}_T$, is closed. Since $\pi(E)$ is dense in $\hat{E}_T$, $\forall y_0 \in \hat{E}_T$, $\exists y_1 \in \pi(E)$ such that $\|y_0 - y_1\| \leqq \epsilon > 0$ given. The set $A = \{x \in E; \|\pi(x) - y_1\| \leqq \epsilon\}$ is a translate of ϵT, and therefore closed. Now if $(x_0, y_0) \notin \Gamma(\pi)$, $\|\pi(x_0) - y_0\| > 2\epsilon$ for some ϵ (which we shall use to pick y_1), and thus $x_0 \in W$, the complement (open) of A in E, and (x_0, y_0) belongs to the open set $W \times B_\epsilon$, where B_ϵ is the open ball $\{y \in \hat{E}_T; \|y - y_1\| < \epsilon\}$; therefore, the complement of $\Gamma(\pi)$ in $E \times \hat{E}_T$ is open, q.e.d.

Remark. Other extensions of both the homomorphism and the closed graph theorem exist, which make use of the notion of duality. We shall state the major ones later (Lecture III). Let us only remark now that, in view of the last lemma, the hypothesis that E is barreled, in the closed graph theorem, cannot be seriously weakened (in the category of l.c.s.).

LECTURE III
SPACES OF LINEAR AND BILINEAR MAPS. EQUICONTINUITY. DUALITY.

A. Topological Vector Spaces of Linear Operators, Equicontinuity, and the Banach-Steinhaus Theorem

1) Topologies on L(E,F). Let E, F be two l.c.s., and Σ a collection of bounded sets of E. We call Σ-topology on L(E,F), and hence $L_\Sigma(E,F)$ the corresponding space, the topology of uniform convergence over elements of Σ. This is a locally convex topology on the vector space L(E,F) that can be defined by the semi-norms (where $B \in \Sigma$ and $p \in$ Spec F are arbitrary)

$$p_B(u) = \sup_{x \in B} p(u(x)),$$

and a base of 0-neighbourhoods of which is given by the sets (where $B \in \Sigma$ and $V \in \mathcal{U}$, a base of 0-neighbourhoods in F, are arbitrary): $M(B,V) = \{u \in L(E,F);\ u(B) \subset V,\ B \in \Sigma,\ V \in \mathcal{U}\}$. The topology is thus separated if F is separated and

$$\bigcup_{B \in \Sigma} B$$

is total in E, and does not change if the B's are replaced by all their homothetics, or their convex closed circled hulls.

Remarks.

a) One can also define Σ-topologies for vector spaces L of mappings from a topological space E into a t.v.s. F, the M(B,V) defining a 0-neighbourhood base (Σ being a priori any family of subsets of E): iff u(B) is bounded in F for all $u \in L$ and $B \in \Sigma$, this topology will be a vector space topology.

b) It can be shown that if E is barreled, F quasi-complete,

$$\bigcup_{B \in \Sigma} B = E,$$

then $L_\Sigma(E,F)$ is also quasi-complete (cf. Schaefer, e.g.).

c) In the literature, the gothic capital $\mathfrak{S}$ is usually used instead of Σ.

2) Major examples.

i) Σ = finite sets. We get the topology of simple (pointwise) convergence, the space being often denoted L_s.

ii) Σ = convex circled compact [resp. precompact] sets: we get the compact [resp. precompact] convergence, and write L_c (in the precompact case].

iii) Σ = all bounded sets: we get the bounded convergence, and write L_b.

[The topology of precompact convergence is preferred to the more usual compact convergence when the space E is not complete].

For instance, in Banach or Hilbert spaces, L_b is the uniform topology, defined by the operator norm, while L_s is the so-called "strong" topology. Using the dual space F' of F, a "weak" topology can also be defined in the general case like in the Hilbert space case (e.g., $u_\alpha \to 0$ iff $f'(u_\alpha(x)) \to 0$ for all $x \in E$ and $f' \in F'$), but it is not (in general) a Σ-topology in the above sense.

Definition. If E is a l.c.s., $E'_\beta = L_b(E, \mathbf{C})$ is called the strong dual, and $E'_\sigma = L_s(E, \mathbf{C})$ the weak dual [do not mix the strong or weak topology on the dual with, e.g., the "strong" operator topology introduced above!].

Lemma. A subset $H \subset L_\Sigma(R, G)$ is bounded iff either for any 0-neighbourhood $V \subset F$,

$$\bigcap_{f \in H} f^{-1}(V)$$

absorbs every $B \in \Sigma$, or $\forall B \in \Sigma$,

$$\bigcup_{f \in H} f(B)$$

is bounded in F.

The proof is trivial (remember that a set is said bounded iff it is absorbed by every 0-neighbourhood, and that

$$\lambda M(B, V) = M(B, \lambda V)).$$

3) Equicontinuous subsets of L(E,F).

Definition. A set H of linear maps from a t.v.s. E into another F is said equicontinuous if $\forall$ 0-neighbourhood V in F, there exists a 0-neighbourhood $U \subset E$ such that $f(U) \subset V$ for all $f \in H$.

Remark. This is the same definition as for equicontinuous sets of real functions. Due to the vector structure, it will be enough here to have equicontinuity at 0. It is also obvious that a subset $H \subset L(E,F)$ is equicontinuous iff for any 0-neighbourhood V in F,

$$\bigcap_{f \in H} f^{-1}(V)$$

is a neighbourhood in E [or, equivalently, there exists a 0-neighbourhood U in E such that

$$\bigcup_{f \in H} f(U) \subset V].$$

Therefore an equicontinuous subset of L(E,F) is bounded for any Σ-topology. The converse is false in general. However:

Theorem. If E and F are two l.c.s., with E barreled, then any simply bounded subset H (i.e., bounded in $L_s(E,F)$) is equicontinuous.

Proof. Let V be any closed circled convex 0-neighbourhood in F;

$$W = \bigcap_{f \in H} f^{-1}(U)$$

is also closed convex circled in E; since H is bounded in $L_s(E,F)$ from the last lemma, W is also absorbing, thus a barrel, and a 0-neighbourhood if E is barreled.

Remark. If E and F are two t.v.s., with E a Baire space, the same result holds. Indeed, let V be any 0-neighbourhood in F and choose V_1 closed balanced 0-neighbourhood such that $V_1 + V_1 \subset V$;

$$W = \bigcap_{f \in H} f^{-1}(V)$$

is closed balanced absorbing, and therefore

$$E = \bigcup_{n=1}^{\infty} nW$$

and (since E is Baire) W has an interior point, whence $U = W + W$ is a 0-neighbourhood and since $f(U) \subset V$ for all $f \in H$, H is equicontinuous. One can show that any Baire l.c.s. is barreled, but the converse is false ((LF) spaces need not be Baire, but are barreled).

4) Corollary (principle of uniform boundedness). Let $H \subset L(E,F)$, E and F normed, and suppose that $\sup\{||f(x)||, f \in H\} < \infty$ for every $x \in M$, some non-meager subset of E. Then $\sup\{||f||, f \in H\} < \infty$.

Indeed, the vector subspace E_M generated by M is a dense Baire subspace of E; therefore, the (bounded in $L_s(E_M,F)$) set H_0 of restrictions of the $f \in H$ to E_M is equicontinuous, and thus bounded in norm (bounded in $L_b(E_M,F)$). But the restriction to E_M

is an isomorphism $L_b(E,F) \to L_b(E_M,F)$ since the unit ball of E_M is dense in that of E, and therefore H is bounded in $L_b(E,F)$.

Banach-Steinhaus theorem. Let E and F be two l.c.s. with E barreled, or t.v.s. with E Baire, and $\mathfrak{F}$ be a filter in $L_s(E,F)$, either with a countable base or containing a bounded subset, and converging pointwise to a map $f_1 \in F^E$ (the set of all maps from the set E to the set F). Then $f_1 \in L(E,F)$ and the convergence is uniform over every precompact subset of E.

Sketch of the proof. Let $\Phi \in \mathfrak{F}$ be bounded in $L_s(E,F)$ and Φ_1 its closure in F^E (endowed with the topology of pointwise convergence). Then Φ is equicontinuous; moreover, it is easily seen that the space of linear maps from E to F (without topology) is closed in F^E; and therefore the elements of Φ_1 are linear and, due to the equicontinuity of the maps $f \to f(x)$ from F^E to $F (x \in E)$, continuous and Φ_1 is equicontinuous in L(E,F). In addition, one can see that on equicontinuous subsets, the topologies induced by $L_c(E,F)$ and $L_s(E,F)$ coincide, whence the theorem in the case of a (simply) bounded filter, and in particular for an elementary filter (section filter of a sequence), and therefore also for a filter with countable base (which is the intersection of all finer elementary filters).

Corollary. If E and F are Banach spaces, $M \subset E$ is not meager, $\{f_n\} \subset L(E,F)$ and $\{f_n(x)\}$ is a Cauchy sequence for all $x \in M$, then f_n converges to some $f \in L(E,F)$ in the topology of $L_c(E,F)$.

Indeed, from the principles of uniform boundedness, $\{f_n\}$ is bounded in $L_b(E,F)$, hence equicontinuous; moreover, if we denote by E_M the linear hull of M (which is a Baire space), from the previous theorem, f_n converges to some $\tilde{f}$ in $L(E_M,F)$ which has a unique extension $f \in L(E,F)$, and $\{f_n, f\}$ is still equicontinuous, whence the result.

B. Bilinear Mappings; Hypocontinuity

1) Definitions. If E_1, E_2, F are three vector spaces, a map $f: E_1 \times E_2 \to F$ is said bilinear whenever the associated maps $E_1 \times \{x_2\} \to F$ and $\{x_1\} \times E_2 \to F$ (for all fixed $x_1 \in E_1$ and $x_2 \in E_2$) are linear. If E_1, E_2, and F are t.v.s., the map f is continuous iff it is continuous at (0,0). It is called separately continuous if all the above-mentioned linear maps are continuous. Obviously, (jointly) continuous bilinear maps are separately continuous, but the converse is false in general [analogy for functions of two real variables: the function $xy/(x^2+y^2)$, defined as 0 in (0,0), is separately continuous but not jointly continuous at this point].

Moreover, this leaves place for notions intermediate between separate and joint continuity, grouped under the name of hypocontinuity.

We shall say that a family Σ of subsets of a l.c.s. is saturated if:

a) $A \in \Sigma$, $B \subset A \Rightarrow B \in \Sigma$, $\lambda A \in \Sigma$ (λ: scalar).

b) $A_1, \ldots, A_n \in \Sigma \Rightarrow$ the closed convex hull of $A_1 \cup \ldots \cup A_n$ belongs to Σ.

c) All one-point subsets $\{x\}$, $x \in E$, belong to Σ.

[Often—but not always—Σ will be composed of bounded sets; the condition c) will ensure us later that separate continuity is included.]

Definition. If Σ_1 and Σ_2 are two (e.g.) saturated families in l.c.s. E_1, E_2 (respectively), we shall say that a bilinear map $f: E_1 \times E_2 \to F$ is (Σ_1, Σ_2) hypocontinuous if its restrictions to $A_1 \times E_2$ and $E_1 \times A_2$ are continuous, for all $A_1 \in \Sigma_1$ and $A_2 \in \Sigma_2$.

2) Examples.

a) Σ_1 and Σ_2 consists of all finite-dimensional bounded sets: we get the separate continuity defined before.

b) Σ_1 or Σ_2 consists of all subsets (the other being arbitrary): this is (joint) continuity.

c) If Σ_2 (e.g.) consists of all finite-dimensional bounded subsets, we shall speak (for simplicity) of Σ_1-hypocontinuity. Then (Σ_1, Σ_2) hypocontinuity is equivalent to Σ_1-hypocontinuity and Σ_2-hypocontinuity.

d) Other important families considered: Σ_1 and Σ_2 are all bounded sets of E_1 and E_2 (respectively); when one speaks of "hypocontinuity" without additional specification, one usually refers to this kind (bi-bounded) of hypocontinuity. One also sometimes takes for Σ_1 and Σ_2 all sets contained in convex circled compact sets (cf. Schwartz, Reference 10), or for Σ_1 all bounded sets and for Σ_2, when E_2 is considered as the dual of some l.c.s., all equicontinuous sets (cf. Bruhat, Reference 11).

3) Remarks.

a) If Σ_1 is a (saturated) family of bounded sets in E_1, a bilinear map $f: E_1 \times E_2 \to F$ is Σ_1-hypocontinuous iff either:

i) $\forall W$, 0-neighbourhood in F, and $\sigma \in \Sigma_1$, $\exists V$, 0-neighbourhood in E_2 such that $f(\sigma \times V) \subset W$; or

ii) $\forall \sigma \in \Sigma_1$, the set of linear maps $\{x \to f_x = (y \to f(x,y));\ x \in \sigma\}$ is an equicontinuous subset of $L(E_2, F)$; or

iii) $y \to f_y = (x \to f(x,y))$ is a continuous linear map $E_2 \to L_{\Sigma_1}(E_1, F)$.

b) Saturated families are not exactly necessary to define hypocontinuity: it is enough to take a "base" of such a family (all "bases" having the same "saturated hull" define the same notion).

c) For a family H of bilinear maps $E_1 \times E_2 \to F$, one can similarly define the notion of (Σ_1, Σ_2) equihypocontinuity (and in particular of separate, or of joint, equicontinuity).

We shall denote by $B(E_1 \times E_2; F)$ the family of all bilinear maps.

4) Theorem. If E_1 and E_2 are metrizable t.v.s., with E_1 Baire or E_1 barreled and F l.c.s., then every separately equicontinuous family $H \subset B(E_1 \times E_2; F)$ is equicontinuous.

(For a proof, c.f. e.g. Schaefer where a proof of the following intuitively obvious extension theorem can also be found):

Extension theorem. If E_i is dense in $\overline{E}_i$, Σ_i is a family of bounded sets in E_i such that $\overline{\Sigma}_i = \{$closures in $\overline{E}_i$ of the $\sigma \in \Sigma_i\}$ covers $\overline{E}_i$, and F is quasi-complete, then every (Σ_1, Σ_2) hypocontinuous map f of $B(E_1 \times E_2; F)$ has a unique extension $\overline{f} \in B(\overline{E}_1 \times \overline{E}_2; F)$ which is (Σ_1, Σ_2) hypocontinuous.

5) Topologies on spaces of bilinear maps. Let G be a subspace of $B(E_1 \times E_2, F)$, Σ_i be a family of bounded subsets of E_i; then G is a t.v.s. under the topology of (Σ_1, Σ_2) convergence (uniform convergence over sets $A_1 \times A_2$, with $A_i \in \Sigma_i$) iff for all $f \in G$, $f(A_1 \times A_2)$ are bounded in F. This is, e.g., the case for the space $\mathcal{B}(E_1 \times E_2, F)$ of continuous bilinear maps. If E_1 and E_2 are l.c.s., Σ_1 a saturated family of bounded subsets of E_1 such that $\overline{A}_1$ is complete for all $A_1 \in \Sigma_1$, Σ_2 a total family of bounded sets in E_2, then the space $\beta(E_1 \times E_2, F)$ of separately continuous bilinear maps is a l.c.s. with the (Σ_1, Σ_2) topology [cf. Schaefer].

C. Duality

1) Vector spaces and duality. Let E be a vector space. Its algebraic dual E^* is the set of all linear maps $E \to \mathbb{C}$, and this is a vector space.

On $E \times E^*$, we have then a canonical bilinear form, namely the map $(x, x^*) \to x^*(x) \in \mathbb{C}$. Of course, $\forall x^* \neq 0$, $\exists x$ such that $x^*(x) \neq 0$; and also for all $x \neq 0$, $\exists x^*$ such that $x^*(x) \neq 0$ (take a basis of E containing x and define, e.g., $x^*(x) = 1$, $x^*(x') = 0$ for any other base element x' of E, and extend by linearity). This leads us to:

Definition. If E and F are two vector spaces and B a bilinear form on $E \times F$, we shall say that they are in duality (relatively to B) if for all $x \neq 0$ in E, there exists $y \in F$ such that $B(x, y) \neq 0$, and if $\forall y \neq 0$ in F, $\exists x \in E$ such that $B(x, y) \neq 0$.

If $\dim E < \infty$ and $F \subset E^*$, we must have $F = E^*$; this is no longer true if $\dim E = \infty$. Via B, F can, however, always be identified with a subspace of E^*.

If (E, F; B) is a dual pair, the polar of a subset $M \subset E$ is defined as the set $M^o = \{y \in F;\ \mathrm{Re}\, B(x, y) < 1 \text{ for all } x \in M\}$. It is a convex subset of F, containing 0. When M is a subspace of E, $\mathrm{Re}\, \lambda B(x, y) < 1$ for all scalars λ and $x \in M$, whence $B(x, y) = 0$; the polar M^o is then the orthogonal (relatively to B) of the subspace M.

E can be endowed with a projective topology, noted $\sigma(E, F)$, namely the coarsest such that all linear forms $x \to B(x, y)$ (for $y \in F$) are continuous. This is a locally convex separated topology. One defines similarly $\sigma(F, E)$ on F. Polars M^o of sets $M \subset E$ are closed in $\sigma(F, E)$, and the bipolar M^{oo} of M is the closed (for $\sigma(E, F)$) convex hull of $M \cup \{0\}$.

2) If E is a t.v.s., its topological dual E' is the subspace of E^* consisting of all continuous linear maps $E \to \mathbf{C}$. It may be reduced to $\{0\}$ (e.g., the metric spaces l^p for $0 < p < 1$; cf. I.B.9). However, if E is a l.c.s., then E' is rich enough to separate points, i.e., $\forall x, y \in E$, $x \neq y$, $\exists f \in E'$ such that $f(x) \neq f(y)$. This is a consequence of the following:

Lemma. If $x_1, \ldots, x_n$ are linearly independent vectors in a l.c.s. E, then there are $f_1, \ldots, f_n$ in E' such that $f_\mu(x_\nu) = \delta_{\mu\nu}$.

Indeed, take the linear forms x_μ^*, on the n-dimensional space generated by the x_μ, defined by $x_\mu^*(x_\nu) = \delta_{\mu\nu}$ (dual basis) and extend them to some f_μ, continuous on E, by the Hahn-Banach theorem.

The topology $\sigma(E, F)$ defined by a (non-empty) subspace F of E^*, the algebraic dual of a given vector space E, will be locally convex iff F separates points in E. Therefore, if E is a l.c.s., the topology $\sigma(E, E')$, often called the weakened topology of E, is locally convex (we shall write E_σ for E endowed with this topology).

On the dual E' of a l.c.s., we can also define the topology $\sigma(E', E)$, called the weak topology of the dual, and we often write E'_σ when we endow E' with this topology. Notice that $E'_\sigma = L_s(E, \mathbf{C})$. Barrels in a l.c.s. E correspond by polarity to closed convex circled bounded subsets of E'_σ, and therefore E is barreled iff every bounded subset of E'_σ is equicontinuous (because, as one can see, the polars of any fundamental family of equicontinuous sets in E' form a neighbourhood base of 0 in E). Each $u \in L(E_\sigma, F_\sigma)$ [in particular each $u \in L(E, F)$] defines canonically a transposed map $u' \in L(F'_\sigma, E'_\sigma)$ by $u'y'(x) = y'(ux)$ when $x \in E$, $y' \in F'$.

3) Compatibility between duality and topology. If $(E, F; B)$ is a dual pair, a l.c. topology $\mathcal{C}$ on E is said compatible (or consistent) with the duality if the subspace $(E, \mathcal{C})'$ of E^* is identical to F (identified, via B, with a subspace of E^*); $\mathcal{C}$ is thus finer than $\sigma(E, F)$, which is the coarsest compatible. From the corollary to the separation theorem (II.B.1.b) we see that the closure of any convex subset of E is the same in all l.c. topologies compatible with the duality; and, from the above-mentioned property of polars of equicontinuous sets, any such topology $\mathcal{C}$ is the topology of uniform convergence on the class of $\mathcal{C}$-equicontinuous subsets of F. Now since $E'_\sigma = L_s(E, \mathbb{C})$, $\sigma(E', E)$ is induced on E' by the product topology $\mathbb{C}^E$; therefore, from the Tychonov theorem (a set in a product space is compact iff each component is compact) and the definition of equicontinuity, the weak closure Φ_1 of any equicontinuous subset $\Phi \subset E'_\sigma$ (which, from the proof of the Banach-Steinhaus theorem (A.4), is equicontinuous and identical with the closure of Φ in $\mathbb{C}^E$) is (weakly) compact—this result is known as the Alaoglu-Bourbaki theorem. Thus, any topology compatible with duality can be considered as that of uniform convergence over a saturated family of weakly relatively compact sets; the converse is also true (cf., e.g., Schaefer for a proof):

> Theorem (Mackey-Arens). A topology $\mathcal{C}$ on E is compatible with the duality $(E, F; B)$ iff it is the Σ-topology for a saturated family Σ, covering F, of $\sigma(F, E)$ relatively compact subsets of F.

The topology $\tau(E, F)$ of uniform convergence on all $\sigma(F, E)$ compact convex circled subsets of F is therefore the finest compatible l.c. topology, and called the Mackey topology. A l.c.s. E is called a Mackey space if its topology is $\tau(E, E')$. It can be seen that barreled and bornological spaces are Mackey spaces, and that the completion, product, l.c. direct sum and inductive limit of Mackey spaces is a Mackey space.

Moreover, it is easily checked that bounded subsets in F'_σ are bounded for the topology of uniform convergence over the family Σ of all convex circled, $\sigma(F, E)$ bounded complete, subsets of F, which contains the convex circled $\sigma(F, E)$-compact subsets of F. That is to say, every $\sigma(E, F)$ bounded subset of E is bounded for $\tau(E, F)$ and thus all (separated) l.c. topologies compatible with duality have the same bounded sets.

4) Reflexivity. The strong dual E'_β of a (separated) l.c.s. E is $L_b(E, \mathbb{C})$ (bounded convergence topology, which may happen to be non-compatible with duality), and the bidual E'' is defined as the dual of E'_β. E can be identified with a vector subspace of E'', but the

strong bidual $(E'_\beta)'_\beta$ does not always induce on E its original topology. The latter is, however, always induced by the so-called natural (and in general coarser) topology on E", $E''_\epsilon = L_e(E'_\beta, \mathbb{C})$, the topology of uniform convergence over the equicontinuous subsets of E' (which is identical with the strong bidual topology iff E is infrabarreled).

Definitions. A l.c.s. is said semi-reflexive if E = E" (algebraically) and reflexive if $E = E''_\beta$ (with identification of the topologies).

5) In E' we have the following inclusions of families of subsets: $Q \subset C \subset B \subset B_\sigma$ where Q = equicontinuous subsets, C = sets with weakly compact closed convex hull, B = strongly bounded sets, and B_σ = weakly bounded sets; and we have the following characterizations:

a) $Q = B_\sigma$ iff E is barreled.

b) Q = B iff E is infrabarreled (this follows from the definition).

c) Q = C iff E is a Mackey space.

Moreover, it is easily seen (using polars) that $B = B_\sigma$ if E is quasi-complete (weakly bounded subsets of E' are then strongly bounded), and from a) it follows that E is reflexive iff it is semi-reflexive and barreled.

A l.c.s. E is a Schwartz space (cf. II.A.1) iff $\forall A \in Q$, $\exists B \in Q$ convex circled weakly closed such that A is relatively compact in the Banach (since B is weakly complete) space E'_B; this implies that E is quasi-normable (cf. Grothendieck), i.e., that E'_B and E'_β define the same topology on A. If E is complete, then $L_c(E, \mathbb{C}) = E'_\gamma = \lim E'_B$ and thus the dual $E'_\beta = E'_\gamma$ is ultrabornological (e.g., the strong dual $\mathcal{D}'$ of $\mathcal{D}$ is ultrabornological). Note that a Schwartz space can also be defined as a quasi-normable space where bounded sets are relatively compact.

6) Theorem. A semi-reflexive space is quasi-complete. A l.c.s. E is semi-reflexive iff every bounded subset of E is relatively compact for the weakened topology $\sigma(E, E')$, and thus reflexive iff the latter holds and, in addition, it is barreled.

[For a proof, which uses standard l.c.s. arguments, cf. Schaefer, e.g.; the last result contains in particular the well-known criterium of reflexivity of a Banach space: that the unit ball be compact in the weakened topology.]

Definition. A Montel space (or (M)-space) is a barreled l.c.s. in which every bounded set is relatively compact.

Equivalent definition. A l.c.s. space is Montel iff it is reflexive and closed bounded sets are compact (for the strong topology)—the equivalence follows from the preceding remarks. It can be seen that the strong dual of a Montel space is still a (M)-space, and that a metrizable (M)-space is separable.

The name "Montel" comes from the fact that the compactness of bounded sets was well-known in classical mathematics for the (FM)-space of entire analytic functions $\mathcal{H}(\mathbb{C})$ [cf. II.B.3 example; one can also consider the space $\mathcal{H}(\Omega)$ of analytic functions in some open domain $\Omega \subset \mathbb{C}^n$, also with the topology of uniform convergence over compact subsets of Ω]: indeed, closed bounded sets in this l.c.s. were called by Paul Montel "normal families of analytic functions," and he proved that every infinite sequence in such a family contains a subsequence converging to some analytic function, namely that in the family any filter (since this space is separable) has a limit point (which means compactness—cf. I.A.6).

7) Definition. A (DF) space is a l.c.s. E which has a countable base of bounded sets, and for which every bounded countable union of equicontinuous sets in E'_β is equicontinuous.

These spaces were introduced by Grothendieck, and have some important properties. For example, the strong dual of any metrizable l.c.s. is a (DF)-space, and the strong dual of a (DF) space is a (F) space. However, not every (DF) space is the dual of some (F). In fact, all normable spaces, and more generally all infrabarreled l.c.s. having a countable base of bounded sets, are (DF) spaces. But Fréchet-Schwartz spaces (FS) and their duals (DFS) are reflexive (they are Montel spaces), the two categories (FS) and (DFS) are dual, and in particular (DFS) spaces are (LF).

Lemma. Let u be a linear map from a (DF) space E into any l.c.s. F. If the restrictions of u to every bounded set of E are continuous, then u is continuous.

8) Duality and inductive or projective limits. One would expect that duality will exchange inductive and projective limits. This indeed is the case in many specific examples. However, in the general case of l.c.s., one has to be a little more careful. We shall here limit ourselves to quote the following result:

Theorem.

a) If $E = \lim \operatorname{ind} E^\alpha$, then $E'_\sigma = \lim \operatorname{proj} (E^\alpha)'_\sigma$ (weak topologies, the defining maps being the adjoint maps).

b) If in a) the E^α are a sequence of reflexive (DF) spaces, the same result holds for the strong duals.,

c) If $E = \lim \operatorname{proj} g^{\beta\alpha} E^\alpha$, the projection of E on each E^α being dense, then (with eventually a slightly more general definition of inductive limits, and using also the adjoint maps but here the Mackey topologies) $E'_\tau = \lim \operatorname{ind} (E^\alpha)'_\tau$.

9) The general closed graph and homomorphism theorems.

Definition. A l.c.s. F is said fully complete (cf. Reference 13), or a Ptak space, when a subspace $A \subset F'_\sigma$ is closed iff $A \cap C$ is (weakly) closed in C for every equicontinuous subset C of F'. It is said almost fully complete, or B_r-complete, when every dense subspace $A \subset F'_\sigma$ such that $A \cap C$ is closed in C for all equicontinuous $C \subset F'$ is identical with E'.

It is trivial that fully complete implies almost fully complete, and one can see that either of them implies complete. Closed subspaces of either of them are of the same kind, and it follows from the next theorem (Part i)) that every separated quotient of fully complete spaces are fully complete (this need not be true for complete spaces; and this holds for every almost fully complete space iff all of them are fully complete).

Theorem. Let E be a barreled space and F a l.c.s., then:

i) If F is fully complete and $u \in L(F, E)$ surjective, u is a topological homomorphism.

ii) If F is almost fully complete, a linear map from E into F is continuous iff its graph is closed.

For a proof, see, e.g., Schaefer. The result of ii) holds also (cf. Reference 13) if we suppose that E has an inductive topology defined by a family of Baire l.c.s. (which implies it is barreled) and F is a Hausdorff inductive limit of a sequence of fully complete spaces.

LECTURE IV
NUCLEARITY AND TOPOLOGICAL TENSOR PRODUCTS. EXAMPLES AND APPLICATIONS.

A. Topological Tensor Products

1) Tensor products. Let E_1 and E_2 be vector spaces. Their tensor product $E_1 \otimes E_2$ is defined by the universal property that any bilinear map from $E_1 \times E_2$ into a vector space F can be factorized as a "canonical" bilinear map χ: $E_1 \times E_2 \rightarrow E_1 \otimes E_2$ composed with a linear map $E_1 \otimes E_2 \rightarrow F$. Now if E_1 and E_2 are (Hausdorff) l.c.s. and Σ_1, Σ_2 saturated families of subsets of E_1, E_2 (respectively), one can put on the vector space $E_1 \otimes E_2$ the finest l.c. topology for which the canonical bilinear map χ is (Σ_1, Σ_2) hypocontinuous (cf. III.B) (it is a projective topology); and then one can define the completion (resp. quasi-completion) of $E_1 \otimes E_2$ endowed with this topology, that is usually denoted by $E_1 \hat{\otimes}_{\Sigma_1\Sigma_2} E_2$ (resp. $E_1 \tilde{\otimes}_{\Sigma_1\Sigma_2} E_2$). The most important of these topologies is the so-called projective tensor product topology π,

for which Σ_1 (or Σ_2) consists of all the subsets; it is the finest l.c. for which χ is continuous.

Other topologies are, e.g., the inductive tensor-product ι, the finest l.c. for which χ is separately continuous (Σ_1 and Σ_2 consist of finite-dimensional bounded sets), and the β (resp. γ) topology, for which Σ_1 and Σ_2 consist of all bounded (resp. subsets of circled convex compact) sets. In order of increasing finess, we have obviously $\pi \leqq \beta \leqq \gamma \leqq \iota$. In the notations of III.B we have the following:

Theorem.

$$(E_1 \otimes_\pi E_2)' = \mathcal{B}(E_1 \times E_2; \mathbb{C}) \text{ and } (E_1 \otimes_\iota E_2)' = \beta(E_1 \times E_2; \mathbb{C}).$$

Another very important tensor product topology, coarser than all the preceding ones, is not of this type. It is the ϵ-topology of biequicontinuous convergence, in which $E_1 \otimes E_2$ is considered as a subspace of $(E_1' \otimes E_2')^*$ by $(x \otimes y)(x' \otimes y') = x'(x)y'(y)$, and endowed with the topology of uniform convergence on subsets $S_1 \otimes S_2$ where S_1 (resp. S_2) is an arbitrary equicontinuous subset of E_1' (resp. E_2'). The l.c.s. $E \otimes_\epsilon F$ is a topological vector subspace of the l.c.s. $E \epsilon F = \beta_e(E_\gamma' \times F_\gamma'; \mathbb{C})$ of bilinear forms over $E_\gamma' \times F_\gamma'$ endowed with the biequicontinuous convergence topology, and (cf. Schwartz) is dense in it for any l.c.s. F iff E possesses the so-called approximation property that can be defined by the property that $E' \otimes E$, identified with the space of finite-rank operators, is dense in $L_c(E)$ (or equivalently that the identity map of E is a limit point of $E' \otimes E$). All known spaces have this property (and the question of the hypothetical existence of a l.c.s. not having this property can be reduced to Banach spaces); in particular, projective limits of Hilbert spaces (and their subspaces) have it.

2) The projective tensor product topology is easily defined in terms of semi-norms. Indeed, if we have $p \in \operatorname{Spec} E$ and $q \in \operatorname{Spec} F$, we define a semi-norm $r = p \otimes q$ on $E \otimes F$ by $r(z) = p(x)q(y)$ when $z = x \otimes y$, $x \in E$ and $y \in F$, and more generally $r(z) = \inf\{\Sigma_i p(x_i) q(y_i);\ z = \Sigma_i x_i \otimes y_i\}$; then $r \in \operatorname{Spec}(E \otimes_\pi F)$, and the set of all such semi-norms $p \otimes q$ generates the π topology on $E \otimes F$ (it is in fact enough to take p and q in some sets generating the topologies of E and F respectively, in which every couple of semi-norms has an upper bound).

Theorem.

a) If E and F are metrizable l.c.s., every element $z \in E \hat{\otimes}_\pi F$ can be written as the sum $z = \Sigma_n \lambda_n (x_n \otimes y_n)$ of an absolutely convergent series where $\Sigma |\lambda_n| < \infty$ (i.e., the sequence $\{\lambda_n\}$ of scalars belongs to the Banach space ℓ^1) and x_n, $x_n \to 0$ in E, F (respectively).

b) If E and F are Banach spaces, with F reflexive, then $E \hat{\otimes}_\pi F'$ can be identified with $(L_c(E, F))'$.

Remark. From the latter theorem follows that

$$L^1(\mathbb{R}^k) \hat{\otimes}_\pi L^1(\mathbb{R}^\ell) \approx L^1(\mathbb{R}^{k+\ell}),$$

where L^1 is the usual Banach space of Lebesgue (absolutely) integrable functions, but that this is false for the L^p spaces, $p>1$, and in particular for the Hilbert spaces L^2. More generally, if H_1 and H_2 are two Hilbert spaces, then the Hilbert (completed) tensor product $H_1 \hat{\otimes} H_2$ is strictly larger than $H_1 \hat{\otimes}_\pi H_2$. Indeed, elements of the former are still sums $z = \Sigma\lambda_n(x_n \otimes y_n)$ but with $\Sigma|\lambda_n|^2 < \infty$; the former corresponds to Hilbert-Schmidt operators between H_1 and H_2, while the latter corresponds to operators of the trace class, which we shall call also nuclear.

3) Integral forms and maps. Let E, F be two l.c.s.; since, on $E \otimes F$, we have $\epsilon \leqq \pi$, the dual I(E, F) of $E \hat{\otimes}_\epsilon F$ is a subspace of the space $\mathcal{B}(E \times F, \mathbb{C})$ of continuous bilinear forms. Its elements were called by Grothendieck integral bilinear forms and can be (cf. Schaefer, e.g.) represented in the form $z \to u(z) = \int z_0(x', y')d\mu(x', y')$ where the integral is taken over the product $S \times T$ of two suitable closed equicontinuous (hence compact) subsets of (respectively) E'_σ, F'_σ, with respect to some measure μ, and where z_0 is the restriction to $S \times T$ of $z \in E \hat{\otimes}_\epsilon F$ considered as a continuous bilinear form over $E'_\sigma \times F'_\sigma$ (the latter is obvious when $z \in E \otimes F$, hence for uniform limits over each product $S \times T$ of elements of $E \otimes F$, i.e., for any $z \in E \hat{\otimes}_\epsilon F$).

Any element $u \in I(E, F)$ defines in the usual manner a map $v \in L(E, F'_\sigma)$, called an integral linear map, that is of the form (integration over $S \times T$):

$$E \ni x \to v(x) = \int x'(x)y' d\mu(x', y') =$$

$$= \left[y \to \int x'(x)y'(y) d\mu(x', y') \right] \in F'_\sigma .$$

B. Nuclear Spaces and Maps

1) Nuclear maps in Banach spaces. As mentioned above, if E and F are l.c.s., any element $g \in E' \otimes F$ defines a $f \in L(E, F)$ by

$$x \to f(x) = \sum_{i=1}^{r} f_i(x) y_i$$

if

$$g = \sum_{i=1}^{r} f_i \otimes y_i;$$

these f are called finite-rank operators (f(E) is finite-dimensional) and the map τ: g → f is injective. Now if E and F are Banach spaces, then τ is continuous on $E'_\beta \otimes_\pi F$ and can therefore be extended to an injection $\tau: E'_\beta \hat{\otimes}_\pi F \to L(E, F)$. The elements of $\operatorname{Im}\tau$ are called <u>nuclear</u>.

<u>Definition</u>. If E and F are any l.c.s., a <u>map</u> $f \in L(E, F)$ is <u>compact</u> (resp. <u>bounded</u>) if there exists some 0-neighbourhood U in E, the image f(U) of which is relatively compact (resp. bounded) in F.

A bounded map f can be decomposed as follows (in the notations of II.A):

$$E \to E_U \xrightarrow{f_0} F_B \to F$$

where B is convex circled bounded and contains f(U), and $f_0 \in L(E_U, F_B)$. If F_B is complete, f_0 can be extended to $\bar{f}_0 \in L(\hat{E}_U, F_B)$ with the same factorization. A nuclear map between Banach spaces is compact (in Banach spaces, compact operators are limits in norm of finite-rank operators). This leads us to the general case:

2) <u>Nuclear maps in l.c.s</u>. The main definition and characterization are (cf., e.g., Schaefer):

<u>Definition</u>. If E and F are l.c.s., a map $f \in L(E, F)$ is nuclear if there exists U, convex circled 0-neighbourhood in E, such that f(U) is contained in a bounded set B with F_B complete, and such that the induced map $\bar{f}_0: \hat{E}_U \to F_B$ is a nuclear map between Banach spaces (we factorize f as above).

<u>Theorem</u>. $f \in L(E, F)$ is nuclear iff it is of the type

$$x \to f(x) = \sum_{n=1}^{\infty} \lambda_n f_n(x) y_n$$

where $\{\lambda_n\} \in \ell^1$, $\{f_n\}$ is equicontinuous in E', and $\{y_n\} \subset B$ convex circled bounded subset of F with F_B complete.

Corollary.

i) Nuclear maps are compact.

ii) If $F = G'_\sigma$ for some l.c.s. G and $\{y_n\}$ is equicontinuous in G', then the nuclear map $f \in L(E, G'_\sigma)$ is integral.

3) Nuclear spaces. These spaces play an important role in several domains of mathematical analysis (closely connected to physics), due, in particular, to the above-mentioned "trace" property $\Sigma |\lambda_n| < \infty$ of nuclear maps. For proofs in the following, we refer to Schaefer and Grothendieck.

Definition. A l.c.s. E is said nuclear if there exists a base $\mathcal{U}$ of convex closed circled 0-neighbourhoods in E such that for each $V \in \mathcal{U}$ the canonical map $E \to E_V$ is nuclear.

Thus, a Banach space is nuclear iff it is finite-dimensional.

Theorem. A l.c.s. E is nuclear iff either of the following (equivalent) conditions holds:

o) Its completion is nuclear.

i) Every linear continuous map from E into any (B) space is nuclear.

ii) $\forall$ convex circled 0-neighbourhood U, $\exists V \subset U$ such that the canonical projection $E_V \to E_U$ is nuclear.

iii) For any l.c.s. F, the canonical injection $E \otimes_\pi F \to \mathcal{B}_e(E'_\sigma \times F'_\sigma; \mathbb{C})$ is a topological homomorphism (in which case the image is dense).

iv) The canonical map $E \hat{\otimes}_\pi \ell^1 \to E \hat{\otimes}_\epsilon \ell^1$ is a topological isomorphism (onto).

As a consequence of condition iv), one can see that a (F) space E is nuclear iff every unconditionally (= for any rearrangement of the terms) convergent series Σx_n is absolutely convergent (i.e., for any $p \in \operatorname{Spec} E$, $\Sigma p(x_n) < \infty$), which generalizes the well-known Dvoretzky-Rogers theorem from (B) spaces to (F) spaces. Condition iii) is in fact the original definition of Grothendieck. It expresses that, on the tensor product $E \otimes F$, the π and ϵ topologies coincide, and therefore (when E is nuclear) $I(E, F) = \mathcal{B}(E \times F; \mathbb{C})$. A nuclear space is also a Schwartz space, the map $E'_A \to E'_B$ mentioned in III.C.5 being nuclear (this is also a characteristic property); thus, a barreled quasi-complete nuclear space is a Montel space (but a complete nuclear space need not be barreled, and can have a non-quasi-complete strong dual). The following structure theorem is of great importance (and has been used as a definition by Gelfand and Vilenkin).

4) Theorem.

i) A Fréchet space E is nuclear iff $E = \text{limproj } g_{mn} H_n$, where the H_n are Hilbert spaces and the maps g_{mn} nuclear for all $m < n$ (it is a "countably Hilbert space").

ii) The topology of every nuclear space E can be generated by an infinite family of semi-norms, each of which originates from a positive semi-definite Hermitian form on $E \times E$ (for which there exists another semi-norm that is nuclear relatively to it, in the sense expressed by condition ii) of the preceding theorem).

iii) In particular, every complete nuclear space is some projective limit of a family of Hilbert spaces.

Permanence properties.

i) Subspaces, separated quotient spaces, products, countable l.c. direct sums, and thus projective and countable inductive limits, of nuclear spaces are nuclear.

ii) If E and F are two nuclear spaces, then $E \hat{\otimes}_\pi E$ $(= E \hat{\otimes}_\epsilon F)$ is nuclear.

iii) If E is a Fréchet space or a complete (DF) space, then E'_β is nuclear iff E is nuclear [but $(\mathbb{C}^d)'_\beta$ is not nuclear when $d > \aleph_0$, namely the condition that E be Fréchet is necessary].

iv) If E is a complete nuclear (DF) space and F a nuclear (F) or (DF) space, then $E \hat{\otimes}_\iota F$ is also nuclear.

v) If F is nuclear and E any l.c.s. [resp., any l.c.s. with E'_β nuclear; or if F is a nuclear (F) space and E a nuclear (DF) space], then $L_s(E, F)$, isomorphic to a subspace of the nuclear space F^E [resp. $L_b(E, F)$; or L_b and its strong dual] is nuclear.

Remarks.

a) If E and F are both Fréchet spaces, or barreled (DF) spaces, the projective and inductive (and thus all intermediate) tensor product topologies on $E \otimes F$ are identical and give barreled metrizable (resp. (DF)) spaces. More generally, if both are (DF) spaces, then $\pi = \beta$ and $E \otimes_\pi F$ is (DF) also, and if both are barreled then $\beta = \gamma = \iota$.

b) If E and F are either (DF) spaces, or (F) spaces with E nuclear, then the strong topology on $(E \otimes_\pi F)' = \mathcal{B}(E \times F; \mathbb{C})$ is is identical to the a priori coarser bibounded topology (uniform convergence over products of bounded sets, in $E \times F$). If E and F are (DF) spaces with E nuclear, then $(E \hat{\otimes}_\pi F)'_\beta \approx E'_\beta \hat{\otimes}_\pi F'_\beta$ and $(E \hat{\otimes}_\pi F)''_\beta \approx E''_\beta \hat{\otimes}_\pi F''_\beta$, and thus if E and F are nuclear (F) spaces, then $(E \hat{\otimes}_\pi F)'_\beta \approx E'_\beta \hat{\otimes}_\pi F'_\beta = L_b(E, F'_\beta)$.

5) We shall end this section by the following abstract form of the famous kernel theorem (in French "théorème des noyaux," which explains the choice of the word "nuclear").

Kernel theorem. If E is nuclear and F a l.c.s., then every continuous bilinear form $f \in \mathcal{B}(E \times F; \mathbb{C})$ originates already from some space $E'_A \hat{\otimes}_\pi F'_B$, where A and B are some equicontinuous subsets of E' and F' (respectively), that is to say can be written as

$$(x, y) \to f(x, y) = \sum_{i=1}^{\infty} \lambda_i x'_i(x) y'_i(y)$$

where $\{\lambda_i\} \in \ell^1$ and $\{x'_i\}$, $\{y'_i\}$ are equicontinuous sequences in E', F', converging to zero in E'_A, F'_B (respectively). In particular (cf. the end of Remark b) above), if $E = \varprojlim H_n$ and $F = \varprojlim F_p$ are (F) spaces, then for every (separately) continuous bilinear map f, there exist integers m and q and a nuclear operator (kernel) $A \in L(H_m, F'_q)$ such that $f(x, y) = (Ax)(y)$; when the F_p are also Hilbert spaces, $x \otimes y$ can be viewed as a finite-rank operator in $L(H'_m, F_q)$ with transposed $(x \otimes y)' \in L(F'_q, H_m)$, and one has then $f(x, y) = \mathrm{Tr}((x \otimes y)'A)$.

C. Examples and Applications

1) Some important function spaces.

a) $\mathcal{C}_K(\mathbb{R}^n)$, Banach space of continuous functions with support in a compact set $K \subset \mathbb{R}^n$, and the norm

$$p_K(f) = \max_{x \in K} |f(x)|.$$

$$\mathcal{C}_0(\mathbb{R}^n) = \operatorname{limind}_K \mathcal{C}_K(\mathbb{R}^n),$$

the (LB) space of continuous functions with compact support, the dual of which is the space of (Radon) measures. This space is not to be mixed with the space $\mathcal{C}(\mathbb{R}^n)$, Fréchet space of continuous functions (with the semi-norms p_K), the dual of which is the space of measures with compact support.

b) We have already seen the spaces $\mathcal{D}^m_\ell$, $\mathcal{D}_\ell$, $\mathcal{D}^m$, $\mathcal{D}$, and their duals. $\mathcal{D}_\ell$ and $\mathcal{D}$, and their duals, are nuclear. $\mathcal{D}'$ (with its strong topology) is the (l.c.) space of distributions. A subspace $\mathcal{H}$ of $\mathcal{D}'$ is said a space of distributions when its topology is finer than that induced by $\mathcal{D}'$, and is said normal if, moreover, $\mathcal{D}$

is a dense subspace of $\mathcal{H}$ with a finer topology. The most important such spaces are $\mathcal{E}'$ and $\mathcal{S}'$, and their duals $\mathcal{E}$ and $\mathcal{S}$ defined as follows.

$\mathcal{E}$ is the nuclear (F) space of C^∞ functions on $\mathbb{R}^n$, with the topology of uniform convergence in all derivatives on any compact and (e.g.) semi-norms $p_{k\ell}(f) = \sup\{|D_m f(t)|;\ |t| < \ell,\ |m| < k\}$ where D_m is a differential monomial (of order $|m| = m_1 + \ldots + m_n$). Its dual $\mathcal{E}'$ is the space of distributions with compact support. [What has been said here on spaces of the type $\mathcal{C}$, $\mathcal{D}$ and $\mathcal{E}$ can be generalized when $\mathbb{R}^n$ is replaced by a manifold.]

$\mathcal{S}$ is the nuclear (F) space of C^∞ functions with rapid decrease at infinity, with the topology defined by (e.g.) the semi-norms $p_{k\ell}(f) = \sup\{|(1+|t|)^\ell D_m f(t)|;\ t \in \mathbb{R}^n,\ |m| < k\}$. Its dual is the space of tempered distributions.

c) We have already defined the nuclear (F) space $\mathcal{H}(\Omega)$ of analytic functions in some domain $\Omega \subset \mathbb{C}^n$, with semi-norms

$$p_K(f) = \sup_K |f| .$$

For instance, $\mathcal{H}(\mathbb{C}^n)$ is a closed subspace of $\mathcal{E}(\mathbb{R}^{2n})$. Its dual is the space $\mathcal{H}'$ of analytic functionals. If K is a compact in $\mathbb{C}^n$, one defines also $\mathcal{H}(K) = \varinjlim \mathcal{H}(\Omega)$ for $\Omega \supset K$; we get a space of type (DF), the dual of which is a nuclear (F) space $\mathcal{H}'(K)$. One can also define the space $\mathcal{H}(\Omega; E)$ of analytic functions f with values in some (quasi-complete) l.c.s. E in the usual manner (derivability or Cauchy integral, e.g.) or equivalently (cf. Grothendieck) by the property that $\forall e' \in E'$, the function $z \to \langle f(z), e' \rangle$ is in $\mathcal{H}(\Omega)$.

2) Vector-valued distribution and the kernel theorem (cf. Schwartz for more details).

a) A distribution with value in a l.c.s. E (usually taken quasi-complete) is an element of the space $L_b(\mathcal{D}, E)$, often noted $\mathcal{D}'(E)$. Notice that $L_b(\mathcal{D}, E) \approx L_c(\mathcal{D}, E)$, and is also isomorphic to $\mathcal{D}' \epsilon E$, and to $\mathcal{D}' \hat{\otimes}_\pi E$ if E is quasi-complete. A distribution of order $\leqq m$ is an element of $L_c(\mathcal{D}^m, E) \approx \mathcal{D}'^m_\gamma \hat{\otimes}_\epsilon E$ (when E is quasi-complete), where $\mathcal{D}'^m_\gamma = L_c(\mathcal{D}^m, \mathbb{C})$. If $\mathcal{H}$ is a distribution space, we denote $\mathcal{H}(E) = \mathcal{H} \epsilon E \approx L_\epsilon(\mathcal{H}'_\gamma, E)$. When $\mathcal{H}$ is normal and has the so-called γ-topology of $(\mathcal{H}'_\gamma)'_\gamma$ (which is the case of most spaces), then $L_c(\mathcal{H}; E) \approx \mathcal{H}'_\gamma(E)$; moreover, if $\mathcal{H}$ is normal and E complete, then $T \in \mathcal{D}'(E)$ is in $L(\mathcal{H}, E)$ if it is continuous on $\mathcal{D}$ endowed with the topology induced by $\mathcal{H}$.

b) One should pay attention to one fact: a distribution $T \in \mathcal{H}(E)$ is scalarly in $\mathcal{H}$, i.e., for every $e' \in E'$, we have $\langle T, e' \rangle \in \mathcal{H}$; but the converse need not be true; it is true for most spaces, but not all (spaces like $\mathcal{D}^m$, with finite m, are counter-examples); there even exist distributions $T \in \mathcal{D}'(E)$ that are scalarly in C and are not defined by a scalarly continuous function having values in E (but rather by some kind of δ-measure for instance); such "pathologies" may be quite common: general systems of imprimitivity, considered as operator-valued distributions, give rise to this kind of phenomenon (cf., e.g., Bruhat).

c) Questions of support give rise also to some pathologies. For instance, the space $\overline{\mathcal{E}}'(E)$ of vector-valued distributions with compact support, that can be endowed with the inductive-limit topology of the $\mathcal{E}'_K(E)$ (distributions with compact support in the compact set K), is in general a subspace with a (strictly) finer topology of $\mathcal{E}'(E)$, though both induce on $\mathcal{E}'_K(E)$ the same topology. When E is a (DF) space (e.g., a Banach space), they are isomorphic; if there exists a continuous norm on E, they coincide algebraically. This is no longer true even when E is an (F) space (not normed); e.g., the identity map of $L_c(\mathcal{E},\mathcal{E}) = \mathcal{E}'(\mathcal{E})$ has support in the whole space, thus is not in $\overline{\mathcal{E}}'(\mathcal{E})$, though scalarly it is a compact-support distribution (the scalarly compact supports "explode," namely their union covers the whole space). Such a phenomenon occurs also when considering general systems of imprimitivity because the space of differentiable vectors in a group representation is at best (when the representation space is Hilbert, or Fréchet) a (F)-space.

d) Similarly, the space $\mathcal{D}(E) = \mathcal{D}\hat{\otimes}_\pi E$ is not in general the space $\overline{\mathcal{D}}(E)$ of C^∞ mappings from $\mathbb{R}^n$ into E, with compact support, endowed with the topology of limind $\mathcal{D}_K(E)$. One can see (cf. Grothendieck) that if $E = \mathcal{D}(\mathbb{R}^\ell)$ then $\mathcal{D}(\mathbb{R}^n \times \mathbb{R}^\ell) = \mathcal{D}(\mathbb{R}^n)\hat{\otimes}_\iota \mathcal{D}(\mathbb{R}^\ell)$ and that if E is a (F) space, then $\mathcal{D}(E) \approx \mathcal{D}\hat{\otimes}_\iota E$ and is a (LF) space. However, the topology of $\mathcal{D}(\mathbb{R}^n, \mathcal{D}(\mathbb{R}^\ell))$ is coarser than that of $\mathcal{D}(\mathbb{R}^n \times \mathbb{R}^\ell)$ on these (algebraically identical) vector spaces. Similarly, one can see that in the general case the topology induced by $\overline{\mathcal{D}}(E)$ on $\mathcal{D} \otimes E$ is finer than π and coarser than the inductive topology ι. The spaces $\mathcal{D}(E)$ and $\overline{\mathcal{D}}(E)$ coincide under the same conditions on E as above. In particular if there exists a continuous (true) norm on a space E containing a non-normable (F) subspace, then $\mathcal{D}(E) = \overline{\mathcal{D}}(E)$ algebraically only and therefore $\mathcal{D}(E)$ is not barreled, and its strong dual is not quasi-complete (cf. Grothendieck, where $\mathcal{D}$ and $\overline{\mathcal{D}}$ are noted $\tilde{\mathcal{D}}$ and $\mathcal{D}$, respectively).

e) One calls <u>E-distribution</u> an element of the dual $(\overline{\mathcal{D}}(E))'$, and more generally <u>E-distribution with values in F</u> an element of $L(\overline{\mathcal{D}}(E), F)$. When E is a Fréchet space, every distribution with values in E' defines canonically an E-distribution, because it defines a separately continuous bilinear form on $\mathcal{D} \times E$, thus a continuous form on $\mathcal{D} \hat{\otimes}_\iota E = \bar{\mathcal{D}}(E)$ in this case. Conversely (in the general case) any E-distribution defines canonically (and obviously) an E'-valued distribution. One should not, however, mix the two notions, which are not completely equivalent (in the general case, an E'-valued distribution defines an E-distribution iff for every compact K, it transforms some 0-neighbourhood in $\mathcal{D}_K$ into an equicontinuous set of E').

f) The kernel theorem, classically can be written:

$$\mathcal{D}'_{x,y} \approx \mathcal{D}'_x(\mathcal{D}'_y) = \mathcal{D}'_x \hat{\otimes}_\epsilon \mathcal{D}'_y = \mathcal{B}_b(\mathcal{D}_x \times \mathcal{D}_y; \mathbf{C}).$$

In particular, the algebraic equality of the first and last terms means that to every separately continuous bilinear form f on $\mathcal{D}_x \times \mathcal{D}_y$ corresponds a distribution-kernel A such that one has (writing formally distributions as functions) $f(\varphi, \Psi) = \int A(x, y)\, \varphi(x)\, \Psi(y)\, dx\, dy$, for all $\varphi \in \mathcal{D}_x$ and $\Psi \in \mathcal{D}_y$. [The last equality of the kernel theorem uses the fact that, as a consequence of the Mackey-Arens theorem, cf. the end of III.C.3, the bounded sets in the reflexive space $\mathcal{D}$ are the same for all topologies compatible with the duality between $\mathcal{D}$ and $\mathcal{D}'$, and therefore

$$\mathcal{D}'_x \hat{\otimes}_\epsilon \mathcal{D}'_y = \mathcal{B}_e((\mathcal{D}'_x)'_\sigma \times (\mathcal{D}'_y)'_\sigma; \mathbf{C}) = \mathcal{B}_b(\mathcal{D}_x \times \mathcal{D}_y; \mathbf{C})$$

since $\mathcal{D}'$ is nuclear and barreled; notice that these separately continuous maps are in fact also hypocontinuous relatively to bounded sets. The proof uses the similar result for $\mathcal{E}'$, which follows from the end of Remark B.4.b; but is a little more complicated due to the fact that $\mathcal{D}$ is not an (F) space].

One has also

$$\mathcal{E}_{x,y} = \mathcal{E}_x \hat{\otimes} \mathcal{E}_y, \quad \mathcal{S}_{x,y} = \mathcal{S}_x \hat{\otimes} \mathcal{S}_y, \quad \mathcal{E}'_{x,y} = \mathcal{E}'_x \hat{\otimes} \mathcal{E}'_y,$$

$$\mathcal{S}'_{x,y} = \mathcal{S}'_x \hat{\otimes} \mathcal{S}'_y$$

(where $\hat{\otimes}$ means any tensor-product topology from ι to ϵ, since they are here all the same) and we have seen that $\mathcal{D}_{x,y} = \mathcal{D}_x \hat{\otimes}_\iota \mathcal{D}_y$ (the $\pi = \epsilon$ topology gives the same vector space, but with a coarser topology); one has also $\bar{\mathcal{E}}'_x(\mathcal{E}'_y) = \mathcal{E}'_x(\mathcal{E}'_y)$ since $\mathcal{E}'$ is a (DF) space.

To end this part, we shall mention the following curiosity: The canonical injection $\mathcal{D} \to \mathcal{D}'$ can be associated to some kernel $\delta(y-x)$ belonging to $\mathcal{D}' \hat{\otimes}_\pi \mathcal{D}$, $\mathcal{E} \hat{\otimes}_\pi \mathcal{E}'$, etc; the map $y \to \delta(y-x)$ is an $\mathcal{E}'$-valued C^∞-function, but also a $\mathcal{D}$-valued distribution.

3) Applications to group theory. We shall mention these very briefly. (We already pointed at some of them in the preceding section.)

If T is a continuous representation of a Lie group G having a (regularly imbedded) normal abelian subgroup Γ in a l.c.s. E, then (provided T is so-called tempered on Γ, e.g., $||T(x)||$ polynomially bounded in $|x|$ when $x \in \Gamma$, if E is a Banach space) one can define for every $\varphi \in \mathcal{S}(\hat{\Gamma})$ the operator $P(\varphi) = \int_\Gamma \hat{\varphi}(x)\, T(x)\, dx$, where $\hat{\Gamma}$ is the Fourier dual of Γ and $\hat{\varphi}$ the Fourier transform of φ, dx the Haar measure of Γ, the integral converging (a priori) in $L_s(E)$. Defining, for $y \in G$, ${}^y\varphi(\xi) = \varphi({}^y\xi)$, where ${}^y\xi(x) = \xi(yxy^{-1})$ with $\xi \in \hat{\Gamma}$ and $x \in \Gamma$ (we write $\xi(x)$ the value in x of the character ξ), we shall have $P(\varphi_1\varphi_2) = P(\varphi_1)\, P(\varphi_2)$ and $T(y)P(\varphi)T(y^{-1}) = P({}^y\varphi)$. The map $(\varphi, a) \to P(\varphi)a$ will also be a continuous map $\mathcal{S}(\hat{\Gamma}) \times E \to E$ and define an $L_b(E)$-valued distribution P. If T is irreducible (there are no non-trivial closed subspaces invariant under T(G)), the support of this distribution, which is a generalized system of imprimitivity, will be just the closure of one orbit of $\hat{\Gamma}$ under G. From this, one can show that T is, in a sense, induced by an irreducible representation of the stabilizer of a point in this orbit (cf. Bruhat); the proof uses heavy machinery of topological tensor products, vector-valued distributions, etc. If we drop the temperedness condition, one can still (under some conditions) give a meaning to the formal Fourier transform $P(\xi) = \int_\Gamma \xi(x)T(x)dx$ even for non-unitary characters ξ; in this case the generalized system of imprimitivity will be some kind of operator-valued analytic functional. [In the classical case of unitary representation, the system of imprimitivity introduced by Mackey was a projector-valued measure over Borel sets in $\hat{\Gamma}$, satisfying similar multiplicative and quasi-invariance conditions.]

4) Nuclear direct sums. It can be shown, using the Hermite transformation (cf., e.g., Schwartz), that the space $\mathcal{S}$ is isomorphic to the space (s) of rapidly decreasing sequences $x = (x_n)$ (i.e., sequences the product of which by any sequence (n^k), for arbitrary fixed k, is bounded); $\mathcal{S}'$ is then isomorphic to the space (s') of slowly increasing sequences. [Notice that the product of any finite number of spaces (s) is still isomorphic to (s).] This space (s) can be viewed as a nuclear direct sum of $\aleph_0$ one-dimensional (and therefore nuclear) spaces, which is larger than the l.c. direct sum (the latter, (φ) in the notations of Köthe, consists of sequences with all but a finite number of terms equal to zero). The (nuclear) tensor product $(s) \otimes_\pi E$, where E is any nuclear space, can thus be seen as a nuclear direct sum of $\aleph_0$ copies of E, larger than the

l.c.-direct sum $(\varphi) \otimes_\pi E$. We can still go further defining nuclear direct sums of countably many nuclear (F) spaces (that need not be the same), by use of their representation as projective limits of Hilbert spaces. This construction, due to Flato and Simon, may be utilized to define a nuclear topology (different from the "natural" one) on the vector space of differentiable vectors for some direct sums of representations of a Lie group, provided the space of differentiable vectors for each component is nuclear (which is the case for irreducible unitary representations of, e.g., semi-simple groups, or Poincaré group for non-zero mass).

Suppose indeed we are given a sequence of nuclear (F) spaces

$$H^i_\infty = \varprojlim g^i_{mn} H^i_n = \bigcap H^i_n$$

where the H^i_n are Hilbert spaces $(i, m, n \in \mathbb{N})$ and the maps g^i_{mn} are Hilbert-Schmidt for $m < n$ (this is enough) namely, denoting by p^i_m the norm in H^i_m and by $\{e^i_{nk}\}$ $(k \in \mathbb{N})$ an orthonormal base of H^i_n, that

$$\sum_k p^i_m (g^i_{mn} e^i_{nk})^2 = (\beta^i_{mn})^2 < \infty .$$

We now define the spaces $H_n = \oplus_i H^i_n$ with the usual norm given by

$$p_n(\varphi_n)^2 = \sum_i p^i_n(\varphi^i_n)^2$$

when $\varphi_n = (\varphi^i_n) \in H_n$, and write $\tilde{H}_n$ for the same space with the "corrected" norm

$$\tilde{p}_n(\varphi_n)^2 = \sum_i (\alpha^i_n)^2 p^i_n(\varphi^i_n)^2 ,$$

where $\alpha^i_n > 0$ with $\sup \alpha^i_n < \infty$ —under the latter norm, the space is pre-Hilbert only, i.e., is not complete (in general). Now $g_{mn} = \oplus_i g^i_{mn}$ will not be Hilbert-Schmidt for $m < n$ as maps from H_n to H_m, but we can choose weights α^i_n so that they will be Hilbert-Schmidt as maps in $L(\tilde{H}_n, \tilde{H}_m)$; the condition for this reads:

$$\sum_{k,i} (\alpha^i_m)^2 p^i_m (g^i_{mn}(e^i_{nk}/\alpha^i_n))^2 = \sum_i (\beta^i_{mn})^2 (\alpha^i_m/\alpha^i_n)^2 < \infty ,$$

when $m < n$. If we write $\beta^i_n = \beta^i_{n-1,n}$, we can thus choose

$\alpha_n^i = i^n \beta_n^i \ldots \beta_1^i \alpha_0^i$, where $\{\alpha_0^i\}$ is a fastly enough decreasing sequence (if all spaces H_∞^i are the same, one can replace all β_n^i by 1 and thus take any element of (s)). The semi-norms $\tilde{p}_n$ will thus define on the vector space $H_\infty = \varinjlim g_{mn} H_n$ a new topology, giving a nuclear ("prefréchet") space $\tilde{H}_\infty$ that is larger than the l.c. direct sum of the H_∞^i. The identity $H_\infty \to \tilde{H}_\infty$ maps H_∞ onto a dense subspace of the completion $\hat{H}_\infty$ of $\tilde{H}_\infty$. Conversely, if θ is an isometry $\hat{H}_0 \to H_0$ between the completion $\hat{H}_0$ of $\tilde{H}_0$ and the Hilbert space H_0, then $\theta(\tilde{H}_\infty)$ is a dense subspace of H_∞ with a finer topology (if we endow it with the topology of $\tilde{H}_\infty$).

Remark. If we replace the direct sums $H_n = \oplus H_n^i$ by a direct integral $H_n = \int H_n(\lambda) d\mu(\lambda)$, the situation is more complicated (the straightforward extension fails) and we know a construction of this type only in the case where all $H_\infty(\lambda)$ are the same.

5) Field theory. We remind the reader that quantized fields are in fact distributions with values in the space of linear operators in some Hilbert space having a common invariant domain, and endowed with the corresponding weak operator topology. The latter operator space being far from normed, the introduction we made here to l.c.s. may be, for this reason also, not so far from physics as one could think from first sight.

References

1. N. Bourbaki. Espaces vectoriels topologiques. Hermann, Paris (1953-55).
2. H. H. Schaefer. Topological vector spaces. Macmillan, New York (1966).
3. H. G. Garnir, M. De Wilde, F. Schmets. Analyse fonctionelle: théorie constructive des espaces linéaires à semi-normes. (Tome I, théorie générale). Birkhäuser, Basel (1968).
4. F. Trèves. Locally convex spaces and linear partial differential equations. Springer, Berlin (1967).
5. A. Grothendieck. Produits tensoriels topologiques et spaces nucléaires. AMS Memoirs, No. 16, Providence, R.I. (1955).
6. A. Grothendieck. Leçons sur les espaces vectoriels topologiques. Sao Paulo (2nd. ed.), (1958).
7. I. M. Gel'fand, G. E. Shilov. Generalized functions (Vol. 2: fundamental spaces). Fizmatgiz, Moscow (1958).
8. I. M. Gel'fand, N. Y. Vilenkin. Generalized functions (Vol. 4: Applications of harmonic analysis). Fizmatgiz, Moscow (1961). (The last two books exist in several translations, including English.)

9. L. Schwartz. Théorie des distributions. Hermann, Paris (re-edition in one augmented volume, 1968).
10. L. Schwartz. Théorie des distributions à valeurs vectorielles. Extrait des tomes 7 et 8 des Annales de l'Institut Fourier, Université de Grenoble (1959).
11. A. Robertson and W. Robertson. Topological vector spaces. Cambridge (1964).
12. F. Bruhat. Sur les représentations induites des groupes de Lie. Bull. Soc. Math. Fr. 84, 97-205 (1956).
13. J. L. Kelley, I. Namioka et al. Linear topological spaces. Van Nostrand, Princeton (1963).
14. A. Grothendieck. Sur certains espaces de fonctions holomorphes. J. Reine u. Angew. Math., 192 (1953), pp. 35-64 and 77-95.
15. G. Köthe. Topological vector spaces I. Springer, Berlin (1969).

CONTENTS

B. The Hahn-Banach theorem and related results
Geometric and analytic forms of the Hahn-Banach theorem; consequences and related results.

C. Linear operators: definitions; the closed graph theorem
Continuity and factorization of linear operators; homomorphism and closed graph theorems; generalizations and connection with barreled spaces.

Lecture III. Spaces of Linear and Bilinear Maps. Equicontinuity. Duality

A. Topological vector spaces of linear operators, equicontinuity, and the Banach-Steinhaus theorem
Σ-topologies on $L(E, F)$ and major examples; equicontinuous and bounded sets; the Banach-Steinhaus theorem and related results.

B. Bilinear mappings, hypocontinuity
Definitions; separate and joint continuity, hypocontinuity; topologies on spaces of bilinear maps.

C. Duality
The notion of duality; weak and weakened topologies; compatibility between duality and topology. Mackey theorem and topology; reflexitivity; Montel and (DF) spaces; duality and inductive or projective limits; fully complete spaces and the closed graph theorem.

Lecture IV. Nuclearity and Topological Tensor Products; Examples and Applications

A. Topological tensor products
Topologies on tensor products; semi-norms; the cases of metrizable and of Hilbert spaces.

B. Nuclear spaces and maps
Nuclear maps in Banach and in general spaces; nuclear spaces; properties and characterizations of nuclear spaces; the abstract kernel theorem.

C. Examples and applications
Some important function spaces; vector-valued distributions and the kernel theorem; applications to group theory; nuclear direct sums; applications to field theory.

LECTURES ON
MATHEMATICAL ASPECTS OF PHASE TRANSITIONS†

M. Kac
The Rockefeller University
New York, New York

Section 1

In these lectures I shall discuss a number of questions relating to the phenomenon of phase transitions concentrating mainly on the mathematical aspects and emphasizing some unsolved problems.

I shall restrict myself to Ising models, and I shall begin by a review of some facts about the most famous of such models, i.e., the two-dimensional Ising model with nearest neighbor interaction. Consider an $N \times M$ square lattice

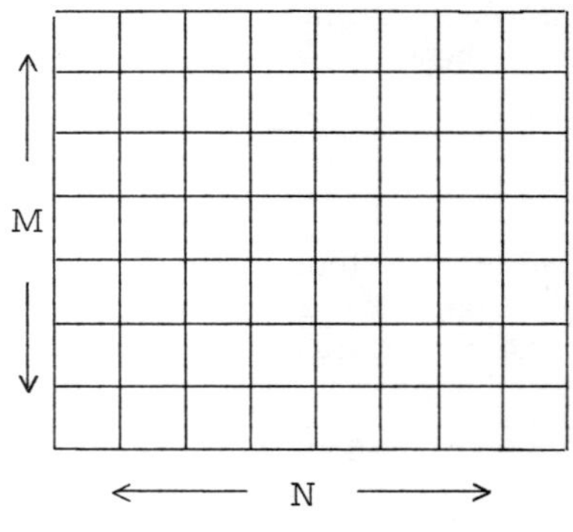

and imagine that on each lattice point P there "sits" a little magnet which can point in one of two directions. Mathematically, it means that with each point P of the lattice we associate a scalar "spin" μ_P which can assume only the values +1 and -1, i.e.,

---

†Presented at NATO Summer School in Mathematical Physics, Istanbul, 1970.

$$\mu_P = \begin{cases} +1 \\ -1 \end{cases}$$

The state of the system is specified by the knowledge of all the "spins" and the interaction energy is assumed to be given by the expression

$$E = -J \sum^{*} \mu_P \mu_Q \tag{1.1}$$

where $J > 0$ is a constant (measuring the strength of the interaction) and the star (*) on the summation sign indicates that only nearest neighbors are taken into account.

In other words,

$$E = -J \sum v(P, Q) \mu_P \mu_Q \tag{1.2}$$

where $v(P, Q)$ is zero (0) if P and Q are not nearest neighbors and one (1) if they are. Also, to simplify matters, the lattice is wrapped on a torus so that each site has four nearest neighbors.

If there is also an external magnetic field of intensity m, the energy is given by the expression

$$E = -J \sum^{*} \mu_P \mu_Q - m \sum \mu_P . \tag{1.3}$$

If the system as described is in thermal equilibrium with a heat bath of absolute temperature T, then it is an axiom of statistical mechanics that the probability of a given state is proportional to

$$e^{-E/kT} \qquad (k \equiv \text{Boltzmann constant})$$

and the thermodynamics of the system can be derived from the partition function

$$Q_{N,M} = \sum e^{-E/kT} , \tag{1.4}$$

the sum being over all the (2^{NM}) states.

Writing

$$Q_{N,M} = e^{-\Psi/kT} ,$$

we are led to the definition of free energy Ψ but in order to have meaningful thermodynamics, Ψ must be an extensive quantity, i.e., it should be proportional to NM when both N and M are large.

Mathematically, the limit

$$\lim_{\substack{N\to\infty \\ M\to\infty}} \frac{\Psi}{NM} = \psi$$

should exist or, in other words, the limit

$$\lim_{\substack{N\to\infty \\ M\to\infty}} \frac{1}{MN} \log Q_{N,M} \tag{1.5}$$

should exist. Denoting this limit by

$$-\frac{\psi}{kT},$$

we have a proper definition of free energy ψ per lattice site. This free energy is a function of

$$\nu = \frac{J}{kT} \tag{1.6}$$

and

$$\omega = \frac{m}{kT}. \tag{1.7}$$

Magnetization is defined as the average of $\sum \mu_P$, i.e.,

$$\frac{\sum \left(\sum_P \mu_P\right) e^{-E/kT}}{\sum e^{-E/kT}} = \frac{\partial}{\partial\omega} \log Q_{N,M}(\nu,\omega)$$

$$= \frac{\partial}{\partial\omega}\left(-\frac{\psi}{kT}\right) = NM \frac{\partial}{\partial\omega}\left(-\frac{\Psi/NM}{kT}\right)$$

Only a little optimism is needed to believe that in the thermodynamic limit (i.e., $N\to\infty$, $M\to\infty$) magnetization per site is given by the formula

$$\zeta = \lim_{\substack{N\to\infty \\ M\to\infty}} \left\langle \frac{1}{NM} \sum_P \mu_P \right\rangle = \lim_{\substack{N\to\infty \\ M\to\infty}} \frac{\sum \left(\frac{1}{NM} \sum_P \mu_P \right) e^{-E/kT}}{\sum e^{-E/kT}}$$

$$= \lim_{\substack{N\to\infty \\ M\to\infty}} \frac{\partial}{\partial\omega} \left(- \frac{\Psi/NM}{kT} \right) = \frac{\partial}{\partial\omega} \left(- \frac{\psi}{kT} \right) .$$

(Only the last step contains elements of danger.)

A word about taking limits. It is actually a bit subtle. (I ask you to believe me that it is all right to first let $N\to\infty$ and then $M\to\infty$ (the order does not matter but coupling of the two limits does introduce a number of spurious difficulties).)

Section 2

To calculate $Q_{N,M}(\nu,\omega)$, we introduce the transfer matrix $L(\vec{\mu}, \vec{\mu}')$. It is $2^M \times 2^M$ matrix indexed by binary vectors $\vec{\mu}$, $\vec{\mu}'$ (i.e., $\vec{\mu}$ and $\vec{\mu}'$ are M-dimensional vectors with components +1 and -1) and defined by the formula

$$L(\vec{\mu}, \vec{\mu}') = e^{\frac{\omega}{2}\sum_1^M \mu_k} e^{\frac{\nu}{2}\sum_1^M \mu_k \mu_{k+1}} e^{\nu \sum_1^M \mu_k \mu_k'} e^{\frac{\nu}{2}\sum_1^M \mu_k' \mu_{k+1}'} e^{\frac{\omega}{2}\sum_1^M \mu_k'} \tag{2.1}$$

where it is understood that $M+1$ is to be replaced by 1.

It is now a simple matter to note that

$$Q_{N,M} = \text{Trace}\,\{L^N\} = \Lambda_1^N + \Lambda_2^N + \cdots + \Lambda_{2M}^N \tag{2.2}$$

where $\Lambda_1, \Lambda_2, \ldots$ are the eigenvalues of L in decreasing order.

It follows at once that

$$-\frac{\psi}{kT} = \lim_{M\to\infty} \frac{1}{M} \log \Lambda_1 . \tag{2.3}$$

If $\varphi_1(\vec{\mu}; \omega)$ is the (normalized) principal eigenvector of (2.1), we obtain by a simple calculation the magnetization

$$\zeta = \frac{\partial}{\partial\omega}\left(-\frac{\psi}{kT}\right) = \lim_{M\to\infty} \sum_{\vec{\mu}} \left(\frac{\sum_1^M \mu_k}{M}\right) \varphi_1^2(\vec{\mu};\omega) = \lim_{M\to\infty} \sum_{\vec{\mu}} \mu_\ell \varphi_1^2(\vec{\mu};\omega). \tag{2.4}$$

(It is immaterial which $\ell (1 \leqq \ell \leqq M)$ we take because of our choice of "periodic boundary condition," i.e., wrapping the lattice on a torus).

If for sufficiently low temperatures one has

$$\lim_{\omega\to 0^+} \zeta \neq 0 \tag{2.5}$$

(i.e., magnetization at field zero is non-vanishing) while for sufficiently high temperatures the limit (2.5) is 0, one says that the system undergoes a phase transition.

Section 3

How can

$$\lim_{\omega\to 0^+} \zeta \neq 0 \,? \tag{3.1}$$

First of all, let us note that without $M\to\infty$ there is no hope for (3.1).

In fact, for fixed M,

$$\lim_{\omega\to 0^+} \varphi_1^2(\vec{\mu};\omega) = \varphi_1^2(\vec{\mu};0) = \varphi_1^2(\vec{\mu}) \tag{3.2}$$

where $\varphi_1(\vec{\mu};0) = \varphi_1(\vec{\mu})$ is the principal eigenvector of

$$L_0(\vec{\mu}, \vec{\mu}\,') = e^{\frac{\nu}{2}\sum_1^M \mu_k\mu_{k+1}}\, e^{\nu\sum_1^M \mu_k\mu_k'}\, e^{\frac{\nu}{2}\sum_1^M \mu_k'\mu_{k+1}'}. \tag{3.3}$$

Since the elements of the matrix L_0 are all positive, we have by a well known theorem of Frobenius that the maximum eigenvalue is non-degenerate and consequently $\varphi_1(\vec{\mu})$ is uniquely defined. It is clear that

$$\varphi_1(\vec{\mu}) = \varphi_1(-\vec{\mu}) \tag{3.4}$$

and hence

$$\lim_{\omega\to 0^+} \zeta = \sum \mu_\ell \varphi_1^2(\vec{\mu}) = 0.$$

We now know that for sufficiently low temperatures

$$\lim_{\omega\to 0^+} \zeta \neq 0 \qquad (3.5)$$

and this is clearly due to taking the limit $M\to\infty$ first, i.e., the model being bona fide two-dimensional.

The rigorous proof of (3.5) is not at all trivial but an intuitive reason is easy to give.

For very low temperatures ($\nu \gg 1$) and $\omega > 0$ the spins tend to align themselves in the direction of the external field and one might expect that

$$f_+(\vec{\mu}) = \begin{cases} 1 & \vec{\mu} = (1, 1, \ldots) \\ 0 & \text{otherwise} \end{cases}$$

might be a reasonable approximation to $\varphi_1(\vec{\mu};\omega)$.

Using f_+ as a trial vector, we obtain an approximation to Λ_1

$$\Lambda_1 \sim (Lf_+, f_+) = e^{M\omega} e^{2M\nu}.$$

Let us now define

$$f_-(\vec{\mu}) = \begin{cases} 1 & \vec{\mu} = (-1, -1, \ldots, -1) \\ 0 & \text{otherwise} \end{cases}$$

and take $af_+ + bf_-$ as a trial vector. We now obtain

$$\Lambda_1 \sim \frac{a^2 + b^2 e^{-2M\omega} + 2ab\, e^{-M(\omega+2\nu)}}{a^2 + b^2} e^{M(\omega+2\nu)}$$

and by choosing a, b so as to maximize the estimate, we obtain

$$\Lambda_1 \sim \frac{(1+e^{-2M\omega}) + \sqrt{(1-e^{-2M\omega})^2 + 4e^{2M(\omega+\nu)}}}{2} e^{M(\omega+2\nu)},$$

and hence it is better to have $b \neq 0$. In fact, one finds that

$$\frac{b}{a} = 2 \frac{e^{-M(\omega + 2\nu)}}{(1 - e^{-2M\omega}) + \sqrt{(1 - e^{-2M\omega})^2 + 4e^{-2M(\omega+\nu)}}}$$

For M fixed if $\omega \to 0^+$, we see that

$$\frac{b}{a} \to 1.$$

But for a fixed $\omega > 0$ and $M \to \infty$

$$\frac{b}{a} \to 0$$

exponentially fast.

For large M and $\nu \gg 1$, f_+ is a "good approximation" to φ_1 and one obtains

$$\zeta(\omega) \sim 1$$

for all $\omega > 0$, which is as it should be.

Section 4

The onset of a phase transition is also connected with the onset of long-range order. To deal with this we need the correlation function $\rho(r)$ defined by the formula

$$\rho(r) = \lim_{\substack{N\to\infty \\ M\to\infty}} \langle \mu_{k,\ell}\mu_{k+\nu,\ell} \rangle = \lim_{\substack{N\to\infty \\ M\to\infty}} \frac{\Sigma\, \mu_{k,\ell}\mu_{k+\nu,\ell}\, e^{-E/kT}}{\Sigma\, e^{-E/kT}} . \quad (4.1)$$

The study of $\rho(r)$ is of special interest in the absence of an external magnetic field and I shall therefore assume now that $\omega = 0$.

It is again quite easy to express $\rho(r)$ in terms of the transfer matrix and one obtains

$$\rho(r) = \lim_{M\to\infty} \rho_M(r) = \lim_{M\to\infty} \sum_{j=1}^{\infty} \left(\frac{\mathring{\Lambda}_j}{\mathring{\Lambda}_1}\right)^r \left(\sum_{\vec{f1}} \mu_\ell \varphi_j(\vec{\mu})\varphi_1(\vec{\mu})\right)^2 . \qquad (4.2)$$

The $\mathring{\Lambda}_j$ and the $\varphi_j(\vec{\mu}) = \varphi_j(\vec{\mu}; 0)$ are, of course, the eigenvalues and the corresponding (normalized) eigenfunctions of L_0 (see (3.3)).

One says that the system exhibits long-range order† if

$$\lim_{r\to\infty} \rho(r) = \rho(\infty) \neq 0. \qquad (4.3)$$

Section 5

But how can (4.3) be valid? Again if M were finite, this would be impossible because by a theorem of Frobenius the maximum eigenvalue of a matrix with positive entries is non-degenerate and hence

$$\mathring{\Lambda}_2 < \mathring{\Lambda}_1$$

implying that

$$\lim_{r\to\infty} \rho_M(r) = 0.$$

The onset of long-range order can therefore take place only if, in the limit as $M\to\infty$, $\mathring{\Lambda}_1$ becomes degenerate in the sense that

$$\lim_{M\to\infty} \frac{\mathring{\Lambda}_2}{\mathring{\Lambda}_1} = 1. \qquad (5.1)$$

Actually (5.1) is not enough to insure long-range order. We must also have

$$\lim_{M\to\infty} \sum \mu_\ell \varphi_2(\vec{\mu})\varphi_1(\vec{\mu}) \neq 0. \qquad (5.2)$$

†Formula (4.2) and the concept of long-range order were introduced first by Adkin and Lamb.

One can define a phase transition as the onset of long-range order and a natural question arises as to the relation of the two definitions.

In all known cases they are equivalent (although a number of points of rigour remain here and there) and, in fact, one has

$$\rho(\infty) = \zeta^2(0+). \tag{5.3}$$

That we are dealing here with an important and highly non-trivial question can be seen by discussing the relation between (5.3) and coexistence of two phases.

Physically one expects that if there is a phase transition, then for sufficiently low temperatures and in the absence of an external field the system contains two phases: one consisting of spins of magnetization $+\zeta(0+)$ and the other consisting of spins of magnetization $-\zeta(0+) = \zeta(0-)$.

In other words, the distribution of

$$\frac{1}{NM} \sum \mu_P$$

in the thermodynamic limit must be

$$\tfrac{1}{2}\delta(\zeta - \zeta(0+)) + \tfrac{1}{2}\delta(\zeta - \zeta(0-)).$$

This implies, in particular, that

$$\lim_{\substack{N\to\infty \\ M\to\infty}} \left\langle \frac{1}{(NM)^2}\left(\sum \mu_P\right)^2 \right\rangle = \zeta^2(0+). \tag{5.4}$$

On the other hand (formally at least),

$$\lim_{\substack{N\to\infty \\ M\to\infty}} \left\langle \frac{1}{(NM)^2}\left(\sum \mu_P\right)^2 \right\rangle = \rho(\infty) \tag{5.5}$$

and (5.3) follows.

Mathematically (5.3) is striking for the reason that $\zeta(0+)$ involves only the principal eigenvector while $\rho(\infty)$ involves also the eigenvector belonging to the eigenvalue next to the largest. In order to have (5.3), there must therefore be a very special relationship between φ_1 and φ_2.

This special relationship is best expressed by saying that φ_2 is (approximately) the "antisymmetric version" of φ_1. The meaning of this becomes clear if one thinks of very low temperatures ($\nu \gg 1$). Then

$$\varphi_1(\vec{\mu}) \sim \begin{cases} 1/\sqrt{2} & \vec{\mu} = (1, 1, \ldots 1) \text{ and } \vec{\mu}(-1, -1, \ldots -1) \\ 0 & \text{otherwise} \end{cases}$$

and

$$\varphi_2(\vec{\mu}) \sim \begin{cases} 1/\sqrt{2} & \vec{\mu} = (1, 1, \ldots, 1) \\ -1/\sqrt{2} & \vec{\mu} = (-1, -1, \ldots, -1) \\ 0 & \text{otherwise.} \end{cases}$$

A little more precisely, we must have

$$\lim_{M\to\infty} \sum \mu_\ell \varphi_2(\vec{\mu})\varphi_1(\vec{\mu}) = \lim_{\omega\to 0^+} \lim_{M\to\infty} \sum \mu_\ell \varphi_1(\vec{\mu};\omega), \qquad (5.6)$$

and this would surely follow if we had

$$\lim_{M\to\infty} \sum_{\frac{\mu_1+\ldots+\mu_M}{M} > 0} \varphi_2(\vec{\mu})\,\varphi_1(\vec{\mu}) = \frac{1}{2} \qquad (5.7a)$$

$$\lim_{\omega\to 0^+} \lim_{M\to\infty} \sum_{\frac{\mu_1+\ldots+\mu_M}{M} > 0} \varphi_1(\vec{\mu};\omega)\varphi_1(\vec{\mu}) = \frac{1}{2}\,. \qquad (5.7b)$$

Both (5.7a) and (5.7b) are clear on intuitive grounds although no rigorous proof of (5.7b) is available.

It should be noted that the original $\frac{1}{2}$ in (5.7b) is in the fact that $\varphi_1(\vec{\mu})$ (in the absence of the field) is even, i.e.,

$$\varphi_1(\vec{\mu}) = \varphi_1(-\vec{\mu}),$$

while $\varphi_1(\vec{\mu};\omega)$ is for $\omega > 0$ very strongly biased in the direction of positive magnetization. In fact, one should have (though again there seems to be no proof available)

$$\lim_{M\to\infty} \sum_{\frac{\mu_1+\ldots+\mu_M}{M}<0} \varphi_1^2(\vec{\mu};\omega) = 0. \tag{5.8}$$

Section 6

It is interesting to record that we have a mathematically analogous situation for a class of models which physically appear to be quite different. I am referring to models with weak long-range interactions and in particular to the one-dimensional model with interaction energy

$$E = -J\gamma \sum_{1\leqq i<j\leqq N} e^{-\gamma|i-j|} \mu_i \mu_j. \tag{6.1}$$

This case has been studied so extensively that only a cursory review will suffice.

The reason why the coice (6.1) is so convenient is that $\exp(-|t|)$ is the covariance function of the so-called Ornstein-Uhlenbeck process $U(\tau)$ and one can write (as before $\nu = J/kT$)

$$\exp\left\{\nu\gamma \sum_{1\leqq i<j\leqq N} \exp(-\gamma|i-j|)\mu_i\mu_j\right\} =$$

$$= \exp(-\tfrac{1}{2}N\nu\gamma)\exp\left\{\tfrac{1}{2}\nu\gamma \sum_{i,j=1}^{N} \exp(-\gamma|i-j|)\mu_i\mu_j\right\},$$

as $\exp(-\frac{1}{2}N\nu\gamma)$ times the mathematical expectation of

$$\exp\left\{\sqrt{\nu\gamma} \sum_{k=1} U(k\gamma)\mu_k\right\}.$$

Because of this the partition function Q_N corresponding to the energy, (6.1) can be written in the form

$$Q_N = \sum \exp(-E/kT) = 2^N \exp(-\tfrac{1}{2}N\nu\gamma)\left\langle \prod_{k=1}^{N} \cosh\sqrt{\nu\gamma}\, U(k\gamma)\right\rangle, \tag{6.2}$$

where I use now (and shall use in the sequel) $\langle\ \rangle$ to denote also mathematical expectations referring to stochastic processes.

Because the Ornstein-Uhlenbeck process is Markoffian (in addition to being Gaussian and stationary), the density $W(u_1, u_2, \ldots u_N)$ of the joint distribution of $U(\gamma), U(2\gamma), \ldots, U(N\gamma)$ factorizes, i.e.,

$$W(u_1, u_2 \ldots, u_N) =$$

$$= W(u_1)P_\gamma(u_1 | u_2)\, P_\gamma(u_2 | u_3) \cdots P_\gamma(u_{N-1} | u_N), \tag{6.3}$$

where

$$W(x) = \frac{1}{\sqrt{2\pi}} \exp(-\tfrac{1}{2}x^2); \tag{6.4}$$

$$P_\gamma(x | y) = \frac{1}{\sqrt{2\pi(1 - \exp(-2\gamma)}} \exp\left\{-\frac{(y - x\exp(-\gamma))^2}{2(1 - \exp(-2\gamma))}\right\}. \tag{6.5}$$

Thus we obtain that

$$Q_N = 2N \exp(-\tfrac{1}{2}N\nu\gamma) \int_{-\infty}^{\infty}\int_{-\infty}^{\infty} \cdots \int_{-\infty}^{\infty} \prod_{k=1} \cosh(\sqrt{\nu\gamma}\, u_k)\, W(u_1)$$
$$\times \prod_{k=1}^{N-1} P_\gamma(u_k | u_{k+1})\, du_1\, du_2 \ldots du_N. \tag{6.6}$$

Those unfamiliar with the rudiments of the theory of stochastic processes need merely verify the identity

$$\exp\left\{\tfrac{1}{2}\nu\gamma \sum_{i,j=1}^{N} \exp(-\gamma|i-j|)\mu_i\mu_j\right\} =$$

$$= \int_{-\infty}^{\infty} \cdots \int_{-\infty}^{\infty} \exp\left(\sqrt{\nu\gamma} \sum_{1}^{N} u_k u_k\right) W(u_1) \tag{6.7}$$
$$\times \prod_{k=1}^{N-1} P_\gamma(u_k | u_{k+1})\, du_1 \ldots du_N$$

(which is not too difficult an exercise in multiple integration) whence (6.6) follows almost at once.

Returning to (6.6) we introduce the symmetric kernel

$$K(x,y) = \{\cosh(\sqrt{\nu\gamma}x)\}^{\frac{1}{2}} \frac{W(x)\,P_\gamma(x\,|\,y)}{(W(x)W(y))^{\frac{1}{2}}} \{\cosh(\sqrt{\nu\gamma}y\}^{\frac{1}{2}} \tag{6.8}$$

and note that

$$Q_N = 2^N \exp(-\tfrac{1}{2}N\nu\gamma) \int_{-\infty}^{\infty} \cdots \int_{-\infty}^{\infty} \{W(u_1)\cosh(\sqrt{\nu\gamma}\,u_1)\}^{\frac{1}{2}} K(u_1,u_2)$$
$$\times\ K(u_2,u_3)\ldots K(u_{N-1},u_N)\,\{W(u_N)\cosh(\sqrt{\nu\gamma}\,u_N)\}^{\frac{1}{2}}\,du_1\ldots du_N. \tag{6.9}$$

Denoting by $\lambda_1, \lambda_2, \ldots$ the eigenvalues (in decreasing order) and by $\phi_1\phi_2\ldots$ the corresponding normalized eigenfunctions of the Kernel $K(x,y)$ we see that

$$Q_N = 2^N \exp(-\tfrac{1}{2}N\nu\gamma) \sum_{j=1}^{\infty} \lambda_j^{N-1} \left(\int_{-\infty}^{\infty} \phi j(x)\,\{W(x)\cosh(\sqrt{\nu\gamma}x)\}^{\frac{1}{2}}\,dx \right)^2, \tag{6.10}$$

and it follows at once that

$$-\frac{\psi}{kT} = \lim_{N\to\infty} \frac{1}{N} \log Q_N = \log 2 - \tfrac{1}{2}\nu\gamma + \log \lambda_1. \tag{6.11}$$

Similarly one can derive a formula for the correlation coefficient

$$\rho(r) = \lim_{N\to\infty} \langle \mu_k \mu_{k+r} \rangle, \tag{6.12}$$

where k must now approach ∞ with N to avoid boundary effects (it is no longer convenient to introduce periodic boundary conditions in this model).

The result is embodied in the formula (valid only for $r > 1$)

$$\rho(r) = \sum_{j=1}^{\infty} \left(\frac{\lambda_j}{\lambda_1}\right)^r \left\{ \int_{-\infty}^{\infty} \phi_j(x)\,\phi_1(x)\,\tanh(\sqrt{\nu\gamma}\,x)\,dx \right\}^2, \tag{6.13}$$

and the analogy with (4.2) is evident. Equally evident is the analogy between (6.11) and (2.3).

Section 7

To study what happens for small γ we start with (6.8) and note that after a few simple algebraic transformations the kernel K can be written in the form

$$K(x,y) = \exp(\tfrac{1}{2}\gamma)\exp(-\tfrac{1}{2}\gamma\, q(x))\frac{\exp\{-((y-x)^2)/4\sinh\gamma)\}}{\sqrt{4\pi\sinh\gamma}}\exp(-\tfrac{1}{2}\gamma q(x)), \tag{7.1}$$

where

$$\gamma q(x) = \tfrac{1}{2}\tanh(\tfrac{1}{2}\gamma x^2) - \log\cosh(\sqrt{\nu\gamma}x). \tag{7.2}$$

It is easily verifiable that for a smooth function f

$$\int_{-\infty}^{\infty} K(x,y)f(y)dy =$$

$$= \exp(\tfrac{1}{2}\gamma)\exp(-\tfrac{1}{2}\gamma q(x))\exp\left\{(\sinh\gamma)\frac{d^2}{dx^2}\right\}\exp(-\tfrac{1}{2}\gamma q(x))f(x), \tag{7.3}$$

and to lowest order in γ the operators

$$A = \gamma q(x),$$

and

$$B = (\sinh\gamma)\frac{d^2}{dx^2}$$

commute. (If one wants to go beyond the lowest order one must use the Baker-Hausdorf formula or follow a perturbation technique similar to that used by Kac, Uhlenbeck and Hemmer).

Thus, to lowest order, the integral operator K can be replaced by the operator

$$\exp(\tfrac{1}{2}\gamma)\exp\left[\sinh\left\{\gamma\frac{d^2}{dx^2}\gamma q(x)\right\}\right],$$

or (replacing $\sinh\gamma$ by γ and $\tanh\gamma/2$ in (7.2) by $\gamma/2$) by the operator

$$\exp(\tfrac{1}{2}\gamma)\exp\left\{\gamma\frac{d^2}{dx^2}-[\tfrac{1}{4}\gamma x^2-\log\cosh(\sqrt{\nu\gamma}x)]\right\}.$$

It follows that (again to order γ)

$$\log\lambda_1 \approx \tfrac{1}{2}\gamma - \gamma E_0, \tag{7.4}$$

where E_0 is the lowest eigenvalue of the Schrödinger equation

$$\frac{d^2\phi}{dx^2} - \left\{\tfrac{1}{4}x^2 - \frac{1}{\gamma}\log\cosh(\sqrt{\nu\gamma}x)\right\}\phi = -E_0 \tag{7.5}$$

and that the eigenfunctions of K can be taken to be those of (7.5).

To discuss the correlation function (6.13) we need, of course, the whole spectrum of the Schrödinger operators in (7.5) as well as its eigenfunctions. What about this spectrum?

Its character undergoes a dramatic change at $2\nu=1$.

In fact, the potential

$$\tfrac{1}{4}x^2 - \frac{1}{\gamma}\log\cosh(\sqrt{\nu\gamma}x) \tag{7.6}$$

has a unique minimum at $x=0$ if $2\nu<1$ but has two (equal) minima if $2\nu>1$.

These two minima occur at

$$x = \pm\frac{\eta_0(\nu)\sqrt{2}}{\sqrt{\gamma}}, \tag{7.7}$$

where $\eta_0(\nu)$ is the positive root of the transcendental equation $\eta = 2\nu\tanh(2\nu\eta)$.

The depth of each minimum is

$$-\frac{1}{\gamma}\kappa(\nu), \tag{7.8}$$

with $\kappa(\nu) = -\tfrac{1}{2}\eta_0^2 + \log\cosh\sqrt{2\nu}\eta_0$.

If $2\nu<1$ we can expand $\log\cosh(\sqrt{\nu\gamma}x)$ in a power series retaining only terms of order γ or lower (i.e., $\log\cosh\sqrt{\nu\gamma}x \approx \tfrac{1}{2}\nu\gamma x^2$) and replace the potential by

$$\tfrac{1}{4}(1-2\nu)x^2.$$

Thus if $2\nu < 1$,

$$E_n = (n+\tfrac{1}{2})\sqrt{1-2\nu} \tag{7.9}$$

and

$$\log \lambda_n \approx \tfrac{1}{2}\gamma\{1-(2n+1)\sqrt{1-2\nu}\}. \tag{7.10}$$

The $\log \lambda_n$ are therefore more or less evenly spaced, the spacing being of order γ.

For $2\nu > 1$ because the potential has two deep minima each eigenvalue and, in particular, the lowest one is almost degenerate.

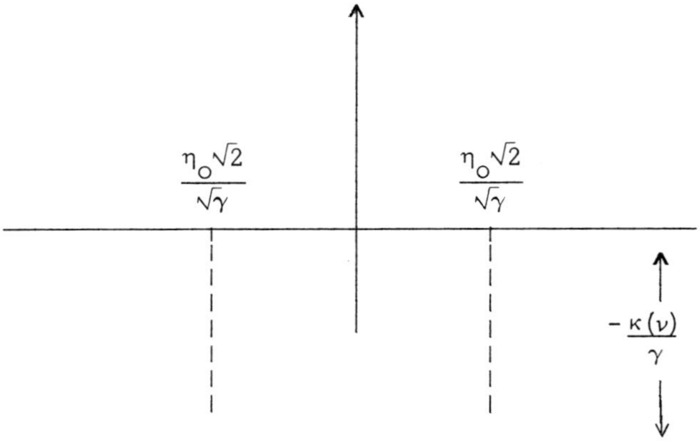

The degree of near-degeneracy can be calculated by a familiar application of the WKB method and it is not too difficult to prove rigorously (see end of §5 of my Brandeis Lectures) that

$$\frac{\lambda_2}{\lambda_1} = 1 - O\{\exp(-c/\gamma)\}. \tag{7.11}$$

A heuristic argument justifying (7.11) is as follows.

We replace each of the potential wells by parabolic wells (it is easily checked that $1-2\nu\,\mathrm{sech}^2(\eta_0\sqrt{2\nu}) > 0$)

$$-\frac{1}{\gamma}\kappa(\nu) + \tfrac{1}{4}(1-2\nu\,\sinh^2(\eta_0\sqrt{2\nu}))\left(x \pm \frac{\eta_0\sqrt{2}}{\sqrt{\gamma}}\right)^2 \tag{7.12}$$

and (setting $\nu' = \nu \operatorname{sech}^2(\eta_o\sqrt{2\nu})$) and $\psi_o(x) = \exp\{-\frac{1}{4}\sqrt{1-2\nu'}x^2\}$ we can take

$$\psi^+ = \psi_o\left(x - \frac{\eta_o\sqrt{2}}{\sqrt{\gamma}}\right) + \psi_o\left(x + \frac{\eta_o\sqrt{2}}{\sqrt{\gamma}}\right)$$

and

$$\psi^- = \psi_o\left(x - \frac{\eta_o\sqrt{2}}{\sqrt{\gamma}}\right) - \psi_o\left(x - \frac{\eta_o\sqrt{2}}{\sqrt{\gamma}}\right)$$

as (un-normalized) trial wave functions corresponding to the lowest and next to the lowest eigenvalues.

A simple calculation based on the Raleigh-Ritz principle, then shows that $E_2 - E_1$ is of the order of the "overlap integral"

$$\int_{-\infty}^{\infty} \psi_o\left(x - \frac{\eta_o\sqrt{2}}{\sqrt{\gamma}}\right) \psi_o\left(x + \frac{\eta_o\sqrt{2}}{\sqrt{\gamma}}\right) dx, \tag{7.13}$$

(divided by the normalizing factor

$$\int_{-\infty}^{\infty} \psi_o^2\left(x - \frac{\eta_o\sqrt{2}}{\sqrt{\gamma}}\right) dx = \int_{-\infty}^{\infty} \psi_o^2\left(x + \frac{\eta_o\sqrt{2}}{\sqrt{\gamma}}\right) dx = \frac{\sqrt{2\pi}}{\sqrt{1-2\nu'}}$$

which is clearly of the order

$$\exp(-c/\gamma). \tag{7.14}$$

Moreover, for $2\nu > 1$ we also have

$$E_1 \approx -\frac{1}{\gamma}\kappa(\nu) + \tfrac{1}{2}\sqrt{1-2\nu'}, \tag{7.15}$$

or, equivalently,

$$\log \lambda_1 \approx \kappa(\nu) + \tfrac{1}{2}\gamma(1 - \sqrt{1-2\nu'}). \tag{7.16}$$

Combining this with the result valid for $2\nu < 1$

$$\log \lambda_1 \approx \tfrac{1}{2}\gamma(1 - \sqrt{1-2\nu'}) \tag{7.17}$$

(see (7.10)), we have by (6.11) the following formula for free energy which is correct to order γ

$$-\frac{\psi}{kT} = \begin{cases} \log 2 - \frac{1}{2}\nu\gamma + \frac{1}{2}\gamma(1 - \sqrt{1 - 2\nu}), & 2\nu < 1, \\ \log 2 + \kappa(\nu) - \frac{1}{2}\nu\gamma + \frac{1}{2}\gamma(1 - \sqrt{1 - 2\nu'}), & 2\nu > 1. \end{cases} \tag{7.18}$$

We recognize at once that to order 1 (i.e., in the limit $\gamma \to 0$) we recover the molecular field theory and that this theory (at least as far as the analytic behavior of thermodynamic functions is concerned) is the result of replacing the lowest eigenvalue E_1 of the Schrödinger equation (7.5) by the minimum of the potential. In other words, the molecular field theory corresponds to the classical approximation which takes no account of the quantum effects caused by the presence of the kinetic energy term d^2/dx^2.

It is, however, the quantum effects that are responsible for the marked change in the behavior of the correlation function as one goes from high $(2\nu < 1)$ to low $(2\nu > 1)$ temperatures.

In fact, for $2\nu < 1$, since

$$\frac{\lambda_1}{\lambda_2} = 1 - O(\gamma),$$

the "correlation length" is of the order $1/\gamma$, i.e., of the same order as the range of the interaction.

For $2\nu > 1$, on the other hand, we have

$$\frac{\lambda_2}{\lambda_1} = 1 - O\{\exp(-c/\gamma)\}$$

and the correlation length becomes of the order $\exp(+c\ \gamma)$ which is incomparably (in fact, enormously!) larger than the interaction length $1/\gamma$.

The decay rate of the correlation function is governed by the rate λ_2/λ_1 which is approximately

$$\exp\{-\gamma(E_1 - E_0)\}.$$

and the quantity $\Delta E = E_2 - E_1$ is par excellence quantum mechanical. In particular, the fact that for $2\nu > 1$, ΔE is of the order $\exp(-c/\gamma)$ is intimately related to the phenomenon of penetration of the barrier.

The remarkable analogy between the limits $M \to \infty$ and $\gamma \to 0$ is reinforced further if one studies our model in an external magnetic field so that the total energy is now

$$E = -J\gamma \sum_{1 \leqq i<j \leqq N} e^{-\gamma|i-j|} \mu_i \mu_j - m \sum_1^N \mu_j . \tag{7.19}$$

Instead of the kernel (6.8), we now have

$$K_\omega(x,y) = \{\cosh(\omega + \sqrt{\nu\gamma}\, x)\}^{\frac{1}{2}} \frac{W(x) P_\gamma(x\,|\,y)}{(W(x)W(y))^{\frac{1}{2}}} \{\cosh(\omega + \sqrt{\nu\gamma}\, y)\}^{\frac{1}{2}} \tag{7.20}$$

and (6.11) remains valid if λ_1 is taken to be the maximum eigenvalue of (7.20).

The discussion for small γ can be carried out as before—the only difference being that the minima in the potential are of unequal depth.

Because of this the principal eigenfunction is centered around $\eta_0(\omega)\sqrt{2}/\gamma$ where η_0 is now the root of

$$\eta = 2\nu \tanh(\omega + 2\nu\eta).$$

We leave it to the reader to verify that for $\nu > \frac{1}{2}$

$$\lim_{\omega \to 0^+} \lim_{\gamma \to 0} \zeta^2(\omega) = \lim_{r \to \infty} \lim_{\gamma \to 0} \rho(r).$$

Section 8

Let me conclude with a brief discussion of certain one-dimensional models to show again how spectral analysis can yield qualitative (non-rigorous so far!) understanding of pretty subtle phenomena.

Consider a one-dimensional Ising model with interaction energy

$$E = -\sum_{1 \leqq i<j \leqq N} v(j-i)\mu_i \mu_j . \tag{8.1}$$

About v I shall assume that

a) $v \geqq 0$, $v(k) = v(-k)$

b) $\sum_{-\infty}^{\infty} v(k) < \infty$ (stability condition)

If, in addition, one assumes that

c) $\sum_{-\infty}^{\infty} |k| v(k) < \infty$

then it will come as no surprise to anyone (though the proof is not at all trivial) that the model will not exhibit a phase transition.

If, on the other hand,

d) $\sum_{-\infty}^{\infty} |k| v(k) = \infty$

the situation is much more interesting and intricate.

At first I conjectured that under condition d) the model will exhibit long-range order for sufficiently low temperatures.

In this I was wrong, and Dyson in a most ingenious way has shown that if

$$\sum_{-N}^{N} |k| v(k) \sim \log \log N,$$

the model will <u>not</u> exhibit phase transition. On the other hand, by an even more ingenious argument, Dyson [1] again has shown that if the divergence in d) is sufficiently strong, long-range order will indeed be present for low enough temperatures.

The case of logarithmic divergence of d) is left open although P. W. Andersen has given a heuristic argument in favor of long-range order.

Can we understand all this "with our fingers"? The answer, I think, is "yes" if we use spectral analysis.

To get spectral analysis going, I shall assume that v(k) is a linear combination of exponentials, i.e.,

$$v(k) = \sum_{\ell=1}^{\infty} a_{\ell} e^{-\sigma_{\ell} |k|}, \quad a_{\ell} \geqq 0, \tag{8.2}$$

and to simplify some calculations I shall also modify slightly the interaction energy and assume it to be

$$E = -\frac{1}{2} \sum_{i,j=1}^{N} v(i-j)(\mu_i + \mu_{i+1})(\mu_j + \mu_{j+1}), \tag{8.3}$$

(with the understanding that the subscript $N+1$ is to be replaced by 1).

Following the usual procedure with minor modifications caused by the modified form of the interaction energy, we arrive at the following result: The correlation

$$R(r) = \lim_{\substack{N\to\infty \\ (k\to\infty)}} \langle \mu_k \mu_{k+r} \rangle \tag{8.4}$$

for $r \geqq 2$ is given by the formula

$$R(r) = \frac{1}{\lambda_1^2} \sum_{j=2}^{\infty} \left(\frac{\lambda_j}{\lambda_1}\right)^{r-1} \left((K_+ - K_-)\phi_j, \phi_1\right)^2 \tag{8.5}$$

where the λ_j and ϕ_j are the eigenvalues (ordered decreasingly) and normalized eigenfunctions of the kernel

$$K = K_+ + K_- \tag{8.6}$$

where

$$K_+ = \prod_k e^{-s_k^2(x_k-b_k)^2} \frac{e^{-((y_k-x_k)^2/2)}}{\sqrt{2\pi}} e^{-s_k^2(y_k-b_k)^2} \tag{8.7a}$$

$$K_- = \prod_k e^{-s_k^2(x_k+b_k)^2} \frac{e^{-((y_k-x_k)^2/2)}}{\sqrt{2\pi}} e^{-s_k^2(y_k+b_k)^2} \tag{8.7b}$$

and†

---

†It should perhaps be mentioned that the free energy ψ is given by the formula

$$-\frac{\psi}{kT} = \log 2 + \tfrac{1}{2}\sum \sigma_k + 2\nu \sum a_k \coth\frac{\sigma_k}{2} + \log \lambda_1 .$$

Convergence of $\Sigma\, a_k \coth(\sigma_k/2)$ is equivalent to the stability condition b) and, to avoid trivial difficulties, we may as well assume that $\Sigma\, \sigma_k$ converges too.

$$s_k = \sinh \frac{k}{2}$$

$$b_k = \frac{\sqrt{a_k} s_k \cosh(\sigma_k/2)}{s_k^2} \sqrt{\nu} \qquad (\nu = 1/kT). \tag{8.8}$$

There appears immediately a difficulty caused by the fact that K_+ and K_- are infinitely-dimensional. The maximum eigenvalues of each are of infinite degeneracy and so is the maximum eigenvalue of $K_+ + K_-$, though this requires a proof.

We are thus forced to interpret $R(r)$ in (8.5) as

$$\lim_{m \to \infty} R_m(r) \tag{8.9}$$

where $R_m(r)$ refers to a model in which $v(k)$ is replaced by $v_m(k)$ which is a sum of only m exponentials, i.e.,

$$v_m(k) = \sum_{\ell=1}^{m} a_\ell e^{-\sigma_\ell |k|} . \tag{8.10}$$

This in turn calls for obvious modifications of the definitions of K_+ and K_-.

For large ν we treat K_+ and K_- as almost non-interacting and approximate the principal eigenfunction by

$$\frac{1}{\sqrt{2}} \{\phi_+ + \phi_-\} \tag{8.11}$$

where

$$\phi_\pm(x) = \prod_k e^{-((\alpha_k^2/4)(x_k \mp b_k)^2)\sqrt{\alpha_k}/\sqrt[4]{2\pi}} \tag{8.12}$$

and

$$\alpha_k = 2\sqrt[4]{s_k^2(1+s_k^2)}. \tag{8.13}$$

Needless to say ϕ_+ is (rigorously) the principal eigenfunction of K_+ and ϕ_- the principal eigenfunction of K_-.

The question now arises what about the eigenfunction corresponding to the next to the highest eigenvalue (before we let $m \to \infty$).

The obvious candidate is the antisymmetric combination

$$\phi_2 \sim \frac{1}{\sqrt{2}} \{\phi_+ - \phi_-\}, \tag{8.14}$$

but if c) is satisfied, one does better with

$$\phi_2 \sim \frac{1}{\sqrt{2}} x_m \{\phi_+ - \phi_-\} \tag{8.15}$$

which yields the value zero (in the limit $m \to \infty$) for the matrix element.

$$((K_+ - K_-)\phi_2, \phi_1)$$

and hence (see (8.5)) no long-range order.

The same conclusion holds if the series in d) diverges sufficiently slowly, and one is lead to the conjecture that long-range order obtains (for sufficiently low temperatures) only if (8.14) is a "better" candidate for ϕ_2 than is (8.15).

In making my original conjecture, I was led astray by the thought that (8.14) is <u>always</u> better than (8.15).

Section 9

If the model is in an external magnetic field, the energy is

$$E = -\sum_{1 \leq i < j \leq N} v(i-j)\mu_i \mu_j - m \sum_{L=1}^{N'} \mu_i , \tag{9.1}$$

and setting

$$\omega = \frac{m}{kT} > 0 \tag{9.2}$$

we find that the magnetization $\zeta(\omega)$ is given by the formula

$$\zeta(\omega) = \frac{1}{\lambda_{max}} \iint \phi_1(\vec{x};\omega)(e^{\omega}K_+ - e^{-\omega}K_-)\phi_1(\vec{y};\omega)d\vec{x}\,d\vec{y}$$

where λ_{max} and ϕ_j are the maximum eigenvalue of the corresponding (normalized) eigenfunction of the kernel

$$e^{\omega}K_+ + e^{-\omega}K_- . \tag{9.3}$$

If

$$\lim_{\omega \to 0} \zeta(\omega) \neq 0 \tag{9.4}$$

for sufficiently low temperatures, one might again say that we have a phase transition. In the lattice gas terminology (9.4) is equivalent to having a flat part in the isotherm.

Let us now notice that the (normalized) eigenfunctions of $K_{\pm}$ are

$$\prod_k \psi_{n_k}\left[\alpha_k(x_k \mp b_k)\right]\sqrt{\alpha_k} \tag{9.5}$$

where

$$\psi_n(x) = \frac{1}{\sqrt[4]{2\pi}}\frac{1}{\sqrt{n!}}e^{-x^2/4}H_n(x) \tag{9.6}$$

are the familiar (normalized) Hermite functions, and α_k, b_k have been defined previously ((8.13) and (8.8)).

Using a well known formula, we have

$$e^{-s^2x^2}\frac{e^{-(y-x)^2/2}}{\sqrt{2\pi}}e^{-s^2y^2} = \sum_{n=0}^{\infty} e^{-(n+\frac{1}{2})\sigma}\alpha\psi_n(\alpha x)\psi_n(\alpha y)$$

$$(s = \sinh \sigma/2, \quad \alpha = 2\sqrt{s^2(1+s^2)} = \sinh\sigma),$$

and it follows that the eigenvalues of K_{+} are

$$e^{-\Sigma(n_k+\frac{1}{2})\sigma_k} \qquad n_k = 0, 1, 2, \ldots .$$

If we define the kernels $K_{\pm}^{(m)}$ in the obvious way (i.e., based on a sum of m exponentials), we see that the eigenvalues of $K_{\pm}^{(m)}$ are

$$e^{-\sum\limits_{k=1}^{m}(n_k+\frac{1}{2})\sigma_k}$$

Assuming σ_r to be decreasing, it follows that the next to the highest eigenvalue $\lambda_2(K_{\pm}^{(m)})$ divided by the highest $\lambda_1(K_{\pm}^{(m)})$ is $\exp(-\sigma_m)$

$$\frac{\lambda_2(K_\pm^{(m)})}{\lambda_1(K_\pm^{(m)})} = e^{-\sigma_m} = 1 - \sigma_m + O(\sigma_m^2).$$

I now conjecture that a similar result holds also for

$$e^{-\omega}K_-^{(m)} + e^{\omega}K_+^{(m)}$$

i.e.,

$$-\alpha\sigma_m < \frac{\lambda_2(e^{-\omega}K_-^{(m)} + e^{\omega}K_+^{(m)})}{\lambda_1(e^{-\omega}K_-^{(m)} + e^{\omega}K_+^{(m)})} < 1 \tag{9.7}$$

where α may depend on ω (but not on m!).

It is also tempting to conjecture that for $\nu \gg 1$ (low temperatures) the function

$$\psi_0^{(m)} = \prod_{k=1}^{m} \frac{\alpha_k}{\sqrt[4]{2\pi}} e^{-(\alpha_k^2/4)(x_k - b_k)^2}$$

is a sensible approximation to the principal eigenfunction of $e^{\omega}K_+^{(m)} + e^{-\omega}K_-$.

"Sensible" is understood to mean that if $\varphi^{(m)}$ is the (exact) principal eigenfunction of $e^{\omega}K_+^{(m)} + e^{-\omega}K_-^{(m)}$, then

$$\lim_{m\to\infty} \int \varphi_1^{(m)} \psi_0^{(m)} dx_1 \ldots dx_m \neq 0. \tag{9.8}$$

Starting with $\psi_0^{(m)}$ as a guess, I iterate it n times, i.e., I consider

$$\left(e^{\omega}K_+^{(m)} + e^{-\omega}K_-^{(m)}\right)^n \psi_0^{(m)}$$

and normalize it.

As $n\to\infty$ I should obtain $\varphi_1^{(m)}$, but if my conjecture (9.7) is right it is sufficient to iterate only

$$n = \frac{A}{\sigma_m}$$

times to obtain the approximate "shape" of $\varphi_0^{(m)}$.

To see this in a little more detail, denote by $\varphi_2^{(m)}$, $\varphi_3^{(m)}$ the eigenfunctions (other than the principal one) of

$e^{\omega}K_+^{(m)} + e^{-\omega}K_-^{(m)}$ and by $\lambda_2^{(m)}, \lambda_3^{(m)}, \ldots$ the corresponding eigen-values.

We then have

$$\left(e^{\omega}K_+^{(m)} + e^{-\omega}K_-^{(m)}\right)^n \psi_0^{(m)} = \sum_j \lambda_j^{(m)n}\left(\varphi_j^{(m)}, \psi_0^{(m)}\right)\varphi_j^{(m)}$$

and hence

$$\left\| \frac{(e^{\omega}K_+^{(m)} + e^{-\omega}K_-^{(m)})^n \psi_0^{(m)}}{\lambda_1^n} - \left(\psi_0^{(m)}, \varphi_1^{(m)}\right)\varphi_1^{(m)} \right\| \leqq \left(\frac{\lambda_2^{(m)}}{\lambda_1^{(m)}}\right)^n$$

and

$$0 \leqq \frac{\|(e^{\omega}K_+^{(m)} + e^{-\omega}K_-^{(m)})^n \psi_0^{(m)}\|^2}{\lambda_1^{(m)2n}} - \left(\psi_0^{(m)}, \varphi_1^{(m)}\right)^2 < \left(\frac{\lambda_2^{(m)}}{\lambda_1^{(m)}}\right)^{2n} .$$

It therefore follows that

$$\left\| \frac{(e^{\omega}K_+^{(m)} + e^{-\omega}K_-^{(m)})^n \psi_0^{(m)}}{\|(e^{\omega}K_+^{(m)} + e^{-\omega}K_-^{(m)})^n \psi_0^{(m)}\|} - \varphi_1^{(m)} \right\| \leqq \text{Const} \frac{(\lambda_2^{(m)}/\lambda_1^{(m)})^n}{(\psi_0^{(m)} \quad \varphi_1^{(m)})} , \tag{9.9}$$

and if both conjectures (9.4) and (9.8) are right we see that the right hand side of (9.9) can be made arbitrarily small provided A in

$$n = \frac{A}{\sigma_m}$$

is chosen sufficiently large.

Section 10

Let us conclude by noting that

$$\left(e^{\omega}K_+^{(m)} + e^{-\omega}K_-^{(m)}\right)^n \psi_0 \tag{10.1}$$

is a linear combination of the 2^n functions

$$\psi_{\epsilon_1,\epsilon_2,\cdots\epsilon_n} = \prod_{k=1}^{m} \frac{\sqrt{\alpha_k}}{\sqrt[4]{2\pi}} e^{-\frac{1}{4}\alpha_k^2[x_k - A_k^{(m)}(\epsilon_1,\cdots\epsilon_n)b_k]^2} \tag{10.2}$$

where

$$A_k^{(m)}(\epsilon_1,\cdots\epsilon_n) = e^{-n\sigma_k} + (1-e^{-\sigma_k})\sum_{r=1}^{n} \epsilon_r e^{-(n-r)\sigma_k}, \tag{10.3}$$

and each ϵ_j can assume only the values ± 1.

Let us now note that if

$$\lim_{m\to\infty} \sum_{k=1}^{m} \alpha_k^2 b_k^2 A_k^{(m)2}(\epsilon_1,\cdots\epsilon_n) = \infty, \tag{10.4}$$

then

$$\lim_{m\to\infty} \int_{\sum_1^m \frac{\sqrt{a_k}}{\sqrt{s_k}} x_k < 0} \psi^2_{\epsilon_1,\epsilon_2,\cdots\epsilon_n} dx_1 \cdots dx_m = 0 \tag{10.5}$$

and if the weight of those $\psi_{\epsilon_1\cdots\epsilon_n}$ in (10.1) for which (10.4) holds were sufficiently large, one could conclude that

$$\lim_{m\to\infty} \int_{\sum \frac{\sqrt{a_k}}{\sqrt{s_k}} x_k < 0} \varphi_1^{(m)2} dx_1 \cdots dx_m = 0 \tag{10.6}$$

for all $\omega > 0$. This in turn would suggest (though not quite rigorously imply) that $\zeta(0+) > 0$.

Now it can be argued (and the argument, I am sure, can be made rigorous) that only for $(\epsilon_1,\epsilon_2,\cdots\epsilon_n)$'s of negligible weight can the $A_k^{(m)}(\epsilon_1,\cdots\epsilon_n)$ be made smaller than some constant times $\exp(-n\sigma_k)$. It would then follow that

$$\lim_{m\to\infty} \sum_{k=1}^{m} \alpha_k^2 b_k^2 e^{-2A\sigma_k/\sigma_m} = \infty \tag{10.7}$$

implies (10.4) and hence the "skewness" of the principal eigenfunction as expressed in (10.6), which persists as $\omega \to 0+$ and which is a deeper indication of the phase transition.

As an example, take

$$\sigma_k = \frac{1}{k^2}, \qquad a_k = \frac{1}{k^{3+\alpha}}, \qquad \alpha > 0$$

and note that for $k \to \infty$ and disregarding constants, we have

$$s_k \sim \frac{1}{k^2}$$

$$b_k \sim k^{(3-\alpha)/2}$$

$$\alpha_k^2 \sim \frac{1}{k^2}.$$

Thus (10.7) is equivalent to

$$\lim_{m\to\infty} \sum_1^m \frac{1}{k^{\alpha-1}} e^{-2A(m/k)^2} = \infty$$

which will be the case for $\alpha < 2$.

For $\alpha > 2$, we actually have

$$\lim_{m\to\infty} \sum_1^m \frac{1}{k^{\alpha-1}} e^{-2A(m/k)^2} = 0,$$

and for $\alpha = 2$ (the subtlest case) the limit is finite (different from 0 and ∞).

If one chooses a_k appropriately, one can make

$$\alpha_k^2 b_k^2 \sim \frac{1}{k \log k}$$

in which case

$$\lim_{m\to\infty} \sum \frac{1}{k \log k} e^{-2A(m/k)^2} = 0,$$

and one would not expect a phase transition. (This corresponds to Dyson's first result quoted in §8).

References

1. Dyson, F. J., Comm. Math. Phys. 12, 91 (1969).
2. Dyson, F. J., Comm. Math. Phys. 12, 212 (1969).
3. Kac, M., 1966 Brandeis University Summer Institute in Theoretical Physics, Vol. I, Gordon & Breach Science Publishers.

THE MATHEMATICAL STRUCTURE OF THE BCS-MODEL AND RELATED MODELS†

W. Thirring
CERN
Geneva

1. Introduction

In systems with infinitely many degrees of freedom one encounters mathematical problems which one is not used to in elementary quantum mechanics. The BCS model is a non-trivial and not too unrealistic example exhibiting features which are typical for many field theoretic problems. It is characterized by a Hamiltonian for a finite volume.

$$H_\Omega = -\epsilon \sum_{p=1}^{\Omega} \left(\sigma_p^{(z)} - 1 \right) - \frac{2T_c}{\Omega} \sum_{p=1}^{\Omega} \sigma_p^- \sum_{p=1}^{\Omega} \sigma_p^+ . \tag{1.1}$$

The $\sigma_p^{(\alpha)}$'s are Pauli matrices acting on the pth mode of a tensor product of two-dimensional Hilbert spaces

$$\mathcal{H} = \mathcal{H}_1 \otimes \mathcal{H}_2 \otimes \ldots \otimes \mathcal{H}_\Omega$$

with the dimension 2^Ω, where Ω is a number proportional to the volume. The $\vec{\sigma}_p$'s have nothing to do with the physical spin of the electrons, but $\sigma_p^+ (\sigma_p^-)$ describes the creation (annihilation) of a Cooper pair, which has the kinetic energy 2ϵ.

We have made the assumption that the kinetic energy of all pairs is the same ("strong coupling limit"). Since there is only an interaction between pairs near the Fermi momentum, this is not completely unreasonable although not very realistic.

The model can be solved exactly if one introduces the operator

†Presented at NATO Summer School in Mathematical Physics, Istanbul, 1970.

$$\vec{S} = \frac{1}{2} \sum_{p=1}^{\Omega} \vec{\sigma}_p \, . \tag{1.2}$$

Then

$$H_\Omega = \epsilon(\Omega - 2S_z) - \frac{2T_c}{\Omega}\left(\vec{S}^2 - S_z(S_z + 1)\right) . \tag{1.3}$$

Defining r, N by

$$S = \frac{\Omega}{2} - r, \quad S_z = \frac{\Omega}{2} - N; \quad 0 \leqq N \leqq \Omega, \quad 0 \leqq r \leqq \mathrm{Min}(N, \Omega - N) \tag{1.4}$$

(N is just the number of pairs), the eigenvalues of H_Ω are

$$E(N,r) = 2\ N + 2T_c(r-N)\left(1 + \frac{1}{\Omega}\right) + \frac{2T_c}{\Omega}(N^2 - r^2) . \tag{1.5}$$

The structure of the lower lying levels is shown in Figure 1 where the energy of the ground state has been normalized to zero.

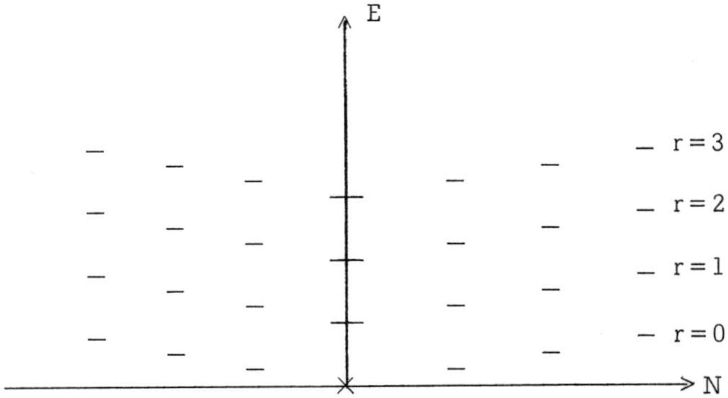

Figure 1

The eigenvalues define two characteristic frequencies which, in suitable units, take the form

$$\begin{aligned} \Delta &= E(N, r+1) - E(N, r) = 2T_c\left(1 - \frac{2r}{\Omega}\right) \\ -2\mu &= E(N+1, r) - E(N, r) = 2\epsilon - 2T_c\left(1 - \frac{2N}{\Omega}\right) \end{aligned} \tag{1.6}$$

where μ is the chemical potential.

Time Development

$$\vec{\sigma}_p \rightsquigarrow \vec{\sigma}_p(t) = e^{iHt}\vec{\sigma}_p e^{-iHt}. \tag{1.7}$$

If one defines

$$\vec{s} = \Omega^{-1} \sum_{p=1}^{\Omega} \vec{\sigma}_p , \tag{1.8}$$

then

$$\begin{aligned} s^{\pm}(t) &= s^{\pm}(0)e^{\mp it(2\epsilon - 2T_c s_z)}, \\ s_z &= \text{const.}, \quad \frac{d}{dt}(\vec{s}\cdot\vec{\sigma}_p) = 0, \end{aligned} \tag{1.9}$$

Equation (1.9) shows that $\vec{s}$ rotates with a frequency 2μ around the z-axis, and one can deduce that the $\vec{\sigma}_p$'s rotate around $\vec{s}$ with a frequency Δ.

In these lectures I shall comment on several of these mathematical questions. Since a good deal of the background material has been covered by previous lectures, I shall not talk about the physical motivations behind the BCS-model and will restrict the mathematical preparation to a discussion of infinite tensor products and traces. Also, I shall not derive this theory in the classical style of deducing lemmas and theorems but only try to convey the ideas behind it. There are many papers in which one can find the necessary theorems but few which explain the heuristics of it.

With these tools I will analyze later the BCS-model and related models and we will find that methods beyond classical mathematics are actually indispensable to understand what is going on.

2. Infinite Tensor Products

2.1. Heuristics.

If one tries to generalize the scalar product for a finite tensor product for a system of N degrees of freedom,

$$\langle x|y\rangle = \prod_{i=1}^{N} (x_i|y_i),$$
$$|x\rangle = |x_1\rangle \otimes |x_2\rangle \otimes \ldots \otimes |x_N\rangle, \tag{2.1}$$

to the case of $N=\infty$ one faces immediately the problem of convergence of the product

$$\prod^{\infty} (x_i|y_i).$$

Even if one is dealing with normalized $|x_i\rangle$, $|y_i\rangle$ such that $|(x_i|y_i)| \leqq 1$, the product need not converge. This would be the case only if all factors were positive. Indeed, for the absolute values there are only the alternatives

$$\text{I:} \quad \prod_{i=1}^{\infty} |(x_i|y_i)| \to c > 0$$
$$\text{II:} \quad \prod_{i=1}^{\infty} |(x_i|y_i)| \to 0. \tag{2.2}$$

Whereas for II we also have

$$\prod_{i=1}^{\infty} (x_i|y_i) \to 0,$$

in the case of I there is the possibility that the phases of $(x_i|y_i)$ spoil the convergence. For instance, if $(x_j|y_j) = e^{i\varphi_j}$, ($\varphi_j$ real), we have

$$\prod_{i=1}^{\infty} (x_i|y_i) = e^{i\sum_{j=1}^{\infty}\varphi_j}$$

and this converges if $\Sigma\phi_i$ does. To cope with this, one adopts the convention*

$$\begin{aligned} I_a \quad &: \quad \prod_{i=1}^{\infty} (x_i | y_i) = c \quad \text{if} \quad \prod_{i=1}^{\infty} (x_i | y_i) \to c \\ I_b,\ II \quad &: \quad \prod_{i=0}^{\infty} (x_i | y_i) = 0 \quad \text{otherwise.} \end{aligned} \tag{2.3}$$

I_b means that if

$$\prod_i |(x_i | y_i)|$$

converges to a value $\neq 0$ but

$$\prod_i (x_i | y_i)$$

does not converge, the scalar product is put equal to zero nevertheless. It turns out that with this definition of the scalar product one can consistently construct a Hilbert space from the linear combinations of these vectors in the usual fashion. Furthermore, it turns out that I_a and I give equivalence relations between vectors unless some factors are zero. To cope with this case, we define class I' if

$$\sum_i ||(x_i | y_i)| - 1| < \infty$$

and I_a' if

$$\sum_i |(x_i | y_i) - 1| < \infty$$

They are equivalence classes since the convergence of

---

*By convergence we always mean unordered convergence since we do not have a natural ordering in the index set. For complex numbers $\prod_i z_i$ is unordered convergent if $\prod_i |z_i|$ converges, and if this is $\neq 0$, then also $\sum_i |\arg z_i|$ has to be $< \infty$.

$$\sum_i |(x_i|y_i) - 1| \quad \text{and} \quad \sum_i |(y_i|z_i) - 1|$$

(or

$$\sum_i ||(x_i|y_i)| - 1| \quad \text{and} \quad \sum_i ||(y_i|z_i)| - 1|)$$

implies the convergence of

$$\sum_i |(x_i|z_i) - 1| \quad \text{(or} \quad \sum_i |(x_i|z_i)| - 1|$$

respectively). In these the infinite tensor product space decays into classes of mutually orthogonal subspaces. The whole space is called C.T.P. (complete tensor product) and classes of the type I_a', strong equivalence classes or I.C.T.P. (incomplete tensor product). Classes of the type I' are called weak equivalence classes. One may picture this as follows:

- C.T.P.
  - weak equiv. class
    - I.C.T.P.
    - I.C.T.P.
    - I.C.T.P.
    - ⋮
  - weak equiv. class
  - ⋮

Of course, this is schematic since there are innumerably many weak equivalence classes and each contains innumerably many I.C.T.P. However, if the individual factors are separable Hilbert spaces then the I.C.T.P.'s are too.

The relevant theorem of this theory states the following. If one takes the algebra $\{\otimes B_i\}''$ of the bounded operators B_i of the individual Hilbert spaces $|\mathfrak{H}_i\rangle$ (defined according to common sense in the C.T.P.) and their weak limits, then they do not lead out of the I.C.T.P. and are there represented irreducibly. In other words the C.T.P. gives a highly reducible representation of the B_i's, the I.C.T.P. being the irreducible sub-spaces. It turns out that within one weak equivalence class the representations are essentially the same—otherwise different. This means that the complication with the phase factors does not lead to anything usefully new and we shall therefore not distinguish between the various I.C.T.P. within one weak equivalence class. Let me illustrate these statements by a simple familiar example.

2.2. The Tensor Product Representations of the Algebra of a Spin Chain.

Let us consider the commutation relations

$$\left[\sigma_j^{(\alpha)}, \sigma_{j'}^{(\alpha')}\right] = 2i\delta_{jj'}\epsilon_{\alpha'\alpha'\alpha'} \, {}_j^{(\alpha'')}$$

where

$$\alpha = 1, 2, 3, \qquad j = 1, 2, \ldots .$$

If we construct the complete tensor product

$$H = \bigotimes_{j=1}^{\infty} H_j$$

(all H_j being two-dimensional), then we obtain a representation π of these commutation relations by

$$\sigma_j^{(\alpha)} \rightsquigarrow 1 \otimes 1 \ldots \otimes \underset{j\text{-th place}}{\sigma^{(\alpha)}} \otimes 1 \otimes \ldots$$

where $\sigma^{(\alpha)}$ is a usual Pauli matrix. This representation decomposes into irreducible sub-representations π_ζ again defined by

$$\sigma_j^{(\alpha)} \rightsquigarrow 1 \otimes 1 \otimes \ldots \otimes \sigma^{(\alpha)} \otimes 1 \otimes \ldots$$

but now the operator on the right hand side acts in an incomplete tensor product, labelled by the index ζ

$$H = \bigotimes_{j=1}^{\infty \, \zeta} H_j .$$

That these representations are irreducible follows from the fact that

$$\pi_\zeta\left(\{\sigma_j^{(\alpha)}\}\right) = \bigotimes_{i=1}^{\infty}{}^{\zeta} B(H_i),$$

by definition of

$$\bigotimes_{j=1}^{\infty}{}^{\zeta} B(H_i)$$

(note that every element $\in B(H_i)$ can be written as a linear combination of the unit matrix and the three Pauli matrices) and this is just

$$B\left(\bigotimes_{i=1}^{\infty}{}^{\zeta} H_i\right).$$

Therefore π as well as all π_ζ are of type I ($\pi = \bigoplus_\zeta \pi_\zeta$).†

If two equivalence classes ζ, ζ' belong to the same weak equivalence class ζ_W, then they are equivalent since there exists a unitary operator

$$U[(z_i)_{i=1,2,\ldots}] \in B\left(\bigotimes_{i=1}^{\infty} H_i\right), \quad (|z_i| = 1 \ \forall\, i),$$

which has the property that

$$U\pi\left(\sigma_j^{(\alpha)}\right)U^* = \pi\left(\sigma_j^{(\alpha)}\right) \quad \forall\ j\ \forall\ \alpha$$

but transforms

$$\bigotimes_{i=1}^{\infty}{}^{\zeta} H_i \quad \text{into} \quad \bigotimes_{i=1}^{\infty}{}^{\zeta'} H_i.$$

In fact, in the natural way of choosing the basis they are even identical.

If ζ, ζ' do not belong to the same weak equivalence class, then π_ζ and $\pi_{\zeta'}$ are inequivalent. This fact will be proved

†In fact, these representations are the usual Fock space representation, and states with finite products of suitable σ^- applied to a "ground state" are dense in this subspace of the C.T.P.

later. From now on, we shall identify all equivalence classes belonging to the same weak equivalence class because they give equivalent representations.

There are many more representations than the above-mentioned ones. Later on we shall, for example, discuss the thermodynamic representations of the BCS-model. But it should be said that, up to now, one cannot construct explicitly all possible representations. However, to facilitate later treatment of the BCS-model we conclude this section with a discussion of certain properties of the representations in our example.

i) Automorphisms. One kind of automorphism is the following:

$$\sigma_j^{(\alpha)} \rightsquigarrow M_{(j)}^{\alpha\beta} \sigma_j^{(\beta)},$$

where $M_{(j)}^{\alpha\beta}$ is an orthogonal (real) (3,3) matrix. Other automorphisms are, for example:

$$\sigma_j^{(\alpha)} \rightsquigarrow \sigma_{P(j)}^{(\alpha)}$$

P being a permutation of the natural members.

For all families of matrices $M_{(j)}$, there is always a unitary operator in the complete tensor product such that

$$U\pi\left(\sigma_j^{(\alpha)}\right)U^* = M_{(j)}^{\alpha\beta}\pi\left(\sigma_j^{(\beta)}\right).$$

But only if U does not lead out of the equivalence classes is $U \in \pi(\{\sigma_j^{(\alpha)}\})''$. However, this is not the case in general.

Let us analyze the situation in an irreducible representation π_ζ. Consider the automorphism

$$\sigma_j^{\pm} \rightsquigarrow e^{\pm i\phi}\sigma_j^{\pm}.$$

If ζ is determined by

$$\begin{pmatrix}1\\0\end{pmatrix} \otimes \begin{pmatrix}1\\0\end{pmatrix} \otimes \cdots,$$

then there exists a

$$U(\phi) \in B\left(\bigotimes_{i=1}^{\infty}{}^{\zeta} H_i\right).$$

The generator is the (unbounded) self-adjoint operator

$$\sum_{j=1}^{\infty} \left(\sigma_j^{(3)} - 1\right)$$

and

$$U(\phi) = \exp i \frac{\phi}{2} \sum_{j=1}^{\infty} \left(\left(\sigma_j^{(3)} - 1\right)\right).$$

If ζ does not contain

$$\begin{pmatrix} 1 \\ 0 \end{pmatrix} \otimes \begin{pmatrix} 1 \\ 0 \end{pmatrix} \otimes \ldots,$$

but another vector, say

$$\begin{pmatrix} a \\ b \end{pmatrix} \otimes \begin{pmatrix} a \\ b \end{pmatrix} \otimes \ldots, \quad (a, b \neq 0),$$

then one could also write formally

$$U_{(\phi)} = \exp\left(i \frac{\phi}{2} \sum_{j=1}^{\infty} (\sigma_j^{(3)} - 1)\right)$$

but neither the left hand side nor the right hand side makes sense, since $U(\phi) \in B(\bigotimes^{\zeta} H_i)$ (i.e., such a U can only be defined in a larger Hilbert space, e.g., the C.T.P.) and $\Sigma(\sigma_j^{(3)} - 1)$ makes a vector of infinite norm out of every non-zero vector.

ii) Isomorphisms of the Representations. Let π_ζ, $\pi_{\zeta'}$ be two representations, such that there exists an isometric operator U fulfilling

$$U\pi_\zeta\left(\sigma_j^{(\alpha)}\right) = M_{(j)}^{\alpha\beta} \pi_{\zeta'}\left(\sigma_j^{(\beta)}\right)U.$$

The two algebras $\pi_\zeta(\{\sigma_j^{(\alpha)}\})''$ and $\pi_{\zeta'}(\{\sigma_j^{(\alpha)}\})''$ are then spatially isomorphic. They are equivalent if $M_j = 1 \ \forall \ j$.

The two C^*-algebras $\pi_\zeta(\{\sigma_j^{(\alpha)}\})$ (the norm closure of all finite linear combinations of finite products of $\pi_\zeta(\sigma_j^{(\alpha)})$' s and $\pi_{\zeta'}(\{\sigma_j^{(\alpha)}\})$ are isomorphic, but generally this isomorphism cannot be extended to the weak closures. Take, for example,

$$\vec{s} = \lim_{N\to\infty} N^{-1} \sum_{j=1}^{N} \pi_\zeta(\vec{\sigma}_j).$$

There are a lot of representations where this limit exists (in the strong operator topology). Define $|n\rangle \in \mathbb{C}^2$ (up to a phase factor) by

$$\vec{n}\vec{\sigma}|\vec{n}\rangle = |\vec{n}\rangle \qquad (\langle \vec{n}|\vec{n}\rangle = 1).$$

Then in all equivalence classes containing

$$|\vec{n}\rangle \otimes |\vec{n}\rangle \otimes \dots ,$$

$\vec{s}$ exists. (Even in many more equivalence classes, but not in all.) A simple calculation yields

$$N^{-1} \sum_{j=1}^{N} \pi_{\vec{n}}(\vec{\sigma}_j) \to \vec{n}$$

($\pi_{\vec{n}}$ being the representation in one of the above equivalence classes), and this limit depends on the representation! If there were an isometric operator U transforming, for example,

$$U\vec{\sigma}_j = M\vec{\sigma}_j U,$$

then consequently

$$U\vec{n} = M\vec{n}U,$$

but $\vec{n}$ is a c-number. Hence this automorphism is incompatible with the weak and strong operator topology; it is only compatible with the norm topology.

Elaborating these ideas, one can easily show that π_ζ is not equivalent to $\pi_{\zeta'}$, unless ζ is in the same equivalence class as ζ'.

iii) Time Development. This is an automorphism depending on t. The simplest case one can consider is spins in a homogeneous external magnetic field $\vec{B}$, which points in the z-direction. Then all spins rotate around this direction. The unitary operator $U(t) \in B^{\#}$ is in suitable units

$$x_1 \otimes x_2 \otimes \dots \rightsquigarrow e^{i(t/2)\sigma^{(3)}} x_1 \otimes e^{i(t/2)\sigma^{(3)}} x_2 \otimes \dots .$$

It describes this rotation, but generally leads out of the I.C.T.P. Consequently, the formal generator

$$\sum_{j=1}^{\infty} \sigma_j^{(3)}$$

does not exist at all (except for $\binom{1}{0} \otimes \binom{1}{0} \otimes \dots$ if a suitable c-number is subtracted). The group $\{U(t)\}$ is weakly measurable—in fact, its matrix elements are for all C_0-vectors of the type

$$\langle a | U(t) | b \rangle = \begin{cases} 1 & \text{if} \quad t = 0 \\ 0 & \text{otherwise} \end{cases}$$

for t sufficiently small—but since the C.T.P. is not separable, Stone's theorem does not apply. (Only in separable Hilbert spaces does weak measurability imply weak continuity of unitary groups.)

Therefore one might argue that it is better not to consider the Schrödinger picture, but the Heisenberg picture. Even so the situation is not much improved; there exists an automorphism

$$\vec{\sigma}_j \rightsquigarrow M(t)\vec{\sigma}_j \doteq \vec{\sigma}_j(t),$$

but this automorphism cannot be extended to the weak closure of the algebra but only to the smallest C^*-algebra containing all $\vec{\sigma}_j$'s.

3. Traces

For the treatment of systems at finite temperatures one works with traces since the thermal expectation value is given by

$$\langle A \rangle = \operatorname{Tr} A e^{-\beta H} / \operatorname{Tr} e^{-\beta H}; \tag{3.1}$$

whereas for finite dimensional spaces there is no difficulty in defining the trace by

$$\mathrm{Tr}\, A = \sum_{i=1}^{N} (e_i | A e_i) , \tag{3.2}$$

the e_i being an orthogonal basis; for $N = \infty$ we run into severe difficulties even for bounded A, not only that

$$\sum_{1}^{\infty} (e_i | A e_i)$$

may not converge, or not absolutely converge, so that it may assume any value in a different basis. Even absolute convergence in one basis does not guarantee the convergence in another basis.

Take

$$\begin{pmatrix} \begin{matrix} 0 & 1 \\ 1 & 0 \end{matrix} & & & \\ & \begin{matrix} 0 & \frac{1}{2} \\ \frac{1}{2} & 0 \end{matrix} & & \\ & & \begin{matrix} 0 & \frac{1}{3} \\ \frac{1}{3} & 0 \end{matrix} & \\ & & & \ddots \end{pmatrix} \tag{3.3}$$

for which the trace is absolutely convergent and zero. In another basis it looks like

$$\begin{pmatrix} \begin{matrix} 1 & \\ & -1 \end{matrix} & & & \\ & \begin{matrix} \frac{1}{2} & \\ & -\frac{1}{2} \end{matrix} & & \\ & & \begin{matrix} \frac{1}{3} & \\ & -\frac{1}{3} \end{matrix} & \\ & & & \ddots \end{pmatrix}$$

for which Tr is only conditionally convergent and may assume any value by a further change of basis.

It turns out that if $(x | Ax) \geqq 0 \ \forall \ x$, the result of (3.2) for $N = \infty$ is independent of the choice of the basis. Thus for positive operators the trace is well defined but may assume the value ∞ (as for $A = 1$, $N = \infty$). We get therefore a finite well-defined trace for linear combinations of positive operators with finite trace (trace-class). It turns out that multiplication by a bounded operator does not spoil this property so that the trace-class is a two-sided ideal.

One may argue that the requirements are too restrictive because the following may happen. Consider $H = C^{(2)} \otimes H_o$ where H_o is ∞-dimensional and consider the operators $\vec{\sigma} \otimes 1$. Since they are of the form

$$\begin{pmatrix} \begin{matrix} 1 & \\ & -1 \end{matrix} & & & \\ & \begin{matrix} 1 & \\ & -1 \end{matrix} & & \\ & & \begin{matrix} 1 & \\ & -1 \end{matrix} & \\ & & & \ddots \end{pmatrix}$$

they do not satisfy the requirements and therefore one cannot define a trace for them. On the other hand, it is clear that this is a trivial situation and it would be silly not to define the trace just as for the σ's in $C^{(2)}$. To cope with this situation, one needs to look more generally for a "relative trace" or "super trace" ϕ which is a positive linear functional over the positive operators of an algebra $\mathcal{A}$ (it may also be infinite) satisfying

$$\begin{aligned} &\text{i)} \quad \phi(\lambda A) = \lambda\phi(A) \qquad A \geqq 0 \quad \lambda \geqq 0 \\ &\text{ii)} \quad \phi(A+B) = \phi(A) + \phi(B) \qquad (3.4) \\ &\text{iii)} \quad \phi(UAU^{-1}) = \phi(A), \quad \forall\, U \text{ unitary}, \in \mathcal{A}. \end{aligned}$$

It turns out that these positive operators for which $\phi(A) < \infty$ are again the positive part of a two-sided ideal. What is more important, for a factor (which is an operator algebra $\mathcal{A}$ such that the only operators $\in \mathcal{A}$ commuting with all other operators of $\mathcal{A}$ are a multiple of unity) ϕ† is unique up to a positive number. Therefore one can use ϕ to classify factors.‡ There are the following categories denoted by III, I and II.

---

†One has to assume in addition to i), ii) and iii) that ϕ is normal, e.g., if A_n is an increasing sequence of operators and A is the supremum of A_n, the $\phi(A_n) \uparrow \phi(A)$.

‡This has some relevance for physics inasmuch as phase transitions will turn out to be accompanied by a change in the type of the representation.

III. The only possibility (except $\phi \equiv 0$) is

$$\phi(A) = \begin{cases} 0 & \text{for } A = 0 \\ \infty & \text{for } A \neq 0 . \end{cases}$$

In this case there is no reasonable linear form with the properties of a trace and we will see that in a way this is the most common situation. (This does not, of course, mean that in the Hilbert space in which the A's are acting one cannot define a trace but the trace-class has an empty intersection with our algebra.)

I. All bounded operators B in a Hilbert space are a factor (they are even irreducible) and because of the uniqueness of the trace ϕ must be of the usual form. These factors are denoted by I_N where N is the dimensionality of the Hilbert space. (Thus for I_∞ if $\phi(I) = \infty$).

II. The remaining cases are denoted by II_1 or II_∞ depending on whether $\phi(1) < \infty$ or $\phi(1) = \infty$. We shall later encounter factors of type III and II_1, (II_∞ is always of the form $II_1 \otimes I_\infty$).

4. The BCS-Model: Ground State

The mathematical structure of the BCS model is instructive since one can treat it exactly in the case of a finite volume of a superconductor as well as in the infinite case.

4.1. The Case $\Omega \to \infty$, Treatment "a la Haag".

We now consider what happens, if $\Omega \to \infty$, in the I.C.T.P. representations.

There are some representations where

$$\vec{s}_\Omega = \Omega^{-1} \sum_{p=1}^{\Omega} \vec{\sigma}_p$$

converges towards a certain c-number $\vec{s}$. For example, the representations characterized by

$$|\{\vec{n}\}\rangle = |\vec{n}\rangle \otimes |\vec{n}\rangle \otimes \ldots \tag{4.1}$$

where $\vec{s}$ converges towards $\vec{n}$ and one might expect that $\vec{s} - \vec{n}$ can be neglected. (Note, however, that in the I.C.T.P. containing

$$\begin{pmatrix}1\\0\end{pmatrix}\otimes\underbrace{\begin{pmatrix}0\\1\end{pmatrix}\otimes\begin{pmatrix}0\\1\end{pmatrix}\otimes\begin{pmatrix}1\\0\end{pmatrix}\otimes\ \cdots}_{4\ \text{times}}\otimes\underbrace{\begin{pmatrix}0\\1\end{pmatrix}\otimes\ \cdots}_{8\ \text{times}}\ \cdots$$

this limit does not exist.)

In these representations, (4.1), Haag's method assumes that

$$H_\Omega = -\epsilon \sum_{p=1}^{\Omega} \sigma_p^{(z)} - \frac{2T_c}{\Omega} \sum_{p,p'} (\sigma_p^+ - n^+)(\sigma_{p'}^- - n^-)$$

$$- 2T_c \sum_p (\sigma_p^+ n^- + \sigma_p^- n^+) + \text{const.} \tag{4.2}$$

converges towards H_B (B stands for Bogoliubov) which is the linear part of H_Ω and has the form

$$H_B = \frac{\Delta}{2} \sum_{p=1}^{\infty} (\vec{n}\vec{\sigma}_p - 1). \tag{4.3}$$

One notices that $H_\Omega - H_B$ is quadratic in $\vec{s} - \vec{n}$ and that H_B and H_Ω have (if $\Omega \to \infty$) the same commutators with all σ_p's. Explicitly,

$$H_B = - \sum_{p=1}^{\infty} \left\{ \epsilon\, \sigma_p^{(z)} + T_c \left(n^{(x)} \sigma_p^{(x)} + n^{(y)} \sigma_p^{(y)} \right) \right\} \tag{4.4}$$

in the I.C.T.P.'s containing $|\{\vec{n}\}\rangle$, and, comparing (4.3) with (4.4), one concludes that the "gap equation"

$$T_c = \Delta/2, \qquad n_z = \cos\theta = \epsilon/T_c \tag{4.5}$$

must hold, i.e., $\mu = 0$, since $n_z = \lim S_z/S$. (The point is that

$$\sum_{p=1}^{\infty} (\vec{n}'\vec{\sigma}_p - \text{const.})$$

can only be defined in the I.C.T.P.'s characterized by $|\{\vec{n}\}\rangle$ if $\vec{n}' = \vec{n}$, otherwise this operator makes a "vector" of indefinite norm out of each non-zero vector.)

4.2. Explicit Treatment.

One can check all assumptions explicitly and obtain the following results

i) H_Ω (or a suitable c-number) converges weakly in the I.C.T.P.'s where

$$\lim \Omega^{-1} \sum_{p=1}^{\Omega} \vec{\sigma}_p = \vec{s} \tag{4.6}$$

exists and the consistency relation (4.5) ("gap equation") is fulfilled which says physically that $\vec{s}$ does not change in time (i.e., there is no rotation around the z-axis). A ground state exists in the I.C.T.P. of $|\{\vec{n}\}\rangle$.

ii) $e^{iH_\Omega t}\vec{\sigma}_p e^{-iH_\Omega t}$ converges strongly if $\vec{s}$ exists, and its limit is

$$e^{iH_B t}\vec{\sigma}_p e^{-iH_B t}$$

if $\vec{s}$ is constant in time. Otherwise the time development of $\vec{\sigma}_p$ consists of a rotation around $\vec{s}$ which itself rotates around the z-axis. In this case there is no Hamiltonian (obtainable either by limiting processes from the H_Ω's, or otherwise) governing the time evolution. For, although a U(t) exists for this motion in the C.T.P., it leaves no vector $\neq 0$ and therefore no I.C.T.P. invariant. Thus U(t) is in none of these representations weakly continuous and can nowhere be written as e^{-iHt}. Furthermore H_B gives the wrong time dependence since it gives the rotation around $\vec{s}$ which is kept constant.

iii) $e^{iH_\Omega t} \nrightarrow e^{-iH_B t}$ since

$$\lim \langle\{\vec{n}\}|e^{iH_\Omega t}|\{\vec{n}\}\rangle = (1 + itT_c)^{-\frac{1}{2}} \tag{4.7}$$

but

$$\langle\{\vec{n}\}|e^{iH_B t}|\{\vec{n}\}\rangle = 1.$$

iv) One can formulate the problem in the language of spin waves, defined by

$$\vec{\sigma}_\lambda = \Omega^{-\frac{1}{2}} \sum_{p=1}^{\Omega} e^{i\lambda p}\vec{\sigma}_p , \tag{4.8}$$

and consider the limit $\Omega \to \infty$. One obtains a different result: in the representation where the gap equation holds, H_Ω converges strongly towards a self-adjoint operator H (on a dense domain) which is not equal to H_B.

$$H = T_c\left[\sum_{\lambda\neq 0} \tau^-_{-\lambda}\tau^+_\lambda + \left(1 - \frac{\omega^2}{T_c^2}\right)p^2\right]$$

$$\tau_0^\pm = (q \pm ip)/\sqrt{2}, \qquad \tau^-_{-\lambda} = (\tau^+_\lambda)^*. \tag{4.9}$$

The $\vec{\tau}$'s are rotated $\vec{\sigma}$'s. If $\Omega = \infty$, they obey pure boson commutation relations

$$\left[\tau^+_\lambda, \tau^-_{\lambda'}\right] = \delta_{\lambda,-\lambda'} \tag{4.10}$$

The spectrum of H is the same as for H_B for almost all degrees of freedom; only one degree of freedom ($\lambda = 0$) has a continuum in its part of the Hamiltonian. (For a free particle this continuum corresponds to states with the same r and different N which, for $\Omega \to \infty$, become degenerate.) It is reminiscent of the rotation of $\vec{s}$ in the sense that in these representations $\vec{s}$ does not rotate but there is also no force which keeps $\vec{s}$ fixed.

5. The BCS-Model: Thermodynamic Limit

We shall now calculate the thermodynamic expectation values

$$\langle A\rangle_\Omega = \mathrm{tr}\left[e^{-H_\Omega/T}A\right]\Big/\mathrm{tr}\left[e^{-H_\Omega/T}\right] \tag{5.1}$$

of elements A belonging to the *-algebra Σ' generated by the $\vec{\sigma}_p$'s. However, we are not really interested in the value of $\langle A\rangle_\Omega$ itself, but rather in the limit $\Omega \to \infty$ which we hope will exist. (The *-algebra Σ' is of infinite dimensions; there is a little problem in defining the trace but one should notice that Σ' consists only of finite linear combinations of elements of the form

$$\sigma_1^{(\alpha_1)} \ldots \sigma_n^{(\alpha_n)},$$

etc., and for these $\langle A\rangle_\Omega$ is well-defined if $\Omega \geqq n$. Hence, the question of the existence of $\lim\langle A\rangle_\Omega$ makes sense for all $A \in \Sigma'$.) The reader should bear in mind that in the expression defining

$\langle A\rangle_\Omega$, S_z, and hence the number of pairs, is not kept fixed so that $\langle A\rangle_\Omega$ corresponds to the grand-canonical expectation value.

Our investigations on the ground state suggest the following conjecture. Splitting H_Ω into two parts,

$$H_\Omega = H_{B,\Omega} + H'_\Omega$$

$$H_{B,\Omega} = -\epsilon \sum_{p=1}^{\Omega} \sigma_p^{(z)} - 2T_c \sum_{p=1}^{\Omega} (\sigma_p^+ \langle\sigma^-\rangle_B + \sigma_p^- \langle\sigma^+\rangle_B)$$

$$= -T\omega \sum_{p=1}^{\Omega} (\vec{\sigma}_p \vec{n}) \tag{5.2}$$

$$H'_\Omega = -\frac{2T_c}{\Omega} \sum_{p=1}^{\Omega} (\sigma_p^+ - \langle\sigma^+\rangle_B) \sum_{p'=1}^{\Omega} (\sigma_{p'}^- - \langle\sigma^-\rangle_B)$$

$$- 2T_c\Omega \langle\sigma^+\rangle_B \langle\sigma^-\rangle_B$$

we know that in the ground state, $H'_\Omega \to 0$ and also does not contribute to the time development of the $\vec{\sigma}_p$'s when $\Omega \to \infty$. Our conjecture is now that H'_Ω also does not contribute to the thermodynamic expectation values when $\Omega \to \infty$, i.e.,

$$\lim_{\Omega\to\infty} \langle A\rangle_{B,\Omega} = \lim_{\Omega\to\infty} \langle A\rangle_\Omega \tag{5.3}$$

with

$$\langle A\rangle_{B,\Omega} = \mathrm{tr}\left[e^{-H_{B,\Omega}/T} A\right] \Big/ \mathrm{tr}\left[e^{-H_{B,\Omega}/T}\right] \tag{5.4}$$

if the self-consistency relations ($\langle\vec{\sigma}\rangle_B = \vec{n}$ th ω where th denotes hyperbolic tangent)

$$\omega = \frac{T_c}{T} \,\mathrm{th}\, \omega, \qquad \cos\theta = \epsilon/T\omega \tag{5.5}$$

are fulfilled. They are the generalization of (4.5) to finite temperatures. But, certainly, the above-stated form of our conjecture cannot be true, since, for example,

$$\langle\sigma^{(\alpha)}\rangle_{\Omega} = 0$$
$$\langle\sigma^{(\alpha)}\rangle_{B,\Omega} = n^{(x)}\ \text{th}\,\omega \neq 0 \qquad (5.6)$$

in general. However, we shall show that the correct thermodynamic limit, which is obtained by using $H_{B,\Omega}$ instead of H_{Ω}

$$\lim_{\Omega\to\infty} \langle A\rangle_{\Omega} = \lim_{\Omega\to\infty} (2\pi)^{-1} \int_0^{2\pi} d\phi \langle A\rangle_{B,\Omega}. \qquad (5.7)$$

(Our self-consistency conditions fix only the angle θ, while ϕ is not determined.)

Before we go over to the proof, two remarks should be added:

i) The energy gap $2T\omega$ depends on the temperature and becomes smaller with increasing temperature corresponding to the decrease of Δ with (r); (4.6).

ii) The first of the equation (5.5) tells us that if $T = T_C$, then $\omega = 0$, as can easily be seen in Figure 2.

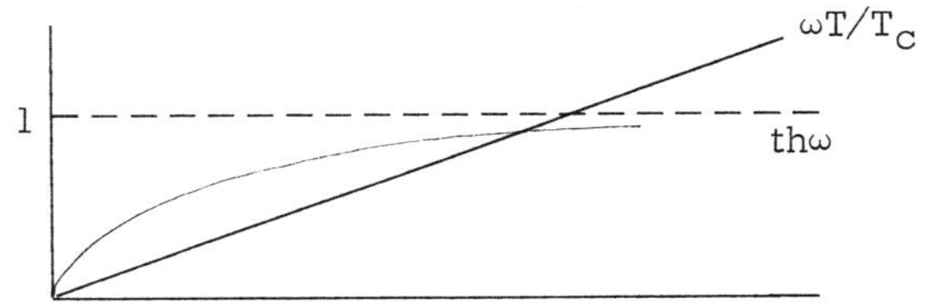

Figure 2

However, from $\cos\theta = \epsilon/T\omega$ we learn that $T\omega > \epsilon$. $\cos\theta$ increases with T until it reaches 1 for T_0 determined by $\epsilon/T_C = \text{th}\,\epsilon/T_0$. Then it remains 1 and the discontinuity of the derivation of θ with respect to T implies a phase transition of second kind.

Now let us indicate how one calculates the thermodynamic limit.

a) The right hand side of (5.7). We shall use a generating functional

$$\hat{A}_\Omega = \exp\left(i \sum_{p=1}^{\Omega} a_p \sigma_p^{(x)}\right) \exp\left(i \sum_{p=1}^{\Omega} b_p \sigma_p^{(y)}\right) \exp\left(i \sum_{p=1}^{\Omega} c_p \sigma_p^{(z)}\right). \qquad (5.8)$$

By differentiating $\langle \hat{A}_\Omega \rangle_\Omega$, one obtains all expectation values. We can even use a simpler functional

$$A_\Omega = \exp\left(i \frac{a}{\Omega} \sum_{p=1}^{\Omega} \sigma_p^{(x)}\right) \exp\left(i \frac{b}{\Omega} \sum_{p=1}^{\Omega} \sigma_p^{(y)}\right) \exp\left(i \frac{c}{\Omega} \sum_{p=1}^{\Omega} \sigma_p^{(z)}\right) \qquad (5.9)$$

because H_Ω as well as $H_{B,\Omega}$ are invariant under all permutations of the indices. Then, for example,

$$\langle \sigma_p^{(x)} \rangle = \frac{\partial}{\Omega(a)} \langle A_\Omega \rangle_\Omega \bigg|_{a=b=c=0} \qquad (5.10)$$

etc. (One might think that one could use a parametrization via the Euler angles, i.e.:

$$A_\Omega = \exp\left(i \frac{a}{\Omega} \sum_{p=1}^{\Omega} \sigma_p^{(z)}\right) \exp\left(i \frac{b}{\Omega} \sum_{p=1}^{\Omega} \sigma_p^{(y)}\right) \exp\left(i \frac{c}{\Omega} \sum_{p=1}^{\Omega} \sigma_p^{(z)}\right) \qquad (5.11)$$

but it turns out that one cannot generate all expectation values by use of this functional.)

Now $\langle A_\Omega \rangle_{B,\Omega}$ can easily be calculated:

$$\langle A_\Omega \rangle_{B,\Omega} = \Big[1 + i\Omega^{-1} \, \mathrm{th}\,\omega (a \sin\theta \cos\phi + b \sin\theta \sin\phi + c \cos\theta + \\ + 0(\Omega^{-2})\Big]^\Omega \rightarrow \exp\Big(i \, \mathrm{th}\,\omega \Big[a \sin\theta \cos\phi + \\ + b \sin\theta \sin\phi + c \cos\theta\Big]\Big) \qquad (5.12)$$

and integration over ϕ yields

$$(2\pi)^{-1} \int_0^{2\pi} d\phi \, \langle A_\Omega \rangle_{B,\Omega} \rightarrow J_0(\mathrm{th}\,\omega \sin\theta \sqrt{a^2+b^2}) \cdot \exp(ic \, \mathrm{th}\,\omega \cos\theta). \qquad (5.13)$$

b) The left hand side of (5.7). The diagonalization of Ω is very simple, and inserting the matrix elements of the rotation group (which in our case are less familiar than the usual matrix elements based on the Euler angles) and paying attention to the degeneracy of the eigenvalues, one obtains

$$\mathrm{tr}\left[e^{-H_\Omega/T} A_\Omega\right] = \sum_{L=0}^{\Omega/2} \sum_{L_z=-L}^{L} \frac{\Omega!\,(2L+1)}{((\Omega/2)-L)!\,((\Omega/2)+L+1)!}$$

$$\cdot \exp\left\{\frac{1}{T}\left[\epsilon\,(2L_z-\Omega) + \frac{2T_c}{\Omega}\,(L(L+1) - L_z(L_z+1))\right]\right\}$$

$$\cdot\, G_\Omega\left(\frac{2L}{\Omega}, \frac{2L_z}{\Omega}; a, b, c\right). \tag{5.14}$$

For the functions G_Ω we find that

$$G_\Omega\left(\frac{2L}{\Omega}, \frac{2L_z}{\Omega}; a,b,c\right) = \langle L, L_z | A_\Omega(a,b,c) | L, L_z\rangle$$

$$\rightarrow\; e^{iwc} J_0\left(\sqrt{(y^2-w^2)(a^2+b^2)}\right)\ (\equiv G_\infty(y,w;a,b,c,)) \tag{5.15}$$

where we introduced the intensive quantities

$$y = 2L/\Omega, \qquad w = 2L_z/\Omega \tag{5.16}$$

which are confined to the triangle in Figure 3.

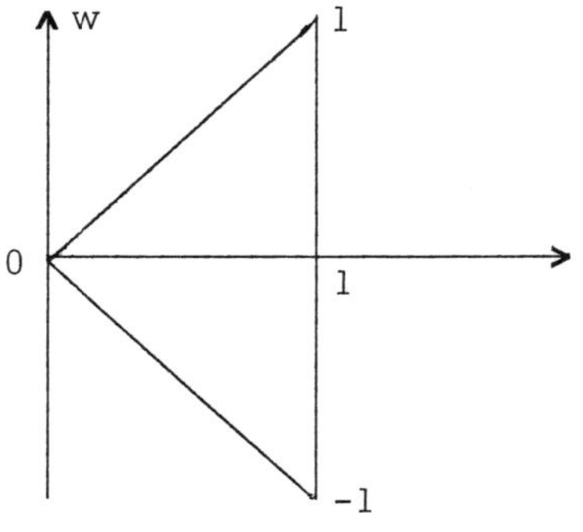

Figure 3

We now have to calculate the limit of the double sum. The usual prescription is to replace the sum by its leading term, and indeed this can be rigorously justified in our case by considering sequences of probability measures on the triangle. What we find is that

$$\sum_{L=0}^{\Omega/2} \sum_{L_z=-L}^{L} \to \Omega^2 \int_0^1 dy \int_{-y}^{y} dw,$$

$$\langle A \rangle_\Omega \to \int_0^1 dy \int_{-y}^{y} dw\, \rho(y, w) G_\infty(y, w; a, b, c) \cdot \left(\int_0^1 dy \int_{-y}^{y} dw\, \rho(y, w) \right)^{-1} \tag{5.17}$$

with

$$\rho(y, w) = \exp\left(-\Omega \left[f(y_0) - f(y) + \frac{T_c}{2T} (w - w_0)^2 \right]\right)$$
$$f(y) = \frac{T_c}{2T} y^2 - \frac{1-y}{2} \ell n(1-y) - \frac{1+y}{2} \ell n(1+y), \tag{5.18}$$

y_0 being determined by

$$f'(y_0) = 0$$

hence

$$y_0 = \mathrm{th} \frac{T_c}{T} y_0, \qquad w_0 = \frac{\epsilon}{T_c} - \frac{1}{\Omega} \tag{5.19}$$

and, by the usual methods,

$$\lim_{\Omega \to \infty} \langle A_\Omega \rangle_\Omega = G_\infty(y_0, w_0; a, b, c). \tag{5.20}$$

The identity of the right hand side and the left hand side of (5.7) is now established if one notices the relations between ω, θ and y_0, w_0:

$$\frac{\omega T}{T_c} = y_o, \qquad w_o = \frac{\epsilon}{T_c} = \cos\theta\, \mathrm{th}\,\omega$$

$$\sin\theta\ \mathrm{th}\,\omega = \sqrt{\left(\frac{T\omega}{T_c}\right)^2 - \frac{\epsilon^2}{T_o^2}} = \sqrt{y_o^2 - w_o^2}\ . \tag{5.21}$$

5.1. The Thermodynamic Representations.

Here we shall list the various factor types which occur. The rather lengthy proofs of these statements can be found in Reference 1.

i) <u>T = 0</u>

$$y_o = 1 = \mathrm{th}\,\omega. \tag{5.22}$$

The thermodynamic representation is an integral over type I factors:

$$\pi = \int^{\oplus} d\phi \cdot \pi_\phi \tag{5.23}$$

π_ϕ is a representation in an I.C.T.P.

$$\overset{\infty}{\underset{p=1}{\bigotimes}}{}^{\zeta} (H_p \otimes H_p), \tag{5.24}$$

all H_p being two-dimensional, such that (in an obvious notation)

$$\vec{\sigma}_p \rightsquigarrow (1 \otimes 1) \otimes \ldots \otimes (\vec{\sigma} \otimes 1) \otimes (1 \otimes 1) \ldots$$

p-th place. (5.25)

The equivalence class is determined by

$$\otimes\ (|\phi\rangle \otimes |\phi\rangle) \tag{5.26}$$

$$(|\phi\rangle\ :\ (\sigma^{(x)} \cos\phi + \sigma^{(y)} \sin\phi)\,|\phi\rangle = |\phi\rangle .$$

Phase factors are irrelevant).

That this is, in fact, the thermodynamic representation can easily be checked by comparison of the expectation values.

Since in this case

$$\bigotimes^{\zeta} (H_p \otimes H_p) = \bigotimes^{\zeta_1} H_p \otimes \bigotimes^{\zeta_2} H_p$$
$$(\zeta_1, \zeta_2 \ni \otimes |\phi\rangle) \tag{5.27}$$

one sees that the thermodynamic representation is just a gauge invariant integral over amplifications of the ground-state representations that we found in the foregoing lecture. (Amplification of an algebra is defined as its tensor product with some algebra of scalars C(H).)

One remark should be added: the operator

$$\vec{s} = \lim \Omega^{-1} \sum_{p=1}^{\Omega} \vec{\sigma}_p \tag{5.28}$$

is no longer a c-number but belongs only to the center. (In fact, $\langle s_x \rangle = 0$ but $\langle s_x^2 \rangle \neq 0$, etc.)

ii) $\underline{0 < T < T_o}$. The critical temperature T_o is determined by

$$\epsilon / T_c = \text{th}\, \epsilon / T_o . \tag{5.29}$$

(Only if $\epsilon = 0$, $T_o = T_c$.)

The thermodynamic representation is now an integral

$$\pi = \int^{\oplus} d\phi \; \pi_\phi$$

over type III-factor representations in I.C.T.P.'s:

$$\bigotimes^{\zeta} (H_p \otimes H_p)$$

as before, but the equivalence class ζ now contains

$$\bigotimes_{p=1}^{\infty} \sqrt{\frac{1+\text{th}\omega}{2}}\, |\vec{n}\rangle \otimes |\vec{n}\rangle + \sqrt{\frac{1-\text{th}\omega}{2}}\, |-\vec{n}\rangle \otimes |-\vec{n}\rangle$$

$$|\vec{n}\rangle : (\sigma\vec{n})| \pm \vec{n}\rangle = \pm|\pm\vec{n}\rangle$$

$$\vec{n} = \begin{bmatrix} \cos\phi \sqrt{1-(\epsilon/T_c\, \text{th}\,\omega)^2} \\ \sin\phi \sqrt{1-(\epsilon/T_c\, \text{th}\,\omega)^2} \\ \epsilon/T_c\, \text{th}\,\omega \end{bmatrix} \tag{5.30}$$

When $\|\vec{s}\| < 1$ one gets a different type of factor since for this ζ the relation (5.27) does not hold any more (i.e., the infinite tensor product is generally not associative.

iii) $\underline{T_0 \leqq T < \infty,\ \epsilon \neq 0}$. In this region our previous calculations have to be modified. So far we have not checked whether (y_0, w_0) is in the triangle at all. Since

$$y_0 = \text{th}\,\frac{T_c}{T}\, y_0, \qquad w_0 = \frac{\epsilon}{T_c} - \frac{1}{\Omega}, \tag{5.31}$$

we find that this is indeed the case as long as $T < T_0$, but not if $T \geqq T_0$; see Figure 4.

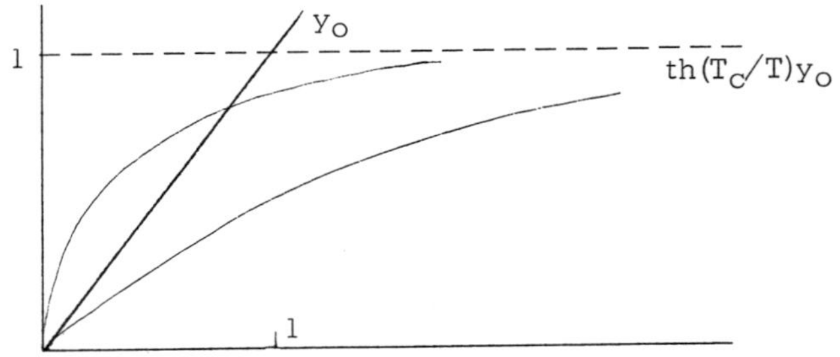

Figure 4

We have $|w_0| \geqq y_0$ in the latter case. The maximum value inside the triangle is attained on the boundary $|w| = y$, and therefore

$$y_0 = \text{th}\,\epsilon/T, \qquad \sin\theta = 0$$
$$\langle A_\Omega \rangle_\Omega \to e^{ic\,\text{th}\,\epsilon/T} \,. \tag{5.32}$$

One sees that there is a phase transition at T_0. The thermodynamic representation is again in an I.C.T.P.

$$\bigotimes_{p=1}^{\infty}{}^{\zeta} (H_p \otimes H_p)$$

but it is now no longer integrated, ($\sin\theta = 0$), but a type III-factor representation. ζ is determined by

$$\bigotimes_{p=1}^{\infty} \left(\sqrt{\frac{1+\text{th}\,\epsilon/T}{2}}\, |\vec{z}\rangle \otimes |\vec{z}\rangle + \sqrt{\frac{1-\text{th}\,\epsilon/T}{2}}\, |-\vec{z}\rangle \otimes |-\vec{z}\rangle \right)$$
$$(|\pm\vec{z}\rangle : \sigma^{(z)}\, |\pm\vec{z}\rangle = \pm\, |\vec{z}\rangle). \tag{5.33}$$

(One could equally use the reference vector

$$\bigotimes_{p=1}^{\infty} \left(\sqrt{\frac{1+\text{th}\,\epsilon/T}{2}}\, |\vec{z}\rangle \otimes |e_1\rangle + \sqrt{\frac{1-\text{th}\,\epsilon/T}{2}}\, |-\vec{z}\rangle \otimes |e_2\rangle \right)$$

$\{|e_1\rangle, |e_2\rangle\}$ being an orthonormal basis, since all these I.C.T.P.'s yield equivalent representations.)

When $\epsilon = 0$, $\langle A_\Omega \rangle_\Omega \to 1$, i.e., there is complete disorder. The thermodynamic representation is of the type stated above, but with reference vector

$$\bigotimes_{p=1}^{\infty} \left(\frac{1}{\sqrt{2}}\, |\vec{z}\rangle \otimes |\vec{z}\rangle + \frac{1}{\sqrt{2}}\, |-\vec{z}\rangle \otimes |-\vec{z}\rangle \right) \tag{5.34}$$

i.e., a type II_1-factor representation. Here we have the remarkable situation that beyond T_c the Hamiltonian vanishes in some sense and a II_1 representation becomes possible.

iv) $\underline{T = \infty}$. We obtain the same type II_1 factor representation as above in both cases $\epsilon = 0$ and $\epsilon \neq 0$.

5.2. An "Isotropic" BCS-Model.

One can apply the above-developed techniques to the following Hamiltonian.

$$H_{\Omega} = \epsilon \sum_{p=1}^{n} \sigma_p^{(z)} - \frac{T_c}{2\Omega} \sum_{p,p'=1}^{\Omega} \vec{\sigma}_p \vec{\sigma}_{p'}, \ (\epsilon > 0). \tag{5.35}$$

We arrive at the equality

$$\text{th}\,\omega = T/T_c\,\omega - \epsilon/T_c \tag{5.36}$$

or ($y_0 = \text{th}\,\omega$)

$$y_0 = \text{th}\,\omega(T_c/T\, y_0 + \epsilon/T) \tag{5.37}$$

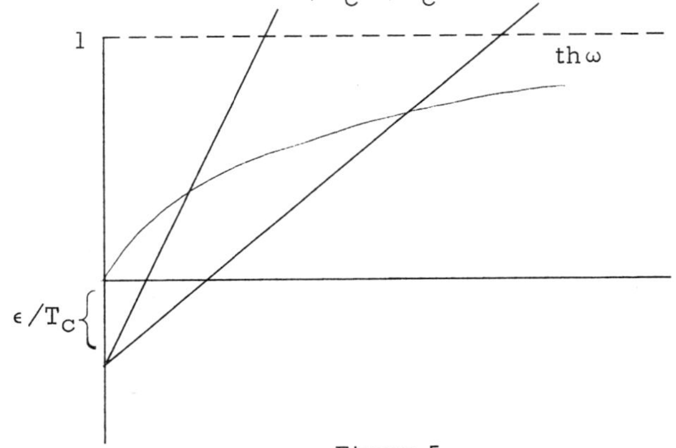

Figure 5

and $\lim\langle A_{\Omega}\rangle_{\Omega} \rightarrow e^{iy_0 c}$. The thermodynamic representation is a
Type I factor representation for $T = 0$
Type III factor representation for $0 < T < \infty$
Type II_1 factor representation for $T = \infty$.

There is no phase transition.

If, however, $\epsilon = 0$, then the calculation yields

$$\langle A_{\Omega}\rangle_{\Omega} \rightarrow \frac{\sin y_0\sqrt{a^2+b^2+c^2}}{y_0\sqrt{a^2+b^2+c^2}} = \lim \frac{1}{4\pi}\int \sin\theta\, d\theta\, d\phi\, \langle A_{\Omega}\rangle_B \tag{5.38}$$

hence the thermodynamic representation is an integral of type III (type I) factor representations if $0 < T < T_0$ $(T = 0)$. At T_0 there is a phase transition, and for $T \geqq T_0$ $\langle A_\Omega \rangle_\Omega \to 1$, hence we have a type II_1 factor representation.

6. Concluding Remarks

I would like to conclude with some remarks about the thermal expectation values. These remarks are not specific for the BCS model. I shall consider the general functional

$$\langle \vec{\sigma}_1 \ldots \vec{\sigma}_j \rangle = \vec{n}_1 \ldots \vec{n}_j \, \eta^j \qquad \vec{n}^2 = 1 \qquad 0 \leqq \eta \leqq 1 \qquad \langle 1 \rangle = 1 .$$

We know that for the BCS model the thermal expectation value is either of this form or an integral over it.

i) The functional is clearly bounded and can therefore be uniquely extended to a bounded linear functional over the norm completion of the *-algebra generated by the $\vec{\sigma}$'s. This completed algebra is a C^*-algebra.

ii) In a particular representation we may form the weak closure of the (representation of the) C^*-algebra. According to the Hahn-Banach theorem, the functional can be extended to a norm-continuous functional over the whole von Neumann algebra. (One can even suppose this extension to be positive, which is a theorem of Krein.) Whether this extension will be normal or not depends on the representation. In the one obtained by the GNS construction, it is of course even weakly continuous since it is a diagonal matrix element. On the other hand, in the irreducible I.C.T.P. representation with "spins up", it is not weakly continuous. Form

$$P_N = \prod_{i=N}^{\infty} \frac{1 + \sigma_i^{(z)}}{2} . \tag{5.39}$$

This product converges strongly towards the projector on the subspace where for $i \geqq N$ all spins are up. $\underset{N \to \infty}{\text{s-lim}} \, P_N = 1$, on the other hand

$$\langle P_N \rangle = \lim_{N' \to \infty} \prod_{i=N}^{N} \langle \frac{1+\sigma_i^{(z)}}{2} \rangle = \lim_{N' \to \infty} \left(\frac{1+\eta}{2} \right)^{N'-N} = 0 \quad \text{if} \quad \eta < 1 . \tag{5.40}$$

$$\Longrightarrow \lim_{N \to \infty} \langle P_N \rangle \neq \langle \lim_{N \to \infty} P_N \rangle .$$

iii) One may raise the question of the existence of a density matrix. We have noted for the ground state that

$$\lim_{\Omega\to\infty} H_{B,\Omega}$$

exists strongly and the same holds in the thermal representation, provided a suitable element from the commutant is subtracted. Thus

$$\lim_{\Omega\to\infty} e^{-\beta H_{B,\Omega}}$$

exists but is not in the trace class, and consequently,

$$\lim_{\Omega\to\infty} e^{-\beta H_{B,\Omega}} / \operatorname{tr} \lim_{\Omega\to\infty} e^{-\beta H_{B,\Omega}} = 0. \tag{5.41}$$

iv) Whether the above-constructed functional can be written as the trace of a density matrix times the element of the algebra depends on the representation. In the thermal representation we have $\langle A\rangle = \operatorname{tr} P_0 A$, P_0 being the projector onto the cyclic vector. However, generally $P_0 \notin$ the von Neumann algebra since we are in a type III factor which does not contain finite-dimensional projectors.

In the representations where the extensions of the functional is not normal, it certainly cannot be written as $\langle A\rangle = \operatorname{tr} \rho A$, since this would imply normality.

v) An exception is the case $\eta = 0$, e.g., $\langle\vec{\sigma}_i\rangle = 0$. This functional is already a trace, and the thermal representation is a II_1 factor.

vi) Although the expectation value cannot be written $\langle A\rangle = \operatorname{tr}(e^{-\beta H}A)/\operatorname{tr}(e^{-\beta H})$, the KMS analyticity property is satisfied in our model.

References

Infinite Tensor Products.

1. J. von Neumann, "On infinite direct products," Comp. Math. 6, 1 (1938).
2. A collection of the main results can be found in M. Guenin, A. Wehrl, W. Thirring, "Introduction to Algebraic Techniques," CERN, 1968-69.

Traces.

3. J. Dixmier, "Des algebres d'opérateurs dans l'espace hilbertien" (Gauthier Villars, Paris, 1957).

The BCS-Model.

Most standard texts on super-conductivity. The calculations mentioned here can be found in more detail in

4. W. Thirring and A. Wehrl, Commun. Math. Phys. 4, 303 (1967), for $T=0$.
5. W. Thirring, Commun. Math. Phys. 7, 181 (1968) for $T\neq 0$; this paper contains also a list of further references.
6. F. Jelinek, Commun. Math. Phys. 9, 169 (1968).
7. Some unpublished results are from F. Jelinek and A. Wehrl.

A SHORT SURVEY OF MODERN STATISTICAL MECHANICS†

Moshé Flato
Department of Mathematical Physics
Royal Institute of Technology
Stockholm, Sweden

Introduction

This series of four lectures was delivered at the NATO Summer School in Mathematical Physics in Istanbul, August, 1970. They did not pretend to cover all the material in this vast domain of modern (or what people usually call rigorous) statistical mechanics. Rather, our aim was to give to non-specialists (by a non-specialist!) a rapid survey as a first contact with this rapidly developing field. We have divided our lectures as follows:

First Lecture. Introduction—Mathematical and physical.
Second Lecture. Thermodynamic limit.
Third Lecture. Low-density expansions, phase transitions.
Fourth Lecture. Invariance of physical states.

[Remark: Since Lectures Two, Three and Four followed the corresponding chapters in the excellent review book by D. Ruelle, we shall reproduce here only the first (and the longest) lecture and refer the interested reader to Ruelle's book.]

* * * *

Section I

What is the place of the modern approach to statistical mechanics in the ensemble of the ancient approaches to this theory?

In order to answer this question, one must first review briefly the ancient points of view: In the particular case of kinetic theory of gases (which is a successful and therefore a good model of how statistical mechanics should look), we have the hypothesis of Molecular Chaos to connect microscopic reversibility with macroscopic irreversibility. This is ad-hoc but still gives a hint for every approach to statistical mechanics.

---

†Presented at the NATO Summer School in Mathematical Physics, Istanbul, 1970.

A good theory should have 1) the least ad-hoc possible connection between microscopic reversibility and macroscopic irreversibility, and 2) a good description relative to the problem of "approach to equilibrium." In order to have this kind of theory, two different approaches were taken by the ancient specialist of statistical mechanics:

a) Ergodicity hypothesis—mainly developed by Boltzman.

b) Master equation—mainly developed by Pauli.

The ergodicity hypothesis states that the time average of a macroscopic quantity is (under equilibrium conditions) equal to the "ensemble-average" of the same quantity. Now one must take an average on infinite-time interval, as the hypothesis was proved for this case. In reality, the time we measure is finite, and in some cases we hope that it can be considered as infinite relative to the molecular-collision time.

The master equation approach will now be explained in quantum statistical mechanics language (though it has a full classical meaning, once the notion of state is well understood!). Write as usual

$$\phi = \sum_n a_n \varphi_n$$

(φ_n orthocomplemented basis). Here ϕ depends on coordinates of the system as well as on those of the external world (the system is not completely isolated!). Suppose that φ_n are stationary wave functions of the system, a_n being wave functions depending only on the coordinates of the external world and on the time. Let O be an observable; then, according to the usual rules, one has for its expectation value in the state ϕ:

$$\langle O \rangle = \frac{(\phi, O\phi)}{(\phi, \phi)} = \frac{\sum_n (a_n, a_m)(\varphi_n, O\varphi_m)}{\sum_n (a_n, a_n)} \; .$$

(The denominator being trivially time independent as the total Hamiltonian is self-adjoint.) (a_n, a_m) is the scalar product in external world variables (integrated) and is a function of time. As a matter of fact, what we really measure is time-average of the former equation (with time longer than molecular collision time but shorter than the resolving time of the apparatus):

$$\langle \overline{O} \rangle = \frac{\sum_n \sum_m \overline{(a_n, a_m)}\, (\varphi_n, O\varphi_m)}{\sum_n (a_n, a_n)} \; .$$

(The bars stand for time-average.) Suppose now that we have a macroscopic system of N particles having the volume V weakly coupled to the external world such that the energy of the system remains between E and E+Δ with $\Delta/E \ll 1$. Choose an orthocomplemented basis of wave function φ_n of N particles in a volume V and let E_n be the corresponding eigenvalues. Then we have $H\varphi_n = E_n\varphi_n$. As statistical mechanics has something special to say about the behaviour of $\overline{(a_n, a_m)}$, we introduce the following postulates:

1) $\overline{(a_n, a_n)} = 1$ $(E \leqq E_n \leqq E + \Delta)$, and $\overline{(a_n, a_n)} = 0$ otherwise. This is an hypothesis of equal a priori probability.

2) For $m \neq n$ $\overline{(a_m, a_n)} = 0$. This is an hypothesis of random phases.

It is evident that the random phase hypothesis implies that our state can be considered as an incoherent superposition of eigenstates. Now, let $P_n(t)$ be the probability that at time t the system is in a state n. Let $T_{mn}(t)$ be the transition probabilities per second from n to m.

The master equation—very well understood intuitively—is:

$$\frac{dP_n(t)}{dt} = \sum_m \left(T_{mn}(t)\, P_m(t) - T_{mn}(t)\, P_n(t) \right).$$

The aim of statistical mechanics is to show that for $t \gg t$ (mol.), we have

$$P_n(t) \to \frac{\overline{(a_n, a_n)}}{\sum_n (a_n, a_n)}$$

with the right hand side satisfying the equal a priori probability postulate.

Now when $t \to \infty$, this last fact can be proved in general. The problem is, therefore, to calculate the molecular time, to prove that

$$P_n(t) \to \frac{\overline{(a_n, a_n)}}{\Sigma (a_n, a_n)}$$

for times much greater than this time, and then to give a <u>rigorous</u> justification of the master equation itself for the considered case. If we now come to compare the ergodicity and master equation approaches, we have to admit that in <u>both</u> attitudes the difficult problem is that of getting the characteristic molecular time—which is, of course, <u>dynamics-dependent</u>.

Though the Pauli approach is more simply workable, both attitudes are similar: Boltzman chaos (for instance, in the case of the transport equation) resembles the Pauli random-phase hypothesis, both being ad-hoc; on the other hand, in both cases (transport and master) the property of solutions when $t \to \infty$ is easily obtainable.

What is the role, then, of rigorous statistical mechanics (=R.M.S.), and which place does it take among these attitudes (=master + ergodicity)? The role, aims and ambitions of R.M.S., as well as its place among ancient approaches, can be summed up by the following statements:

1) To utilize completely rigorously the mathematical technique for open (and half-solved!) problems in statistical mechanics.

2) To make use of more modern mathematical tools (like topology, analysis, ergodicity and measure theory, group theory, C^* and von Neumann algebras, etc.) so as to unify statistical mechanics techniques with those of field theory, to unify the formalisms of classical and quantum statistical mechanics and, hopefully, to solve new problems by the aid of these techniques.

3) Being abstract mathematically, R.S.M. favorizes the (vague) Boltzman ergodicity upon Pauli's master equation approach.

4) Its original ambition was to solve unsolved problems in the field. However, we shall see that it was more successful in reformulation and in making rigorous old results, rather than in solving old (or new) completely unsolved problems.

* * * *

Section II

Let us now pass to some mathematical tools. To begin with, we present some basic facts of life concerning C^*-algebras.

A) Let $\mathcal{A}$ be an algebra over $\mathbf{C}$ (= complex field). A mapping $a \to a^*$ ($a \in \mathcal{A}$) of $\mathcal{A}$ into itself will be called an involution if

1) $(a^*)^* = a \quad \forall a \in \mathcal{A}$.

2) $(a_1 + \lambda a_2)^* = a_1^* + \bar{\lambda} a_2^*$.

3) $(a_1 a_2)^* = a_2^* a_1^* \quad \forall a_1, a_2 \in \mathcal{A}$.

$\mathcal{A}$ is a C^* algebra if it is with involution and with a norm satisfying

1) $\|a^*\| = \|a\| \quad \forall a \in \mathcal{A}$.

2) $\|a_1 a_2\| \leq \|a_1\| \, \|a_2\|$ (triangular inequality) $\forall a_1, a_2 \in \mathcal{A}$.

3) $\mathcal{A}$ is <u>complete</u> with respect to the norm (Banach algebra) which means that every Cauchy sequence converges to a limit belonging to $\mathcal{A}$.

4) $\|a^*a\| = \|a\|^2 \quad \forall a \in \mathcal{A}$.

It is trivial to note that the four axioms are not independent; for instance, 4 + 2 implies 1.

<u>Examples</u>:

1) L is a locally compact topological space (locally compact means that every point has a compact neighbourhood). $\mathcal{C}_0(L)$ space of complex continuous functions on L vanishing at infinity. * is complex conjugation, and $\|\varphi\| = \sup_{X \in L} \varphi(X)$. One checks easily that $\mathcal{C}_0(L)$ is an abelian C^*-algebra.

2) <u>Classical C^*-algebras</u>. H is a complex Hilbert space; B(H) is the algebra of all bounded operators in H with coefficients in $\mathbb{C}$. If $A \in H$, let A^* be the adjoint operator (namely, the operator that for every $\varphi, \phi \in H$ satisfies $(\varphi, A\phi) = (A^*\varphi, \phi)$). Let the norm be defined by $\|A\| = \sup \|A\varphi\|$ $(\|\varphi\| \leqq 1)$. A subalgebra $X \subset B(H)$ which is self-adjoint ($X = X^*$, namely, with every operator belonging to X, its adjoint also belongs to X) and norm-closed is a classical C^*-algebra. Every C^*-algebra is isomorphic to a classical C^*-algebra on some H, and therefore the distinction between C^* and classical C^* is only academic.

A C^*-algebra can either have an identity or not. For instance, $\mathcal{C}_0(L)$ has an identity if and only if (= iff) L is compact (the last fact being trivial in both directions). Suppose $\mathcal{A}$ is a C^* with identity I. $U \in \mathcal{A}$ is said unitary if and only if $UU^* = U^*U = I$. One can prove that the unitaries generate $\mathcal{A}$ by linear combinations (exactly as in the finite-dimensional H case). If we have an algebra $\mathcal{A}$ without identity, we can imbed it in a bigger algebra possessing identity by couples-calculus thus: We consider couples (λ, a) with $\lambda \in \mathbb{C}$ and $a \in \mathcal{A}$, with the axioms:

1) $(\lambda_1, a_2) + (\lambda_2 a_2) = (\lambda_1 + \lambda_2, a_1 + a_2)$

2) $\lambda(\lambda_1, a_1) = (\lambda\lambda_1, \lambda a_1)$

3) $(\lambda_1, a_1)(\lambda_2, a_2) = (\lambda_1\lambda_2, \lambda_1 a_2 + \lambda_2 a_1 + a_1 a_2)$.

In the couples-algebra the elements $(0, a)$ are isomorphic to $\mathcal{A}$, and the identity is the element $(1, 0)$. One can prove that the same procedure holds for C^*-algebras. Thus: every C^*-algebra can be imbedded in a bigger C^*-algebra possessing identity by this last couples-calculus.

B) Definition: The spectrum of an element a of a C^*-algebra with identity is the set: spec (a) = $\{\lambda \in \mathbf{C}$ such that $(a - \lambda I)$ has no inverse in the algebra$\}$. Once more, if $\mathcal{a}$ does not have identity, we imbed it in the couples-algebra which has identity and then define the spectrum of its elements.

Definition: Let $\mathcal{a}$ be a C^*-algebra. We define the spectral norm of the element $a \in \mathcal{a}$ by:

$$\|a\|_{\text{spec.}} = \sup_{z \in \text{Spec}'(a)} |z|$$

where spec'(a) is defined by spec'(a) = spec(0, a) in the imbedding algebra. If a is a self-adjoint $(a = a^*)$, one has $\|a\| = \|a\|_{\text{spec.}}$ (This last fact is trivially seen in the case of classical C^*-algebras in virtue of the spectral theory.) Therefore (as (a^*a) is always self-adjoint), one has $\|a^*a\| = \|a\|^2 = \|a^*a\|_{\text{spec}}$.

Definition: A morphism of a C^*-algebra $\mathcal{a}_1$ in a C^*-algebra $\mathcal{a}_2$ is a (*)-homomorphism of $\mathcal{a}_1$ into $\mathcal{a}_2$. Namely, a morphism is an application π:

$$\mathcal{a}_1 \xrightarrow{\pi} \mathcal{a}_2$$

such that $\pi(a \dotplus b) = \pi(a) \dotplus \pi(b)$ and $\pi(a^*) = (\pi(a))^*$ for each $a, b \in \mathcal{a}_1$.

One can prove that a morphism is injective (one-to-one correspondence) if and only if it is norm-preserving. Moreover, it can be shown that $\|\pi(a)\| \leqq \|a\|$, a thing which shows that every morphism is continuous. The kernel of a morphism (elements of $\mathcal{a}_1$ which go to 0 in $\mathcal{a}_2$) is a closed two-sided ideal in $\mathcal{a}_1$ (closed as continuity implies that the inverse image of a closed is closed).

We now introduce the notion of a state on a C^*.

Definition: $a \in C^*$ is positive if either $a = a^*$ and spec' (a) $\geqq 0$ or equivalently $\exists\, b \in C^*$ such that $a = b^*b$. This induces a partial ordering on the C^*, the positive elements being a closed convex cone in the C^*. (It is convex because a convex combination of positives is positive, and cone here means that multiplication by a real positive number transforms positive elements into positive elements.)

Definition: A state on a C^*-algebra is a continuous linear functional on the C^*, positive on positive elements, having the norm 1. The set of states E is therefore a convex subset of the unit ball of the dual of our C^*. If the C^* has an identity then, of course, $\rho(1) = 1$ where ρ is a state. In this case (as the unit ball of the dual of a C^* is weakly compact) the intersection of this unit ball with the plane $\{\rho \in (C^*)', \rho(1) = 1\}$ is weakly compact. Therefore, if the C^* has an identity, E is weakly compact. If we do not have identity in the C^*, we can utilize the notion of approximate identity which exists in every C^*-algebra!

Definition: Approximate identity is

1) Ordered sets of indices $\{\alpha\}$ such that given $\alpha_1, \alpha_2, \exists\ \alpha > \alpha_1, \alpha_2$.

2) Set of elements U_α of the C^* such that $U_\alpha \geqq 0$ and $\| U_\alpha \| \leqq 1\ \forall \alpha$.

3) $\alpha_1 \leqq \alpha_2 \Longrightarrow U_{\alpha 1} \leqq U_{\alpha 2}$

4) $\lim_\alpha \| U_\alpha a - a \| = \lim_\alpha \| a U_\alpha - a \| = 0 \quad \forall a \in C^*$

Evidently for a state ρ the generalization of $\rho(1) = 1$ in the case of identity will be $\lim_\alpha \rho(U_\alpha) = 1$ for the case of approximate identity.

Definition: A state on a C^*-algebra is said to be pure if it cannot be written as $\alpha\rho_1 + \beta\rho_2$ where $\rho_1 \neq \rho_2$ are two states and $\alpha + \beta = 1$, $\alpha\beta > 0$. We denote by $\mathcal{E}(E)$ the pure states of E (extremal points of E). By applying on E a classical theorem by M. G. Krein, one immediately determines that, if the C^* has an identity, then finite linear-convex combinations of pure states (elements of $\mathcal{E}(E)$) are weakly dense in the set of all physical states E.

* * *

C) We continue with the important notion of representation of a C^*-algebra.

Definition: A representation of a C^*-algebra is a pair (H, π) of a complex Hilbert space H and a morphism (= * homomorphism) π of the C^* into the algebra B(H) of bounded operators on H:

$$C^* \xrightarrow{\pi} B(H).$$

Definition: A representation (H, π) is non-degenerate if the equations $\pi(a)\varphi = 0$ for some $\varphi \in H$ and all $a \in C^*$ imply that $\varphi = 0$ (the only vector annihilated by the representation is 0).

Definition: A representation (H, π) is said to be irreducible if one of the two equivalent criteria holds:

1) Schur irreducibility: Every bounded operator on H which commutes with the representation is a multiple of the identity operator.

2) Topological irreducibility: The only invariant subspaces of H under the action of the representation are 0 and H itself. [Remark: The same equivalence of types of irreducibility exists for unitary representations of groups.† This, however, is not true for Lie algebras representations where, for differential representations, we always have invariant (closed) subspaces different from 0 and H.]

Definition: A cyclic representation of a C^*-algebra is a triplet (H, π, Ω) such that (H, π) is a representation and there exists $\Omega \in H$ fulfilling:

1) $\|\Omega\| = 1$, and

2) $\pi(a)\Omega$ for all $a \in C^*$ is dense in H.

It is very easy to verify that if (H, π, Ω) is a cyclic representation, then the mapping from the C^* to the complex field $a \to (\Omega, \pi(a)\Omega)$ is a state.

Demonstration: Suppose for simplicity that the C^* has an identity, then $\|\rho\| = \rho(1) = 1$. Moreover, for positive elements $x = b^*b$ and

$$(\Omega, \pi(b^*b)\Omega) = (\Omega, \pi(b^*)\pi(b)\Omega) = (\Omega, \pi^*(b)\, \pi(b)\Omega)$$
$$= (\pi(b)\Omega, \pi(b)\Omega) = \|\pi(b)\Omega\|^2 \geqq 0, \quad \text{Q.E.D.}$$

Conversely: If ρ is a state on a C^*-algebra, there exists a cyclic representation $(H_\rho, \pi_\rho, \Omega_\rho)$ such that $\rho(a) = (\Omega_\rho, \pi_\rho(a)\Omega_\rho)$ for all $a \in C^*$. The construction of such a representation is the G.N.S. (= Gelfand, Naimark, Segal) Construction. The same kind of construction is also utilized in the main-reconstruction-theorem in the axiomatic Wightman field theory. It should be remarked that such a kind of construction is unique up to unitary equivalence.

We shall sketch the G.N.S. construction for the case of a C^* with identity (although by adjunction one can make it for any C^*). It is clear that ρ defines a scalar product in the C^*. We take elements $a \in C^*$ having zero norm with respect to this scalar product: $R = \{a \in C^*, \rho(a^*a) = 0\}$.

Claim: R is a left ideal of the C^*.

---

†In this case, however, we have to assume "the only closed invariant". . . .

Demonstration: We first prove that $|\rho(y^*xy)| \leqq \|x\| \rho(y^*y)$:

We look at ρ as functional in x, then $\|\rho(y^*xy)\| = \rho(y^*y)$ (since it is $\rho(y^*1 \cdot y)$). On the other hand, from the definition of the norm, we have

$$\|\rho(y^*xy)\| \geqq \frac{|\rho(y^*xy)|}{\|x\|}$$

and therefore

$$\rho(y^*y) \cdot \|x\| \geqq |\rho(y^*xy)| .$$

Now we prove that if $a \in R$ and $b \in C^* \Rightarrow ba \in R$.

$$|\rho((ba)^*(ba))| = |\rho(a^*b^*ba)| \leqq \|b^*b\| \rho(a^*a) = 0 .$$

Therefore, $ba \in R$.

Now we prove that if $a \in R$ and $b \in R$, then $a+b \in R$. This follows from the Minkowski inequality which states:

$$|\rho((a+b)^*(a+B))|^{\frac{1}{2}} \leqq |\rho(a^*a)|^{\frac{1}{2}} + |\rho(b^*b)| .$$

We have therefore proved our claim.

We now construct the C^*/R. It can be proved that this pre-Hilbert space is also Hausdorff (= separated). Therefore, the completion $\overline{C^*/R}$ of C^*/R in the scalar product topology will be our Hilbert space $H_\rho : H_\rho = \overline{C^*/R}$. Now the mapping of $C^*/R \to C^*/R$ defined by $b \to ab$ with $a, b \in C^*$ can be extended by continuity to a mapping $H_\rho \to H_\rho$. This mapping of $H_\rho \to H_\rho$ will, of course, be $\pi_\rho(a)$. One then must determine that π_ρ is a morphism of the C^* into B(H). Ω_ρ is then defined as the class of 1 in C^*/R. Finally, having constructed our H_ρ, Ω_ρ and π_ρ, one can easily prove that they really satisfy $\rho(a) = (\Omega_\rho, \pi_\rho(a)\Omega_\rho)$ for every $a \in C^*$ by construction.

This completes the scheme of the G.N.S. construction. We now state some theorems connecting notions that we have previously seen.

Theorem: Let ρ be a state on a C^*-algebra. Let $(H_\rho, \pi_\rho, \Omega_\rho)$ be the corresponding G.N.S. cyclic representation. Then the following conditions are equivalent:

1) ρ is a pure state.

2) (H_ρ, π_ρ) is irreducible.

3) The left ideal $R = \{a \in C^*, \rho(a^*a) = 0\}$ is a maximal regular ideal (regular means that there exists $U \in C^*$ such that for every $a \in C^*$ we have $Ua - a \in R$).

[Remark: Under each one of these conditions, H_ρ of the G.N.S. construction simply equals to C^*/R with the quotient norm.]

Theorem: Suppose we have an abelian C^*-algebra. Then there is a one-to-one correspondence between a) the pure states, b) all homomorphisms of the C^* into the complex field $\mathbf{C}$, and c) the maximal regular ideals of the C^*.

Definition: $\mathcal{E}(E)$ (the pure states) with the weak topology is defined as the spectrum of the abelian C^*-algebra. We have already seen that $\mathcal{E}(E)$ is a locally-compact space. We have also seen that $\mathcal{E}(E)$ is compact if and only if the C^* has an identity.

* * *

D) The most important instrument concerning abelian C^*-algebras is the Gelfand isomorphism theorem. It is important as a result and as a typical technique of demonstration concerning abelian C^* algebras. Many of the results in the field are either deduced by it or proved by similar techniques. It is therefore worthwhile to give a demonstration of this theorem.

Definition (generalization of the former one): The spectrum of a commutative Banach algebra B is the set of all characters, endowed with the weak topology.

[Remarks: 1) A character is a homomorphism (necessarily continuous here) of the abelian B into $\mathbf{C}$. 2) The weak topology on characters ρ_n is the topology for which we say that $\rho_n \to \rho$ if and only if $\rho_n(a) \to \rho(a)$ (for every $a \in B$) as complex numbers. We can now formulate the isomorphism theorem.]

Isomorphism Theorem (Gelfand): A commutative C^*-algebra $\mathcal{a}$ with identity is isometrically *isomorphic with the algebra of all complex continuous functions on its compact spectrum S.

Lemma 1: $\|x^2\| = \|x\|^2$ for all $x \in \mathcal{a}$.

Proof:

$$\|x^2\|^2 = \|(x^2)^* x^2\| = \|(x^*)^2 x^2\| = \|(xx^*)^*(xx^*)\|$$

$$= \|xx^*\|^2 = \|x\|^4$$

and therefore

$$\|x^2\| = \|x\|^2 \quad \text{Q.E.D.}$$

Lemma 2:

$$\|x\|_{\text{spec}} = \lim_{n\to\infty} \|x^n\|^{1/n} .$$

(It also states that the limit exists!)

Proof: This is similar to the simple case of operators (see Reference 6, p. 864).

Corollary: $\|x\| = \|x\|_{spec}$ (by taking $n=2^m$ and utilizing the first lemma).

[Remark: This corollary has already appeared for the case of self-adjoint elements.]

Lemma 3: If μ's are characters, then

$$\sup_{\mu \in S} |\mu(x)| = \|x\|$$

Proof: If $\mu \in S$, $\mu(\mu(x)e - x) = 0$. Therefore $\mu(x)e - x \in \text{Ker}\,\mu$, and therefore $\mu(x) \in$ spectrum of x and $|\mu(x)| \leqq \|x\|$ for all μ by the corollary. Moreover, if $\lambda \in$ spectrum of x, then $\lambda e - x$ is singular and thus is contained in a maximal ideal J (Zorn Lemma). Then $\mathcal{A}/J$ is a field (having no non-trivial ideals). By a theorem from algebra we have even that $\mathcal{A}/J \cong \mathbf{C}$. The mapping

$$\mathcal{A} \xrightarrow{\mu} \frac{\mathcal{A}}{J}$$

is a homomorphism and thus a character μ. Moreover,

$$\mu(\lambda e - x) = 0 \Rightarrow \lambda = \mu(x)$$

for this μ. Therefore,

$$\sup_{\mu \in S} |\mu(x)| \geqq \sup_{\lambda \in \text{spec}(x)} |\lambda| .$$

But from the first part, $\sup|\mu(x)| \leqq \sup|\lambda|$, and therefore

$$\sup_{\mu \in S} |\mu(x)| = \|x\| \quad \text{Q.E.D.}$$

Lemma 4: μ is a character $\Rightarrow \mu(x^*) = \overline{\mu(x)}$ (the bar stands for complex conjugation).

Proof: $\mu(x) = (\alpha + i\beta)$, $\mu(x^*) = (\gamma + i\delta)$. We prove first that $\beta + \delta = 0$. Suppose $\beta + \delta \neq 0$. We set

$$y = \frac{x + x^* - (\alpha+\gamma)e}{\beta + \delta} .$$

It is clear that $y = y^*$ and $\mu(y) = i$. Therefore, $\mu(y + iNe) = i(1+N)$, and

$$\|y+iNe\| \geqq |1+N| \Rightarrow (1+N)^2 \leqq \|(y+iNe)(y+iNe)^*\|$$

$$= \|y^2+N^2e\| \leqq \|y^2\| + N^2,$$

for positive N^2, at least.

Choose now $N = \|y^2\|$, and we get

$$(1+\|y^2\|)^2 \leqq \|y^2\| + \|y^2\|^2,$$

namely, a contradiction. Therefore, $\beta+\delta=0$. By replacing x by ix, we prove in the same way that $\alpha-\gamma=0$. Therefore, $\mu(x^*)=\overline{\mu(x)}$ Q.E.D.

[Remark: We have thus proved, until now, that the map from the abelian C^* to the complex continuous functions $x \to$ the functions $(\mu \to \mu(x)) = x(\mu)$ on the spectrum of the C^* is a homomorphism (μ being characters) which is isometric (by Lemma 3)—and therefore one-to-one—and * homomorphism (by Lemma 4).]

To show the theorem, we have to prove that the spectrum S is compact and that the mapping is surjective (onto!).

Proposition 1: The spectrum of a commutative Banach algebra with identity is compact (which here means Hausdorff + Heine-Borel).

Hint of the Proof: S is homeomorphic to a closed subspace of the Cartesian product:

$$Q = \prod_{x \in \mathcal{A}} \{\lambda : |\lambda| \leqq \|x\|\}$$

which is compact in virtue of the Tychonoff Theorem. As a matter of fact, each $\mu \in S$ defines a point $q \in Q$ such that $q(x) = \mu(x)$, and two different characters define different points. The surjectivity is now a trivial consequence of the Stone-Weierstrass theorem.

Stone-Weierstrass Theorem: Let S be a compact space, $\mathcal{C}(S)$ the algebra of all complex continuous functions on S. Let $\mathcal{A}$ be a closed subalgebra of $\mathcal{C}(S)$ containing the unit function e and, together with each f, the conjugate $\bar{f}$ defined by $\bar{f}(s)=\overline{f(s)}$. Then $\mathcal{A} = \mathcal{C}(S)$ if and only if $\mathcal{A}$ separates points in S.

The proof of the isomorphism theorem of Gelfand follows now completely as the S.W. property (separation) is fulfilled by characters by definition. Our proof is then complete.

[Remark: In the case of a C^* without identity, the same theorem holds if one just takes, instead of complex continuous functions on the compact spectrum, the complex continuous functions $\mathcal{C}_0(S)$ on the locally-compact spectrum which vanish at infinity. To demonstrate it, one has to take away the trivial character.]

Corollaries:

1) From what we saw before for abelian C^*-algebras, the spectrum can be identified with the set of all pure states $\mathcal{E}(E)$. Therefore, the Gelfand isomorphism is an isometrically * isomorphism between the abelian C^* and the algebra of complex continuous functions $\mathcal{C}_0(\mathcal{E}(E))$—on the pure states—which vanish at infinity.

2) By duality, it gives a one-to-one correspondence between the physical states (which were elements on the unit sphere of the dual of our C^*) and the probability measures on the pure states $\mathcal{E}(E)$ (which are elements of the dual of the continuous functions $\mathcal{C}_0(\mathcal{E}(E))$.

Example: Suppose we have an abelian C^* with identity. Let $\rho \in E$ be a state, and μ the corresponding measure on $\mathcal{E}(E)$. Now let $(H_\rho, \pi_\rho, \Omega_\rho)$ be the G.N.S. triplet corresponding to the state. Then we identify $H_\rho = L^2(\mu)$. π_ρ: $a \rightarrow \tilde{a} \in B(L^2(\mu))$ for every $a \in C^*$, where $\tilde{a}$ is the image of a in $\mathcal{C}_0(\mathcal{E}(E))$ by the Gelfand isomorphism, and of course $\Omega_\rho = 1 \in L^2(\mu)$ in our case.

* * * *

Section III

We shall now begin with another object which will be very important for our purpose in statistical mechanics—Ergodicity.

Definition: Suppose we are given a topological space. The family formed from open sets by taking all denumerable intersections and complementaries is the Borel tribe of measurable sets.

Definition: A measure μ on such a space is an application of the elements of the Borel tribe into $\mathbb{C}$ such that:

1) If $\{B_j\}$ is a denumerable set of elements of the Borel tribe and $B_i \cap B_k = \emptyset$ for all $i \neq k$, then

$$\mu(\bigcup_i B_i) = \sum_i \mu(B_i) \quad (\sigma\text{-additivity}).$$

2) There exists at least one element B in the tribe such that $|\mu(B)| < \infty$ (so as to avoid trivial cases).

In the coming section we shall suppose that our measures are positive—as one does classically!

* * * *

A) We can now come to the important notion of flow.

Definition: Let (X, m) be a locally-compact measure space with finite total measure $|m(X)| < \infty$. A measure-preserving flow

(and this will be the only kind of flows in which we shall be interested) is a mapping from the real line $\mathbb{R}$ to the one-parameter group of measurable maps of (X, m) into itself: $\mathbb{R} \ni t \to \varphi_t$ such that $m(\varphi_t(S)) = m(S)$ for every S in the Borel-tribe and all $t \in \mathbb{R}$.
[Remark: This is a natural generalization of Hamilton formalism in which, during the motion (here t is the physical time), we have the Liouville theorem for volume in phase-space.]

Definition: A flow (measure preserving!) is called ergodic if $\varphi_t(S) = S$ for all $t \in \mathbb{R}$ implies $m(S) \cdot m(X - S) = 0$.
[Remark: This means that ergodicity is a kind of irreducibility: The only invariant measurable sets under ergodic flow have either measure 0, or their complementary has the measure 0.]

Lemma: φ_t is ergodic if and only if for every two elements A and B of the Borel tribe the Cesàro means:

$$0 = \lim_{T \to \infty} \frac{1}{T} \int_0^T \left\{ m(\varphi_t(A) \cap B) - \frac{m(A) \cdot m(B)}{m(X)} \right\} dt,$$

converges to 0.
[Remark: This last lemma (which we do not prove here) explains clearly how the notion of ergodic flow is connected with Boltzman's ergodicity which we discussed in the beginning.]

Definition: If the last integral converges absolutely to 0, we say that the flow is weakly mixing.
[Remark: Already from the definition of ergodic flow it becomes clear that "non-trivial measurable sets" cannot stay invariant under the flow. Therefore, even ergodic flow has some mixing properties.]

Definition: If

$$\lim_{t \to \infty} m(\varphi_t(A) \cap B) = \frac{m(A) \cdot m(B)}{m(X)},$$

we say that φ_t is strongly mixing.
[Remark: One can give a stronger definition of ergodicity, which is that every almost-everywhere invariant measurable set under the flow ($\varphi_t(S) = S$ only almost-everywhere) satisfies $m(S) \cdot m(X - S) = 0$. However, under fairly general conditions on the measures, the two definitions of ergodicity do coincide!]

B) Our discussion will not be complete if we do not say something about complex measures, which were introduced in the beginning of our discussion.

A question of interest is the following: "How do complex ergodic measures look?"

One can prove that under quite general conditions every complex ergodic measure is simply a complex scalar times a usual positive ergodic measure, which means that the generalization from positive to complex ergodic measures is essentially trivial. In other words, ergodicity implies positivity and therefore probabilistic interpretation.

One should mention that if one generalizes also to the non-abelian case by C^*-algebra techniques (as we have already begun to do and will continue to do), then the same result is valid for C^*-algebras with much <u>weaker</u> properties of commutativity. However, for general C^*-algebras, this result does not hold and one can construct non-trivial "complex-ergodic-functionals."

* * * *

C) We now pass back to the classical case of positive-measures.

Suppose that $f \in L^2(X,m)$, the Hilbert space of square integrable functions on X with positive measure m. The flow defines (by standard arguments) a one-parameter group of unitary operators on $L^2(X,m)$: $U_t f(X) = f(\varphi_t(X))$. By spectral theory, we know that

$$U_t = e^{itA} = \int_{-\infty}^{\infty} e^{i\lambda t} dE(\lambda)$$

with self-adjoint operator A. (Stone Theorem.)

Definition: The spectrum of A will be defined as the spectrum of the flow.

Theorem: φ_t is an ergodic flow if and only if the only invariant measurable functions under it in $L^2(X,m)$ are constants (almost everywhere!).

Proof: First suppose that the only invariant functions in $L^2(X,m)$ are constants (the fact that $|m(X)| < \infty$ enters here, as otherwise constants will not belong to $L^2(X,m)$). For every measurable S we look at its characteristic function (namely, the function which is 1 on S and 0 on $(X-S)$). Now S is invariant under the flow almost everywhere if and only if its characteristic function is invariant almost everywhere. But the only invariant functions in $L^2(X,m)$ are constants, and therefore the only almost everywhere invariant S under the flow are either sets having $C \neq 0$ as their non-normalized characteristic function (and therefore $m(X-S)=0$) or sets having $C=0$ as their characteristic function (and therefore $m(S)=0$). Therefore, the flow is ergodic.

Now conversely: Suppose that the flow is ergodic. Let $f \in L^2(X,m)$ be an invariant function under the flow (almost everywhere!). We divide the complex field into half-open squares with side 2a having as centers the points $2a(n+im)$ in $\mathbf{C}$. Call B_{nk}

those sets of (X, m) such that for them the values of $f(x)$ fall in the square having as center $2a(n+ik)$ and sides $2a$. Now f is invariant under φ_t. This means that the B_{nk} are invariant under φ_t. Since f is measurable, we know that the B_{nk} are measurable in (X, m). By definition they are also disjoint. We therefore decompose

$$(X, m) = \bigcup_{n,k} B_{nk}.$$

(We have, of course,

$$m(X) = \sum_{n,k} m(B_{nk})\Big).$$

Now our flow is ergodic and this means (since the B_{nk} are invariant under it) that only one of the B_{nk}—say $B_{n^*k^*}$—carries the measure of the whole space, all the other B_{nk} having the measure 0. This means—since a can be chosen arbitrarily small—that $f(x)$ is constant almost everywhere, Q.E.D.

Corollary: φ_t is ergodic if and only if $\lambda = 0$ has the multiplicity one in its point spectrum.

Proof: We saw that φ_t is ergodic if and only if the only f, satisfying

$$f(\varphi_t(x)) = U_t f(x) = f(x),$$

are constants. This means that $e^{itA}f(x) = f(x)$ which implies (by differentiation with respect to t) $Af(x) = 0$. From this point, the rest of the proof is straightforward.

* * *

One can also prove that φ_t is weakly-mixing if and only if the point-spectrum of the corresponding A consists of 0 alone, which is of multiplicity 1. If, in addition, the spectral-measure of the continuous spectrum of the flow is absolutely continuous with respect to the Lebesgue measure (which means that the null-sets of second are necessarily null-sets of the first), then one can prove that the flow is strongly-mixing.

For the sake of completeness, we close this small section which dealt with ergodicity with the following definition:

Definition: A flow φ_t is said to be minimal if and only if, for every $x \in X$, the orbit $\varphi_t \cdot x \quad (t \in \mathbb{R})$ is dense in X. This notion of minimality is a general topological notion for flows and, contrary to the notions that were utilized before, does not make use of the measure m on X.

We thus complete our short survey about flows, measures and ergodicity.

* * *

Section IV

We shall now introduce some basic notions concerning integral representations of convex compact sets—a tool which will help us later for the study of decomposition of invariant states into invariant-ergodic states.

A) Let F be a locally-convex topological-vector-space. Let K be a convex compact subset of F. Let $\mathcal{C}(K)$ be the complex-valued continuous functions on K, $\mathcal{C}'(K)$ its dual, namely the measures (Radon measures) on K. We denote $\mathcal{C}'(K) = M$ and by M_+ the positive measures; M_1 will be the positive measures of norm 1. M_1, of course, can be identified with the states on $\mathcal{C}(K)$.

Proposition: For every $\mu \in M_1$ there exists $\rho \in K$ such that for all $f \in F'$ we have

$$f(\rho) = \int_K f(\sigma) d\mu(\sigma).$$

(F' is the dual of F.)

[Remark: By

$$\int_K f(\sigma) d\mu(\sigma),$$

we mean $\mu(f)$ where f is a continuous complex-valued linear function on K.]

Definition: ρ of the last proposition is called the resultant of μ.

Proposition: If $\mu \in M_1$ has a resultant ρ, it can be weakly approximated by measures $\mu' \in M_1$ having the same resultant and finite support:

$$\mu' = \sum_{i=1}^{n} \lambda_i \delta_{\rho_i}, \quad \lambda_i \geqq 0, \quad \sum_{i=1}^{n} \lambda_i = 1, \quad \rho_i \in K$$

and

$$\sum_i \lambda_i \rho_i = \rho.$$

(By weakly approximated we mean that a sequence of measures μ_n is said to converge weakly to a given measure μ if and only if for every continuous function f, the complex numbers $\mu_n(f)$ converge to the complex number $\mu(f)$ in the complex-plane.)

Proposition (Choquet and Meyer): Let $\mu \in M_1$ and μ have a resultant ρ—and let f be an affine upper semi-continuous function on K—then $\mu(f) = f(\rho)$.
[Remark: f is said to be affine if and only if, for any $x_1, x_2 \in K$ and $\alpha \in [0,1]$, we have

$$f(\alpha x_1 + (1-\alpha)x_2) = \alpha f(x_1) + (1-\alpha)f(x_2).$$

f is said upper semi-continuous if and only if it is real-valued and for every real a the set $\{X \in K,\ f(x) < a\}$ is an open set.

* * * *

B) Let $S \subset \mathcal{C}(K)$ be the convex cone of (real-valued) convex continuous functions on K. (It is also clear that $F' \subset S$ if we consider F here as a real vector-space.) Now on M_+ we introduce an order: We say that $\mu_1 < \mu_2$ if and only if, for all $f \in S$, $\mu_1(f) \leqq \mu_2(f)$. (We precisely built this S so as to introduce an order $<$ on M_+. It is clear that if $\mu_1 < \mu_2$ and f is an affine-continuous function, then $\mu_1(f) = \mu_2(f)$ (as in this case, both f and -f are convex).

One can prove the following proposition (which we just mention).

Proposition: Suppose that $\mu \in M_1$ and $\rho \in K$. Then μ has a resultant ρ if and only if $\mu > \delta\rho$.

Definition: $\mu \in M_+$ is said to be maximal if and only if it is maximal with respect to the order $<$ (namely, if there does not exist $\nu \in M_+$ such that $\nu > \mu$). By properly utilizing Zorn's lemma, one can prove that if $\mu \in M_+$, there exists a maximal measure $> \mu$. In particular, for every $\rho \in K$ there exists a maximal measure with resultant ρ.

At this point, we stop our mathematical preparation for the theory of integral representations of convex compact sets. We shall come back, and in more detail, to this topic when we discuss the problem of decomposition of invariant-states into ergodic-states.

* * * *

Section V

We pass now to the mathematical preliminaries needed for the theory of group-invariance of physical states.

A) Traditionally, a state of infinite system was correlation function in classical statistical mechanics and reduced density matrix in quantum statistical mechanics. Our notion of a state, namely the notion of linear continuous functional on a C^*-algebra which is positive on positive elements and which has the norm 1, unifies both former definitions. We shall be interested in states which are invariant under groups. In order to be sure that such states exist, we must impose on the group some restrictions. This brings us to the important notion of groups with an <u>invariant mean</u>. Let G be a locally compact group, $g, h \in G$. Let $\mathcal{C}_B(G)$ be the abelian C^*-algebra of bounded complex continuous functions on G. We define $f_h(g) = f(gh)$ and ${}_hf(g) = f(h^{-1}g)$ as, respectively, the right and left translates of f (by h).

<u>Definition</u>: A state τ on $\mathcal{C}_B(G)$ is a right invariant mean if and only if, for all $h \in G$, we have $\tau(f_h) = \tau(f)$. In a similar way we define the notion of left invariant mean and, therefore, of a two-sided invariant mean. It is also trivial to see that the existence of a left invariant mean is ensured if and only if the existence of a right invariant mean is ensured and then of course (if and only if) the existence of a two-sided invariant mean is ensured.

<u>Definition</u>: If this is the situation (namely, e.g., there exists a left invariant mean), we say that the group has an <u>invariant mean</u>.

<u>Examples of Groups with Invariant Mean</u>:

a) Compact groups.

b) Abelian groups.

c) Semi-direct products of the first by the second.

d) Connected semi-simple groups will have an invariant mean if and only if they are compact.

<u>Corollary of d)</u> (important for physics): The Lorentz group $SO_0(3,1)$ does not have an invariant mean.

We close this subject by stating a necessary-and-sufficient condition for a group G to have an invariant mean. This condition is the condition of existence of a <u>net</u>.

<u>Definition</u>: G (with the former hypothesis) has a net if there exist on G functions $\chi_\alpha(g)$ (real-valued) such that

1) The indices α are ordered and for any $\alpha_1, \alpha_2 \ \exists \alpha \geqq \alpha_1, \alpha_2$.

2) $\chi_\alpha \geqq 0$ and

$$\int_G dg\ \chi_\alpha(g) = 1$$

(dg is the right Haar-measure).

3)

$$\lim_\alpha \int_G dg\ |\chi_\alpha(gh) - \chi_\alpha(g)| = 0 \quad \text{for every } h \in G.$$

[<u>Remark</u>: It is condition 3), of course, which is connected with invariance under G. To simplify, if the set of indices contains only one element, condition 3) will mean that we have on the group a positive function of (right-Haar) volume 1 such that almost everywhere this function is right-invariant under G.]

B) We now pass to the study of group invariance of C^*-algebraically defined physical states. Let $\mathcal{A}$ be a C^*-algebra, and let $\mathrm{Aut}(\mathcal{A})$ be its group of automorphisms. The basic notion in the theory of invariant-states is the notion of a triplet.

<u>Definition</u>: (A, G, τ) forms a <u>triplet</u> if and only if $\mathcal{A}$ is a C^*-algebra, G a group and τ a group-homomorphism of G into $\mathrm{Aut}(\mathcal{A})$:

$$G \xrightarrow{\tau} \mathrm{Aut}(\mathcal{A}).$$

We then say simply that we have a C^*-algebra with group of automorphisms,and symmetry G. It is easy (by duality) to define a linear operator on $\mathcal{A}'$ (the dual of $\mathcal{A}$). We define the linear operator τ' on $\mathcal{A}'$ by

$$\tau_g' f(a) = f(\tau_{g^{-1}} a) \text{ for all } g \in G, \ a \in \mathcal{A} \text{ and } f \in \mathcal{A}'.$$

Denote by $\mathcal{L}_G$ the subspace of $\mathcal{A}$ generated by elements of the form

$$a - \tau_g a.$$

<u>Definition</u>: The <u>G-invariant</u> continuous-linear-forms on $\mathcal{A}$ are those $f \in \mathcal{A}'$ such that for all $a \in \mathcal{L}_G$ we have $f(a) = 0$. We denote them by $\mathcal{L}_G^+$. It is evident that τ_g maps $\mathcal{L}_G^+$ onto itself.

<u>Definition</u>: If E is the set of states on $\mathcal{A}$, then $E \cap \mathcal{L}_G^+$ will be defined as the set of <u>G-invariant</u> states on $\mathcal{A}$. Again, one can prove that our definitions are such that G is unitarily implementable on the cyclic-representation of $\mathcal{A}$ corresponding to the G-invariant state ρ.

<u>Lemma</u> (of I. E. Segal): Let $(\mathcal{A}, G, \tau)$ be a triplet. Let ρ be a G-invariant state $(\rho \in E \cap \mathcal{L}_G^+)$, and let $(H_\rho, \pi_\rho, \Omega_\rho)$ be the cyclic representation of $\mathcal{A}$ corresponding to ρ. Then there exists a unique unitary representation (of G) U_ρ in H_ρ such that

$$U_\rho(g)\pi_\rho(a)U_\rho^+(g) = \pi(\tau_g a) \quad \text{and} \quad U_\rho(g)\Omega_\rho = \Omega_\rho.$$

If, in addition, G is a topological group and the application $g \to \rho(a^* \tau_g a)$ is continuous, then U_ρ is also <u>strongly-continuous</u>. [<u>Remark</u>: In the case of abelian C^*-algebra (with identity), by virtue of Gelfand's theorem, the study of invariant states is the study of invariant measures, which is a part of the classical ergodic theory.]

C) Suppose now that we are given a triplet $(\mathcal{A}, G, \tau)$ and let $\rho \in E \cap \mathcal{L}_G^+$ be a G-invariant state. Let $(H_\rho, \pi_\rho, \Omega_\rho)$ be the cyclic representation of $\mathcal{A}$ corresponding to ρ, and let U_ρ be a unitary representation of G in H_ρ. Denote by P_ρ the projector in H_ρ on the subspace generated by vectors invariant under U_ρ. (Think of the case in which G is the charge-group SO(2): In this case P_ρ will be the projection on subspace generated by uncharged states!)

Definition: $\mathcal{A}$ is a G-abelian C*-algebra if and only if, for all $\rho \in E \cap \mathcal{L}_G^+$ and $a_1, a_2 \in \mathcal{A}$, we have

$$[P_\rho \pi_\rho (a_1) P_\rho, P_\rho \pi_\rho (a_2) P_\rho] = 0.$$

[Remark: Evidently if $G = I$ is the group containing one element, the notion of G-abelian will imply that for every $\rho \in E \cap \mathcal{L}_G^+$ the representation itself is abelian—namely, commutativity of the algebra.]

Definition: Given a triplet $(\mathcal{A}, G, \tau)$ and given the set of G-invariant states $E \cap \mathcal{L}_G^+$, we look at the set of extremal points $\mathcal{E}(E \cap \mathcal{L}_G^+)$ of the G-invariant states. We call $\mathcal{E}(E \cap \mathcal{L}_G^+)$ the G-ergodic states.

[Remark 1: This definition just means that G-ergodic states are in a way "irreducible" G-invariant states, or "pure" G-invariant states.]

[Remark 2: It is important to remark here that the interest of groups with invariant-mean is that if this is the case, and if in addition the homomorphism τ is continuous in some sense, we will see that $E \cap \mathcal{L}_G^+$ is not empty and that P_ρ is explicitly obtainable in terms of U_ρ.]

* * * *

Section VI

We have thus finished our mathematical and physical preliminaries to the subject of rigorous-statistical-mechanics. As was mentioned at the beginning of this lecture, and as will become even more clear in the coming lectures, this new approach to statistical mechanics (very fashionable in the last few years) did not perform any miracles. In particular, I do not think that there were any new physical ideas nor were any unsolved problems solved. However, it brought some level of rigour to a field where it was needed. It also brought new techniques and methods of work and a new confidence in some results; maybe these methods will, at least in the future, prove useful in the formulation of new ideas.

References

For those who would like to study the subject more fundamentally, we have suggested the following references (and also references quoted therein).

1. Arnold, V. I., and A. Avez. Problèmes ergodiques de la mécanique classique. Paris: Gauthier-Villars (1967).
2. Bourbaki, N. Eléments de mathématique (1). Intégration. Paris: Hermann (1965).
3. Choquet, G., and P. A. Meyer. Ann. Inst. Fourier 13 (1963), 139-154.
4. Dixmier, Jacques. Les algèbres d'opérateurs dans l'espace hilbertien. Paris: Gauthier-Villars (1957).
5. Dixmier, Jacques. Les C*-algèbres et leurs représentations. Paris: Gauthier-Villars (1964).
6. Dunford, N., and J. T. Schwartz. Linear operators I, II. New York: Interscience (1958-1963).
7. Flato, M., B. Nagel and D. Sternheimer. Flots ergodiques et mesures complexes. C. R. Acad. Sc. Paris 272 (1971), 892-895.
8. Nagel, B. Some results in non-commutative ergodic theory. K.T.H. Preprint (October 1971).
9. Ruelle, David. Statistical mechanics. Benjamin (1969).

GENERALIZED EIGENVECTORS IN GROUP REPRESENTATIONS†

Bengt Nagel
Royal Institute of Technology
Stockholm

I. Physical Background

The "conventional" framework of quantum mechanics is:

Space of physical states = Hilbert space $\mathcal{H}$

Physical observables = all self-adjoint (s.a.) operators in $\mathcal{H}$ (if there are no superselection rules).

It is, however, clear that "some observables are more physical than others," and also some states in $\mathcal{H}$ are physically realizable only in a limiting sense—take, e.g., states of "infinite energy" (vectors outside the domain of definition $\mathcal{D}(H)$ of the Hamiltonian H, supposed unbounded). In general certain important observables are unbounded; very often these are infinitesimal generators of a symmetry group of the system (as, e.g., H in the example above).

We can then suppose that we are given a "natural set of observables," which for mathematical commodity is closed algebraically to an algebra $\mathcal{a}$; in general $\mathcal{a}$ will contain unbounded operators. We then have to tackle domain questions, and the best we can hope to get is to have a linear subset $\phi \subset \mathcal{H}$, dense in $\mathcal{H}$, on which $\mathcal{a}$ can be applied. It is also convenient to have a topology on ϕ, finer than the one induced from $\mathcal{H}$, such that all operators in $\mathcal{a}$ are continuous in ϕ. We can then extend $\mathcal{a}$ to an isomorphic algebra $\mathcal{a}'$ acting in ϕ' where ϕ' is the "dual" of ϕ (the set of continuous linear functionals on ϕ [linearity in ϕ' is defined by $(\alpha x' + \beta y')(x) = \bar{\alpha}x'(x) + \bar{\beta}y'(x)$] or, equivalently, the set of continuous anti-linear functionals on ϕ). We have $\phi \subset \mathcal{H} \subset \phi'$.

A second reason (the first reason being the existence of unbounded operators in $\mathcal{a}$) for letting $\mathcal{a}$ act only in a subspace of $\mathcal{H}$ is the fact that some operators in $\mathcal{a}$ may have continuous parts in their spectrum. The corresponding "eigenvectors" (formal solutions of $Ax = \lambda x$, $\lambda \in$ continuous spectrum of A) are then not in $\mathcal{H}$, but can be accommodated in an extended space. It is known that if ϕ is

†Presented at the NATO Summer School in Mathematical Physics, Istanbul, 1970.

a nuclear space, then there are "enough" generalized eigenvectors of $A \in \mathcal{A}$ (A s.a. in $\mathcal{H}$) in the dual ϕ'.

(The standard example here is $A = -i\, d/dx$ in $\mathcal{H} = L^2(-\infty, \infty)$. Taking $\mathcal{S} \subset \mathcal{H} \subset \mathcal{S}'$, $\mathcal{S}$ the set of C^∞ functions of rapid decrease (which is nuclear), the "complete set of generalized eigenvectors" of A, $\{\exp(i\lambda x), -\infty < \lambda < \infty\}$, is contained in $\mathcal{S}'$.)

We shall study these problems for the special but important case where $\mathcal{A}$ is the representation $dU(\mathcal{E}(L))$ of the enveloping algebra $\mathcal{E}(L)$ of a Lie algebra L (representation $dU(L)$), derived from a continuous UR (unitary representation) $U(G)$ in $\mathcal{H}$ of a corresponding connected Lie group G.

We shall see that the set $\mathcal{D}$ of differentiable vectors of $U(G)$ (which is dense in $\mathcal{H}$), with a natural topology making it a countably-Hilbert space, is the maximal domain, with the coarsest topology, of $dU(\mathcal{E}(L))$. $U(G)$ and $dU(\mathcal{E}(L))$ are then extended to $U'(G)$ and $dU'(\mathcal{E}(L))$ acting in $\mathcal{D}'$.

If $U(G)$ is a UIR (unitary irreducible), then for many important cases $\mathcal{D}$ is nuclear, so that $\mathcal{D}'$ is a suitable (minimal) domain for working with algebraic "infinitesimal" methods in cases where operators with continuous spectra should be diagonalized.

We shall also try to find convenient concrete realizations of $\mathcal{D}$ and $\mathcal{D}'$ in various cases.

(By adding to $A = -i(d/dx)$ above, $B = x$, $C = I$, we get the infinitesimal generators of a UIR of the three-dimensional, non-abelian nilpotent group. Here $dU(\mathcal{E}(L))$ is evidently the polynomial algebra $\mathcal{P}(x, -i(d/dx))$, with $\mathcal{S}$ being the largest possible domain (in $\mathcal{H}$). The topology in $\mathcal{S}$—which is the standard topology—can be defined (in a way made precise in Sections II and III) by the harmonic oscillator Hamiltonian $\Delta = A^2 + B^2 + C^2$).

When solving the eigenvalue equation $B'x' = \lambda x'$ in ϕ' for an s.a. B, we might get more than "enough" generalized eigenvectors (i.e., corresponding to λ not in the spectrum Sp B, either real or complex, or higher multiplicity than we should have; a priori a B with a purely discrete spectrum, i.e., all eigenvectors in $\mathcal{H}$, might have generalized eigenvectors in $\phi' - \mathcal{H}$ corresponding to a continuous part). A more general corresponding situation is when we "reduce" a $U'(H)$ in $\mathcal{D}'$ (i.e., we seek irreducible components of the restriction of $U'(G)$ to a subgroup $H \subset G$); it turns out that we can get IR's (non-unitary or unitary) of H which are not contained in $U(H)$ ($U(G)$ restricted to H in $\mathcal{H}$). In formal work applying infinitesimal methods, this phenomenon has appeared in the form that starting from a state in a "continuous basis" with a non-compact subgroup H "diagonal" (expliciting the reduction of $U(H)$ in UIR's of H) application of a generator of G outside H transforms the state to a state in a "non-U" IR of H, which is then not contained in U(H). Well-known examples here are $G = SU(1,1)$ ($H = \{\exp(-itJ_2),\ -\infty < t < \infty\}$, non-compact hyperbolic subgroup), and Poincaré group $\mathcal{P}_+^\uparrow$ (H = Lorentz group $\mathcal{L}_+^\uparrow$). We shall see that this thing will, in general, not occur if H is compact.

II. Mathematical Preliminaries

1. Countably-Hilbert Space Generated by a Positive s.a. Operator.

Given $\mathcal{H}$, $(\cdot,\cdot)$, $\|\cdot\|$, s.a. operator A, $(x,Ax) \geqq (x,x)$, $x \in \mathcal{D}(A)$. Define for $\alpha \geqq 0$

$$\mathcal{H}_\alpha = \{x \in \mathcal{H};\ (x,x)_\alpha \equiv (A^\alpha x, A^\alpha x) < \infty\},$$

and for $\alpha \leqq 0$

$$\mathcal{H}_\alpha = \text{Completion of } \mathcal{H} \text{ with respect to scalar product } (x,y)_\alpha.$$

All $\mathcal{H}_\alpha$ are Hilbert spaces, and for $\alpha > \beta$, $\mathcal{H}_\alpha$ is dense linear subset of $\mathcal{H}_\beta$. If $\alpha \geqq 0$, and $x \in \mathcal{H}_\alpha$, then $y \in \mathcal{H}$ defines by $f_y(x) = (y,x)$ an element in $(\mathcal{H}_\alpha)'$; as

$$|(y,x)| = |(A^{-\alpha}y, A^\alpha x)| \leqq p_{-\alpha}(y)\cdot p_\alpha(x) \quad [p_\alpha(x) = [(x,x)_\alpha]^{\frac{1}{2}}],$$

we find by completion that we can identify $(\mathcal{H}_\alpha)' = \mathcal{H}_{-\alpha}$; this is true also for $\alpha < 0$.

Putting

$$\mathcal{D} = \bigcap_1^\infty \mathcal{H}_n$$

(projective limit) and

$$\mathcal{D}' = \bigcup_1^\infty \mathcal{H}_{-n}$$

(inductive limit), we have the chain $\mathcal{D} \subset \ldots \mathcal{H}_n \subset \ldots \mathcal{H} \subset \ldots \mathcal{H}_{-n} \subset \ldots \mathcal{D}'$. Evidently $\mathcal{H}_\alpha (\alpha > 0) = \mathcal{D}(A^\alpha)$ and

$$\mathcal{D} = \bigcap_1^\infty \mathcal{D}(A^n).$$

We topologize $\mathcal{D}$ by the countable set of norms $\{p_n\}_0^\infty$, and $\mathcal{D}$ is then called a countably-Hilbert space. It is easily seen to be complete (i.e., Fréchet), dense in $\mathcal{H}$, with dual $\mathcal{D}'$. $\mathcal{D}$ is reflexive, i.e., $(\mathcal{D}')' = \mathcal{D}$ (with the strong topology on $\mathcal{D}'$), and $\mathcal{H}$ (and thus $\mathcal{D}$) is dense in $\mathcal{D}'$. For $x \in \mathcal{D}$, $y' \in \mathcal{D}'$ (x, y') and (y', x) make sense.

If B is an operator in $\mathcal{H}$ (bounded or not) such that its adjoint B^+ (supposed to exist) is continuous $\mathcal{D} \to \mathcal{D}$, we can define the extension B' of B to $\mathcal{D}'$ by $(x, B'y') = (B^+x, y')$. B' is continuous $\mathcal{D}' \to \mathcal{D}'$, and B' is actually the closure by continuity of B in $\mathcal{D}'$.

If $\mathcal{a}$ is a symmetric algebra (i.e., $B \in \mathcal{a} \Rightarrow B^+ \in \mathcal{a}$) of continuous operators in $\mathcal{D}$, we can extend it to an algebra $\mathcal{a}'$ in $\mathcal{D}'$.

Finally we remark that for any real α and β $(A')^{\alpha-\beta}$ $(=(A^{\alpha-\beta})'$, if $\alpha-\beta$ not integer) is an isometry of $\mathcal{H}_\alpha$ onto $\mathcal{H}_\beta$.

2. When is $\mathcal{D}$ a nuclear space?

Definition 1. A linear map $C: \mathcal{H}_{(1)} \to \mathcal{H}_{(2)}$ is Hilbert-Schmidt (H.S.) if for one (and then for all) complete orthonormal set $\{e_n\}$ in $\mathcal{H}_{(1)}$ we have

$$\sum \|Ce_n\|_{(2)}^2 < \infty .$$

Definition 2. A countably-Hilbert space

$$\mathcal{D} = \bigcap_1^\infty \mathcal{H}_n$$

is nuclear, if for every n there is $m \geqq n$ such that the canonical embedding of $\mathcal{H}_m$ in $\mathcal{H}_n$ ($\mathcal{H}_m \ni x \to x \in \mathcal{H}_n$) is H.S. (This means geometrically that the unit ball $\{x;\, p_m(x) \leqq 1\}$ in $\mathcal{H}_m$ is in $\mathcal{H}_n$ an ellipsoid with half axis a_i, such that

$$\sum a_i^2 < \infty;$$

"the topology gets rapidly finer as we go to higher n").

In our case, this means A^{n-m} is H.S.

Thus: $\mathcal{D}$ nuclear $\Leftrightarrow$ for some $M \geqq 0$, A^{-M} is H.S. This means that A has a discrete spectrum $1 \leqq \lambda_1 \leqq \lambda_2 \cdots$, $\lambda_n \to \infty$ (corresponding complete orthonormal set $\{e_n\}$), such that

$$\sum \left(\frac{1}{\lambda_n}\right)^{2M} < \infty .$$

Especially A^{-1} is compact.

Using $\{e_n\}$ as basis in $\mathcal{H}$ (which is then obviously separable) we get

$$\mathcal{H} = \left\{x = \sum x_n e_n;\ \sum |x_n|^2 < \infty\right\}$$

$$\mathcal{H}_\alpha = \left\{x' = \sum x_n' e_n;\ \sum \lambda_n^{2\alpha} |x_n'|^2 < \infty\right\}$$

$$\mathcal{D} = \left\{x;\ \sum \lambda_n^{2M} |x_n|^2 < \infty,\ \text{all } M\right\}$$ "rapidly decreasing sequences"

$$\mathcal{D}' = \left\{x';\ \sum \lambda_n^{-2M} |x_n'|^2 < \infty,\ \text{some } M = M(x')\right\}$$ "slowly increasing sequences."

If $\mathcal{D}$ is nuclear, then so is $\mathcal{D}'$ (in a more general sense than given by Definition 2), and $\mathcal{D}$ and $\mathcal{D}'$ are Montel spaces.

3. Nuclear Spectral Theorem.

Assumptions:

$$\mathcal{D} \subset \mathcal{H}_{(1)} \subset \mathcal{H} \subset \mathcal{H}'_{(1)} \subset \mathcal{D}'$$

$$\underset{\text{cont.}}{\longrightarrow} \quad \underset{\text{H.S.}}{\longrightarrow}$$

[$\mathcal{H}_{(1)} \to \mathcal{H}$ H.S. implies $\mathcal{D}$ nuclear; if we know that $\mathcal{D}$ is nuclear, then a possible choice of $\mathcal{H}_{(1)}$ is $\mathcal{H}_M$.]

B s.a. in $\mathcal{H}$, continuous $\mathcal{D} \to \mathcal{D}$; extend to B' in $\mathcal{D}'$.

Statement: $\mathcal{H}'_{(1)}$ contains a complete set of generalized eigenvectors of B, i.e., there exists $\{e'(\lambda, i_\lambda);\ \lambda \in \text{Sp B},\ i_\lambda = 1, \ldots n_\lambda \leqq \infty,\ n_\lambda \text{ multiplicity of } \lambda\} \subset \mathcal{H}'_{(1)}$, such that

1. $B'e'(\lambda, i_\lambda) = \lambda e'(\lambda, i_\lambda)$, $i_\lambda = 1, \ldots n_\lambda$
   $\{e'(\lambda, i_\lambda)\}_1^{n_\lambda}$ linearly independent.
2. For some positive measure μ on Sp B (defined uniquely up to equivalence), any $x, y \in \mathcal{H}_{(1)}$, and spectral projector $E(\Delta)$ of B, we have the completeness relation

$$(x, E(\Delta)y) = \int_\Delta \sum_{i_\lambda=1}^{n_\lambda} (x, e'(\lambda, i_\lambda)) \cdot (e'(\lambda, i_\lambda), y)\, d\mu(\lambda).$$

[Obvious generalizations to case when B is normal, or we have a finite set $\{B_i\}$ of commuting s.a. (or normal) operators.]

There are two different ways of deriving this theorem: 1) differentiation of the spectral measure of B (Gel'fand, Kostiuchenko, Berezanskij a.a.)[1] and, 2) starting from von Neumann's direct integral form of the spectral theorem for B (Gårding, Maurin a.a.).[2]

III. Differentiable Vectors and Generalized Eigenvectors

1. Set $\mathcal{D}$ of Differentiable Vectors.

We now apply the construction in Section II.1 to the case of U(G) (corresponding dU(L), dU($\mathcal{E}$(L))) in $\mathcal{H} = \{x, y \ldots\}$. Pick a basis $l_1, \ldots l_n$ in L, and define the infinitesimal generators $J_\nu = i \cdot dU(l_\nu)$, $\nu = 1, \ldots, n$. J_ν are symmetric, even essentially self-adjoint (e.s.a.) on, and leave invariant, the Gårding domain, which is dense in $\mathcal{H}$. Define the Nelson operator (corresponding to the choice of basis in L)

$$\Delta = \sum_1^n J_\nu^2 \in dU(\mathcal{E}(L));$$

it is e.s.a. on the Gårding domain[3] with closure $\bar{\Delta}$. Put $A = \bar{\Delta} + 1$ in II.1; then we have

$$\mathcal{D} = \bigcap_1^\infty \mathcal{D}(\bar{\Delta}^n),$$

and $\mathcal{D}$ is topologized by $\{p_n\}$, $p_n(x) = \| (\bar{\Delta} + 1)^n x \|$.

Definition: $x \in \mathcal{H}$ is a differentiable (analytic) vector of U(G) if the map $G \ni g \to U(g)\, x \in \mathcal{H}$ is C^∞ (analytic).

Denote provisionally by $\mathcal{D}_d$ the set of all differentiable vectors. It is not hard to show that an "infinitesimal" characterization of $\mathcal{D}_d$ is: the largest common invariant domain of $J_1 \ldots J_n$, or, the largest domain of the whole dU($\mathcal{E}$(L)). From this follows, of course, $\mathcal{D} \supset \mathcal{D}_d$. Actually, we have the important result: $\mathcal{D} = \mathcal{D}_d$.[4] Henceforth we use $\mathcal{D}$ for the set of differentiable vectors.

Although the norm p_n in $\mathcal{H}_n$ depends on the choice of basis in L, it can be shown that $\mathcal{H}_n$ (and hence $\mathcal{D}$) and the topology on $\mathcal{H}_n$ (and on $\mathcal{D}$) is independent of this choice.

One shows easily that the topology $\{p_n\}$ on $\mathcal{D}$ is actually the coarsest topology on $\mathcal{D}$ making all maps $\mathcal{D} \to \mathcal{H}$ (and $\mathcal{D} \to \mathcal{D}$) in dU($\mathcal{E}$(L)) continuous. (By a well-known theorem in TVS it is then the unique Fréchet topology making these maps continuous.) Also, one can show that all U(g) are continuous in $\mathcal{D}$. We can then extend U(G)/$\mathcal{D}$ (U(G) restricted to $\mathcal{D}$) and dU($\mathcal{E}$(L)) to U'(G) and dU'($\mathcal{E}$(L)) in $\mathcal{D}'$; U'(G) is a continuous, even differentiable, representation of G in $\mathcal{D}'$ with "differential" dU'(L).

Thus we have, starting from UR U(G) in $\mathcal{H}$: Global representation

$$U(G)/D \sim U(G) \sim U'(G)$$

enveloping "differential"

$$dU(\mathcal{E}(L)) \sim \text{——} \sim dU'(\mathcal{E}(L))$$

in

$$\mathcal{D} \subset \mathcal{H} \subset \mathcal{D}'.$$

If we define irreducibility as topological irreducibility, i.e., non-existence of non-trivial proper invariant closed subspace, one finds: One of

U(G)/D, U(G), U'(G) irreducible $\Rightarrow$ the other two irreducible.

((There is also the simple result:

$dU(\mathcal{E}(L))$ irreducible ($\Longleftrightarrow$ $dU'(\mathcal{E}(L))$ irreducible) $\Rightarrow$ U(G) irreducible,

but it seems very hard to find an example—except the trivial one with finite-dimensional $\mathcal{H}$—where the assumption applies. In fact, one is tempted to make the following (stronger) counter-conjecture:

Given a UIR U(G) in infinite-dimensional $\mathcal{H}$, there always exist non-trivial subspaces of $\mathcal{D}$, invariant under $dU(\mathcal{E}(L))$ and not dense in $\mathcal{H}$.

In the case where U(G) is induced from a UIR in $\mathcal{H}_H$ of a subgroup H, with the coset space G/H a manifold of dimension $\geqq 1$, this conjecture seems to hold, as in this case $\mathcal{H} = L^2(G/H; \mathcal{H}_H)$, and $\mathcal{D}$ will be C^∞ functions from G/H (with certain boundary conditions) to some dense subset of $\mathcal{H}_H$; restriction to C^∞ functions with support inside a compact proper subset gives the desired non-dense invariant subspace, since the operators in $dU(\mathcal{E}(L))$ will take the form of differential operators on G/H. Now for certain classes of groups (e.g., nilpotent, or of the form $A \otimes H$, A abelian invariant subgroup), all UIR's can be expressed as induced representations. For the class of semi-simple groups certain types of UIR's, i.e., those of the discrete series, cannot be directly expressed as induced in the above sense. However, even here the conjecture seems to apply: In the simplest case of G = SU(1,1), the discrete series D^{+}_{-1} representation can be realized in the space

$$\mathcal{H} = e^{i\varphi}\left\{\sum_{0}^{\infty} a_n e^{in\varphi}, \quad \sum |a_n|^2 < \infty\right\}$$

or, equivalently, $z \cdot H^2$ (Hardy space H^2 = set of functions holomorphic in $|z| < 1$, such that

$$\int_0^{2\pi} |f(re^{i\varphi})|^2 d\varphi$$

is bounded as $r \to 1$, i.e., the boundary values of f(z) on $|z| = 1$ exist in L^2-sense). $\mathcal{D}$ is here $z \cdot$ {functions in H^2 which are C^∞ on the boundary}.

It is easily seen that the set

$$z \cdot \left\{ \mathcal{P}\left(z, \frac{1}{\sqrt{1-z}}\right) \cdot \exp\left(-\frac{1+z}{1-z} - \frac{1}{\sqrt{1-z}}\right)\right\} \subset \mathcal{D},$$

and is invariant under z and d/dz (which implies invariance under $dU(\mathcal{E}(L))$). A somewhat deeper analysis shows that {...} is not dense in H^2 (e.g., the constant function 1 cannot be in the closure of {...}). (Example is due mainly to Flato and Sternheimer.)))

2. Some Properties of $dU(\mathcal{E}(L))$ in $\mathcal{D}$.

(Selection of results obtained by Nelson-Stinespring,[3] Nelson[4] and also Segal; also, see Maurin.[2])

H denotes a Lie subgroup of G, with Lie algebra L_H; H = G is allowed, of course.

<u>E.s.a. on $\mathcal{D}$ are</u>:

1. Infinitesimal generators $J_1 \ldots J_n$.
2. Symmetric elliptic operators (especially Nelson operators) and symmetric central elements ("Casimir operators") of $dU(\mathcal{E}(L_H))$.
3. Any symmetric element in $dU(\mathcal{E}(L_H))$, if H is compact.

(If, and only if, A is e.s.a. on $\mathcal{D}$ is its spectral representation uniquely fixed by the action of A in $\mathcal{D}$.)

For the property of <u>strong commutation</u> (spectral projectors commute) an important criterion is due to Nelson: Suppose A and B are s.a. in $\mathcal{H}$ and $\mathcal{D}_1$ is a dense linear subset in $\mathcal{H}$, contained in the domain of A, B, A^2, B^2, AB, and BA, and AB = BA on $\mathcal{D}_1$. Then, if $A^2 + B^2$ is e.s.a. on $\mathcal{D}_1$, we have: $A/\mathcal{D}_1$ and $B/\mathcal{D}_1$ are e.s.a., and A and B are strongly commuting.

From this one can deduce, e.g.:

Strongly commuting are:

a) Any set of symmetric central elements in $dU(\mathcal{E}(L_H))$, with each other and with $i \cdot dU(L_H)$ (and hence $U(H)$).

b) Any commuting set of symmetric elements in $dU(\mathcal{E}(L_H))$, if H is compact.

c) Any infinitesimal generator J_k in $dU(L_H)$ with a symmetric elliptic operator (especially a Nelson operator) in $dU(\mathcal{E}(L_H))$ with which it commutes.

3. Analytic Vectors.

Similar to the characterization of $\mathcal{D}$ as the set

$$\bigcap_{1}^{\infty} \mathcal{D}(\overline{\Delta}^{n})$$

(differentiable vectors of $\overline{\Delta}$) one can show that the set $\mathcal{A}(\subset \mathcal{D})$ of analytic vectors of $U(G)$ is the set of analytic vectors of $\overline{\Delta}^{\frac{1}{2}}$:

$$\mathcal{A} = \mathcal{A}_{\overline{\Delta}^{\frac{1}{2}}} = \Big\{ x \in \mathcal{D};\ \exists a(x) \text{ so that } p_n(x) \leqq [a(x)]^{2n}(2n)!,\ n = 1, 2, \ldots \Big\} .$$

From this one can show that $\mathcal{A}$ is dense in $\mathcal{D}$.

4. When is $\mathcal{D}$ Nuclear?

We have thus in $dU'(\mathcal{E}(L))$ and $U'(G)$ in $\mathcal{D}'$ found a natural minimal extension of $dU(\mathcal{E}(L))$ and $U(G)$. If $\mathcal{D}$ is nuclear, $\mathcal{D}'$ will be the "natural" extended space in which to solve eigenvalue equations, and will thus enable us to give a rigorous foundation for the application of "infinitesimal" methods to problems involving continuous spectra. Such problems almost always occur when reducing a given UIR of a group G after a non-compact subgroup (special case: reduction of the tensor product of two UIR's of a non-compact G).

If we put $G = SO(2)$, denote by $D^{(m)}$ the one-dimensional UIR $\varphi \to e^{im\varphi}$, and take the direct sum

$$D = \sum_{1}^{\infty} \oplus\ N(m) D^{(m)},$$

it is seen that if the finite multiplicity $N(m)$ of $D^{(m)}$ is taken to increase faster than any power of m (e.g., $\sim e^{m}$), then, although $(\overline{\Delta}+1)^{-1}$ is compact, no $(\overline{\Delta}+1)^{-M}$ can be H.S. Thus a non-I UR will have a nuclear $\mathcal{D}$ only under very special conditions (never if we have a direct integral decomposition with continuous contribution, and direct sum only if each IR component enters with finite, not to fast increasing multiplicity; assuming, of course, that the corresponding UIR's have nuclear $\mathcal{D}$). Hence we restrict ourselves to UIR's.

We shall show: $\mathcal{D}$ is nuclear if U(G) UIR

1. of G, semi-simple with finite center;

2. of $G = A \otimes K$, semi-direct product, A abelian invariant subgroup, K compact subgroup; example Euclidean groups);

3. of G, nilpotent;

4. with $m^2 > 0$ of $\overline{\mathcal{P}_+^\uparrow} = R^4 \otimes SL(2,C)$ [$= R^4 \otimes \overline{SO_0(3,1)}$].

We shall also find convenient concrete realizations of $\mathcal{D}$ and $\mathcal{D}'$ in the various cases.

In connection with 4. we shall see that the cases with $m = 0$ (but $p \not\equiv 0$) give non-nuclear $\mathcal{D}$, whereas $m^2 < 0$ probably corresponds to nuclear $\mathcal{D}$. Some comments on how to obtain a smaller, nuclear space $\mathcal{D}_1 \subset \mathcal{D}$ in the case $m = 0$ will also be given.

Proof of 1: For semi-simple G, "maximal compact" subgroup K, we can choose a basis of L diagonalizing the Killing form. We then have infinitesimal generators $J_1 \ldots J_\ell$ (spanning $i \cdot dU(L_K)$), $J_{\ell+1} \ldots J_n$, with Casimir operator

$$C_2 = \sum_1^\ell J_i^2 - \sum_{\ell+1}^n J_i^2 = -\Delta + 2\Delta_K$$

(Δ_K Nelson operator of U(K), restriction of U(G) to K). C_2, Δ, Δ_K are e.s.a. on $\mathcal{D}$, commuting, hence strongly commuting. One finds

$$\mathcal{D} = \mathcal{D}_K \cap \left(\bigcap_1^\infty \mathcal{D}(\overline{C}_2^n) \right),$$

with

$$\mathcal{D}_K = \bigcap_1^\infty \mathcal{D}(\overline{\Delta}_K^n),$$

the set of differentiable vectors of U(K). For a UIR we have $\overline{C}_2 = c_2 \cdot I$, hence $\mathcal{D} = \mathcal{D}_K$, and instead of $\{p_n\}$ we can use $\{p_n'\}$, $p_n'(x) = \| (\overline{\Delta}_K + 1)^n x \|$. Since a UIR of semi-simple G is often given in a discrete basis with K diagonal, $\{p_n'\}$ is a natural set of norms, and $\mathcal{D}$ will be identified as the set of "rapidly decreasing" sequences in this basis, and $\mathcal{D}'$ the set of "slowly increasing" sequences. (For UIR U(G) one can show similarly that $\mathcal{a} = \mathcal{a}_K$.)

Now we assume G has finite center ($\Longleftrightarrow$ K is compact). Then the nuclearity of $\mathcal{D}(=\mathcal{D}_K)$ follows from a combination of the two results:

a) For a UIR U(G) of G, semi-simple with finite center, there is an integer $N \geqq 1$, such that U(K) is contained in N times the regular representation of K (Harish-Chandra; under the stronger assumption that G is a semi-simple matrix group we can take $N=1$; Godement 1952).

b) If U(K) (= any UR of a compact Lie group K) is contained in N times the regular representation, then $(\overline{\Delta}_K+1)^{-m}$ is H.S., where integer $m>\frac{\ell}{4}$, $\ell=\dim K$ (Nelson-Stinespring).

Proof of 2: $G=A \circledS K$. Similar to 1, we can write $\Delta=\Delta_A+\Delta_K$, and Δ_A is central, so

$$\mathcal{D}=\mathcal{D}_K \cap \left(\bigcap_1^\infty \mathcal{D}(\overline{\Delta}_A^n)\right);$$

for a UIR $\mathcal{D}=\mathcal{D}_K$. The statement corresponding to a) for $G=A \circledS K$, has been shown to hold (with $N=1$) by Godement. Part b), of course, can be applied directly.

Also in this case we have realizations of $\mathcal{H}$, $\mathcal{D}$, and $\mathcal{D}'$ in terms of a discrete basis diagonalizing K. Alternatively, constructing UIR's using the method of induced representations, we have $\mathcal{H}=L^2(\Omega;\mathcal{H}_H) \cong L^2(\Omega)\otimes\mathcal{H}_H$, where $\mathcal{H}_H$ is the finite-dimensional representation space of a UIR of the "little group" $H \subset K$, and $\Omega\,(\sim K/H)$ is the orbit of H in the dual $\hat{A}$ of A. Although I have not verified this generally, it seems plausible to conjecture that the action of the infinitesimal generators "separate" in such a way that we get $\mathcal{D}=\mathcal{S}(\Omega)\otimes\mathcal{H}_H$, where $\mathcal{S}(\Omega)$ is the set of C^∞ functions on Ω (with the obvious topology). (This is true, certainly, in the case $G=R^3 \circledS SO(3)$, three-dimensional Euclidean group).

Proof of 3. G nilpotent. A. A. Kirillov[5)] has shown that $\mathcal{H}$ can be realized as the space $L^2(R^m)$, for some integer m, with $\mathcal{D}$ equal to $\mathcal{S}(R^m)$ (which is nuclear), and $dU(\mathcal{E}(L))$ the polynomial algebra of $x_1,\dots x_m$, $\partial/\partial x_1,\dots\partial/\partial x_m$. [Our example in Section I corresponds to $m=1$.]

Proof of 4: ($m>0$, s) (real mass >0, spin s $[=0,\frac{1}{2},\dots]$)-representation of $\mathcal{P}_+^\uparrow$. Here $\mathcal{H}=L^2(\Omega_m)\otimes\mathcal{H}_{SU(2)}$, $\Omega_m=\{p;\, p^2\equiv p_0^2-\mathbf{p}^2=m^2\}$, and $\mathcal{H}_{SU(2)}$ the 2s+1-dimensional representation space of a UIR $D^{(s)}$ of the little group SU(2). In this realization the ten generators $(P_\mu, \overline{M}, \overline{N})$ of $\mathcal{P}_+^\uparrow$ take the form

$$(P_\mu) = (p_0 = \sqrt{p^2+m^2}, \overline{p}), \quad \overline{M} = -i\overline{p} \times \nabla_{\overline{p}} + \overline{S},$$

$$\overline{N} = ip_0 \nabla_{\overline{p}} - \frac{\overline{p}\times\overline{S}}{(p_0+m)} .$$

One finds, e.g., that $p_0(p_0+m)^2\nabla_{\overline{p}} \in dU(\mathcal{E}(L))$, and concludes that $\mathcal{D} = \mathcal{S}(\Omega_m) \otimes \mathcal{H}_{SU(2)}$, which is nuclear.

In the case of $m^2 \equiv -\kappa^2 < 0$-representations, where

$$\mathcal{H} = L^2(\Omega_{i\kappa}; \mathcal{H}_{SU(1.1)}) \cong L^2(\Omega_{i\kappa}) \overline{\otimes} \mathcal{H}_{SU(1.1)}$$

(in this case the completed Hilbert space tensor product $\overline{\otimes}$ is needed, since $\mathcal{H}_{SU(1,1)}$ is infinite-dimensional, in general), the situation is more complicated, since the expressions for the infinitesimal generators given, e.g., by Shirokov[6] become singular in the "plane" $p_3 = -\kappa$, except in the case where the inducing representation of SU(1,1) is the trivial one; in this case we get, as in the real mass case, $\mathcal{D} = \mathcal{S}(\Omega_{i\kappa})$, and thus nuclear. What form $\mathcal{D}$ takes in the general case (or, perhaps rather, how one should realize the induced representation to get a neat form of $\mathcal{D}$) should be investigated more in detail; however, it is reasonable to conjecture that $\mathcal{D}$ will be nuclear also in this case.

This is in contrast to the case of $m^2 = 0$ (but $p \neq 0$) representations. Also in this case we get for "helicity" $\lambda \neq 0$ (or infinite spin) singularities in the infinitesimal operators for $p_1 = p_2 = 0$, $p_3 < 0$ (here the orbit is the forward lightcone $p_0 = |\overline{p}|$, $p_0 > 0$.[7]) But in this case one can show for finite λ (also for $\lambda = 0$) that $(\overline{\Delta}+1)^{-1}$ is not a compact operator (thus no $(\overline{\Delta}+1)^{-m}$ can be H.S.), so that $\mathcal{D}$ cannot be nuclear. This property of $\overline{\Delta}$ is connected with the fact that for these representations of $\mathcal{P}_+^\uparrow$ the dilatations, which induce a one-parameter group of automorphisms of $\mathcal{P}_+^\uparrow$, can be (non-trivially) represented in the given UIR of $\mathcal{P}_+^\uparrow$. (This, of course, is not true for the $m^2 \neq 0$ representations.) We have $\Delta = \overline{M}^2 + \overline{N}^2 + 2\overline{p}^2$; a dilatation $\overline{p} \to \kappa\overline{p}$ leaves $\overline{M}$ and $\overline{N}$ invariant, so we get $\Delta \to \Delta_\kappa = \overline{M}^2 + \overline{N}^2 + 2\kappa^2\overline{p}^2$. Hence, if—which would be the case if $(\overline{\Delta}+1)^{-1}$ were compact—$\overline{\Delta}$ had a smallest (discrete) eigenvalue λ_1, with normalized eigenfunction $\varphi(\overline{p})$, then the "dilatated" function $\varphi_\kappa(\overline{p}) = N_\kappa\varphi(\kappa\overline{p})$ (N_κ chosen so that φ_κ is also normalized) would satisfy $\overline{\Delta}_\kappa\varphi_\kappa = \lambda_1\varphi_\kappa$, from which we deduce

$$(\varphi_\kappa, \bar{\Delta}\varphi_\kappa) = \lambda_1 - 2(\kappa^2-1)(\varphi_\kappa, \bar{p}^2\varphi_\kappa),$$

which for $\kappa > 1$ gives a contradiction.

It is known that (at least for finite λ) the given UIR can actually be extended to a UIR of the fifteen-dimensional conformal group (Lie algebra so(4,2)); since this group is semi-simple with finite center, we get a nuclear space $\mathcal{D}_1 \subset \mathcal{D}$ for the UIR of $\bar{P}_+^\uparrow$ by using $\Delta_{so(4,2)}$ corresponding to this representation of the conformal group. It is not obvious that this is the "simplest" solution; in the case $\lambda = 0$ one could conjecture that a natural choice would be $\mathcal{D}_1 = \mathcal{S}_0(\bar{p})$, rapidly decreasing C^∞ functions in $\bar{p}$-space with an infinite zero in $\bar{p}=0$, corresponding to $\Delta \to \Delta + 1/|\bar{p}|$ in the definition of $\mathcal{D}$. This $\mathcal{D}_1$ is certainly invariant under $dU(\mathcal{E}(L))$ and $U(G)$, dense in $\mathcal{H}$, and nuclear (as a subspace of a nuclear space); whether it has other desirable properties, e.g., essential self-adjointness of various operators in du $(\mathcal{E}(L))$, etc., remains to be investigated. In the case $\lambda \neq 0$ the situation is obviously more complicated.

Evidently more work should be done to clarify the situation for the case $m^2 = 0$ as well as $m^2 < 0$.

$P_+^\uparrow$ is one of the simplest examples of $G = A \otimes H$, with H non-compact, semi-simple. Probably in the general case UIR's of G induced from a maximal compact $K \subset H$, being realized as L^2 functions on the symmetric space H/K will not offer any problem but give a nuclear $\mathcal{D} = \mathcal{S}(H/K) \otimes \mathcal{H}_K$, whereas the other types of representations will give more trouble—especially the "cone type" representations will have a non-nuclear $\mathcal{D}$.

A final, disconnected remark under the heading $\mathcal{D}$ nuclear: Using results in Nelson-Stinespring, one can show that if $\mathcal{D}$ is nuclear, say $(\bar{\Delta}+1)^{-M}$ H.S., where 2M is an integer, then for $\varphi \in C_0^{4M}(G)$ (the set of 4M times continuously differentiable functions with compact support on G), the operator

$$U_\varphi = \int_G U(g)\varphi(g)\,dg$$

has an <u>absolutely convergent trace</u>, i.e.,

$$\sum |(e_n, U_\varphi e_n)| < \infty,$$

$\{e_n\}$ any complete orthonormal set in $\mathcal{H}$. Since $C_0^{4M}(G) \supset C_0^\infty(G)$, the mapping

$$C_0^\infty(G) \ni \varphi \to \sum (e_n, U_\varphi e_n) \in C$$

(complex numbers) (which is independent of the choice of $\{e_n\}$) gives a linear functional on $C_0^\infty(G)$, which can be shown to be continuous in the "Schwartz topology" of $C_0^\infty(G)$; this mapping is the character of the given UR U(G) of G.

5. "Redundant" IR's in the Restriction U'(H) of Subgroup H.

As mentioned before, when "diagonalizing" a non-compact subgroup H (i.e., reducing U(H)) one is in general forced to work in $\mathcal{D}'$; the obvious "infinitesimal" procedure is to find simultaneous eigenvectors (in $\mathcal{D}'$) of a "complete commuting set of s.a. operators in $dU(\mathcal{E}(L_H))$," consisting of Casimir operators plus other operators (possibly even operators outside $dU(\mathcal{E}(L_H))$, when certain UIR's of H occur with multiplicity > 1). Each common eigenspace of the Casimirs will carry a (sometime irreducible or else reducible) representation of H in $\mathcal{D}'$, a component of U'(H). If $\mathcal{D}$ is nuclear, the nuclear spectral theorem ensures us to find in this way all UIR's occurring in the Hilbert space decomposition of U(H). But, as mentioned before, also others will appear, in general. Of course, all non-UIR's occurring (whether corresponding to real or complex eigenvalues of the Casimirs) are excluded; but, as we shall see, there can also appear UIR's which are "redundant." Here one can apply certain criteria derivable from the completeness relation—which criteria we shall not state here—showing that a continuous superposition of vectors in such UIR's cannot give vectors in $\mathcal{H}$.

But the redundant IR's—both unitary and non-unitary ones—have an interest of their own, both from the more mathematical point of view as being in some sense naturally connected with the given U(G) and also from the physical point of view as having a possible application in connection with, e.g., so-called generalized partial wave analysis of matrix elements of a relativistically invariant S-operator.

We shall give below some simple examples of the possibilities that can occur, for $G = SU(1,1)$ and various one-parameter subgroups, and for a $(m > 0,\ s = 0)$ UIR of $G = \mathcal{P}_+^\uparrow$, with $H = \mathcal{L}_+^\uparrow$. Finally we show that for H compact no redundant IR's appear.

Example 1. $G = SU(1,1)$. The Iwasawa decomposition is $G = KAN$ where K ("elliptic subgroup"), A ("hyperbolic") and N ("parabolic") have the infinitesimal generators J_0, J_2, and $K_+ = J_0 + J_1$, respectively. The problem here is simply to diagonalize these generators in $\mathcal{D}'$ (conveniently in the discrete J_0-basis). This has been done,[8] and the result is indicated below (for the continuous principal series; for the other series analogous results hold; R and C are the real and complex numbers, respectively).

	J_0	J_2	K_+
Spectrum:	all integers	$2 \times R$	R
Generalized Eigenvalues:	all integers	$2 \times C$	R

The difference in behaviour of J_2 (A) and K_+ (N) is reflected in the properties of the element of $\mathcal{D}$ expressed in a J_2-basis and a K_+-basis; in the first case they can be extended to pairs of functions meromorphic in the whole plane, whereas in the second case they are, in general, just C^∞ functions on $R-\{0\}$.

Example 2. $G= \mathcal{P}_+^\uparrow = R^4 \otimes SO_0(3,1)$; $H=SO_0(3,1)$. $U(G)=(m>0, s=0)$. $\mathcal{H} = L^2(\Omega_m)$, $\mathcal{D} = \mathcal{S}(\overline{p})$, $\mathcal{D}' = \mathcal{S}'(\overline{p})$. Here the IR-content of U'(H) can essentially be obtained, e.g., from the treatment given by Joos,[9] with some "rigorification" and minor extensions.

The two Casimirs of H are $\overline{M}\cdot\overline{N} = 0$ here and $\overline{N}^2 - \overline{M}^2$, which is seen to be a second order elliptic differential operator; hence all solutions of $(\overline{N}^2-\overline{M}^2)'x'=\mu x'$ in $\mathcal{D}'$ are actually ordinary C^∞ solutions ($\mu = 1+\lambda^2$ in Joos' notation).

One then finds that the possible μ-values cover the whole complex plane; each μ carries an ∞-dimensional representation of H; this representation is irreducible and can be denoted $(j_0=0,\lambda)$ (j_0 is the smallest "spin" that occurs in the decomposition after SO(3)) except for $\mu=-(n^2-1)$, $n = 1, 2, \ldots$ ($\lambda=in$) where the representation splits into a direct sum of a finite-dimensional non-unitary IR, standard notation $\mathcal{D}^{((n-1)/2,(n-1)/2)}$, and an infinite-dimensional UIR (principal series $j_0=n$, $\lambda=0$). Apart from these latter IR's, also those corresponding to real $\mu \geqq 1$ (principal series $j_0=0$), and $0<\mu<1$ (supplementary series, $0<-i\lambda<1$) are unitary; the rest are non-unitary. The spectrum of $\overline{N}^2-\overline{M}^2$ is $[1,\infty)$, hence only the principal series, $j_0=0$, UIR's enter into the decomposition of U(H).

Actually, U'(H) thus contains, each exactly once, all the $\overline{M}\cdot\overline{N}=0$ "completely irreducible representations" obtained by Naimark (see, e.g., his book on Lorentz group representations). It might be pointed out that the IR's obtained by Naimark can also be characterized as the set of irreducible local representations of the Lorentz algebra so(3,1) such that the restriction to so(3) gives a direct sum of the standard finite-dimensional (integrable) representation of so(3), each occurring at most once; such a local representation of so(3,1) is hence always integrable.

To show that if H=K, compact, U(K) and U'(K) have the same IR content, we use the following simple

Lemma: Assume $H\subset G$, and $[U(H), A] = 0$, where s.a. A (e.g., $\overline{\Delta}+1$) is used to define $\mathcal{D}$. Then

1. Every $\mathcal{H}_\alpha$ is invariant under U'(H).

2. $U'(H)/\mathcal{H}_\alpha$ is unitary and unitary equivalent to U(H).

The proof follows from the fact that $(A')^{-\alpha}$ [= $(A^{-\alpha})'$] is an isometry from $\mathcal{H}$ onto $\mathcal{H}_\alpha$. $[U(H), A] = 0$ implies $[U'(H), (A')^{-\alpha}] = 0$, whence 1. and 2. follow; especially the last part of 2. is a consequence of $(A')^{\alpha}(U'(h)/\mathcal{H}_\alpha)(A')^{-\alpha} = U(h)$, considered as operator equality in $\mathcal{H}$; $h \in H$.

Now (as remarked by Dixmier) the assumption of the lemma is fulfilled if $H = K$, compact, and $A = \overline{\Delta} + 1$, Δ suitably chosen: AdK acting in the Lie algebra L of G gives a real finite-dimensional representation of K, which is hence equivalent to an orthogonal representation; a corresponding orthonormal basis in L will give a Nelson operator Δ commuting with U(H), whence $[U(H), \overline{\Delta}] = 0$. As an illustration, take $G = \mathcal{P}_+^\uparrow$ and $H = SU(2)$; evidently $\Delta = \overline{M}^2 + \overline{N}^2 + P_0^2 + P^2$ is rotation invariant.

To conclude the argument, assume V(K) is a component of U'(K) acting in finite-dimensional subspace $\mathcal{L}$ of $\mathcal{D}'$. Then $\mathcal{L} \subset \mathcal{H}_{-n}$ for some n, hence V(K) is a subrepresentation of the UR $U'(K)/\mathcal{H}_{-n}$, thus equivalent to a subrepresentation of U(K). This shows that U(K) and U'(K) contain every IR of K with the same multiplicity (finite or infinite). Actually, if V(K) contains only IR's of K occurring with finite multiplicity in U(K), we have $\mathcal{L} \subset \mathcal{D}$. (Proof: $(A')^{-(n+m)}\mathcal{L}$ is a finite-dimensional subspace of $\mathcal{H}_m$, and must be contained in the unique finite-dimensional subspace of $\mathcal{H}_m$ transforming as a direct sum of maximal (but finite) multiples of the IR's entering into V(K). This subspace is invariant under A (this is true in general only if the subspace is finite-dimensional!), hence we get $\mathcal{L}$ $(= A^{n+m}(A')^{-(n+m)}\mathcal{L}) \subset \mathcal{H}_m$. Since this holds for all m, $\mathcal{L} \subset \mathcal{D}$.) On the other hand, if every IR in V(K) has multiplicity ∞ in U(K), we might even have $\mathcal{L} \cap \mathcal{H} = \{0\}$.

IV. Product Representations

1. Tensor Product of Countably-Hilbert Spaces.

Given two countably-Hilbert spaces

$$\mathcal{D}^{(i)} \subset \ldots \mathcal{H}_n^{(i)} \subset \ldots \mathcal{H}^{(i)} \subset \ldots \mathcal{H}_{-n}^{(i)} \subset \ldots \mathcal{D}^{(i)'},$$

$$i = 1, 2,$$

defined by s.a. $A^{(i)}$.

From these spaces we can form the countably-Hilbert space

$$\left(\bigcap_{n=1}^{\infty} \mathcal{H}_n = \right)\mathcal{D} \subset \ldots \mathcal{H}_n \subset \ldots \mathcal{H} \subset \ldots \mathcal{H}_{-n} \subset \ldots \mathcal{D}'\left(= \bigcup_{n=1}^{\infty} \mathcal{H}_{-n}\right)$$

where $\mathcal{H}_n = \mathcal{H}_n^{(1)} \bar{\otimes} \mathcal{H}_n^{(2)}$ ($\bar{\otimes}$ denotes completed Hilbert space tensor product); the norm in $\mathcal{H}_n$ is p_n, where

$$p_n(x) = \| (A^{(1)})^n (A^{(2)})^n x \| ;$$

we use $A^{(1)}A^{(2)}$ and $A^{(1)} + A^{(2)}$, etc., instead of $A^{(1)} \otimes A^{(2)}$ and $A^{(1)} \otimes 1 + 1 \otimes A^{(2)}$.

Evidently $\mathcal{D}^{(1)} \otimes \mathcal{D}^{(2)}$ (algebraic tensor product) is a linear subset of $\mathcal{D}$, and it is not hard to see that it is dense in $\mathcal{D}$, hence $\mathcal{D} = \mathcal{D}^{(1)} \hat{\otimes} \mathcal{D}^{(2)}$; similarly $\mathcal{D}' = \mathcal{D}^{(1)'} \hat{\otimes} \mathcal{D}^{(2)'}$ (strong dual topology).

If we denote by A the s.a. operator defined in $\mathcal{H}$ by $A^{(1)} + A^{(2)}$, we can form Hilbert subspaces of $\mathcal{H}$.

$$\mathcal{H}_{[n]} = \left\{ x \in \mathcal{H};\ p_{[n]}(x) = \| A^n x \| < \infty \right\} .$$

It is not hard to verify that

$$\mathcal{H}_{2n} \underset{\text{cont.}}{\overset{\subset}{\longrightarrow}} \mathcal{H}_{[2n]} \underset{\text{cont.}}{\overset{\subset}{\longrightarrow}} \mathcal{H}_n ;$$

hence we also have

$$\mathcal{D} = \bigcap_1^{\infty} \mathcal{H}_{[n]} ,$$

and the set $\{p_{[n]}\}$ can be used to define the topology of $\mathcal{D}$.

From the relation between A and $A^{(1)}$, $A^{(2)}$ it is easy to deduce:

$$\mathcal{D} \text{ is nuclear } \Longleftrightarrow \mathcal{D}^{(i)} \text{ nuclear}, \quad i = 1, 2.$$

If $\mathcal{D}$ is nuclear, and

$$\left\{ e_n^{(i)} \right\}_{n=1}^{\infty}$$

are the $A^{(i)}$-bases in $\mathcal{H}^{(i)}$, then

$$\left\{e_m^{(1)} \otimes e_n^{(2)}\right\}_{m,n=1}^{\infty}$$

is evidently an A-basis in $\mathcal{H}$, and, e.g.,

$$\mathcal{D} = \left\{x = \sum_{m,n} x_{mn} e_m^{(1)} \otimes e_n^{(2)};\ \sum \left(\lambda_m^{(1)} \lambda_n^{(2)}\right)^{2N} |x_{mn}|^2 < \infty, \text{ all } N;\right.$$

or

$$\left.\sum \left(\lambda_m^{(1)} + \lambda_n^{(2)}\right)^{2N} |x_{mn}|^2 < \infty, \text{ all } N\right\}.$$

We can also remark that, according to well-known results for nuclear spaces, if at least one $\mathcal{D}^{(i)}$ is nuclear, then the projective, bi-equicontinuous, and countably-Hilbert topologies (the last can be shown to lie between the two first topologies, generally) on $\mathcal{D}^{(1)} \otimes \mathcal{D}^{(2)}$ agree. Also, under the same assumption,

$$\mathcal{D}' = \mathcal{B}(\mathcal{D}^{(1)}, \mathcal{D}^{(2)}),$$

the space of (separately) continuous bilinear forms on the product set $\mathcal{D}^{(1)} \times \mathcal{D}^{(2)}$.

2. Tensor Product of Representations of G.

Given two UR's $U^{(i)}(G)$ of G in $\mathcal{H}^{(i)}$, $i = 1, 2$ (correspondingly $\Delta^{(i)}$, $\mathcal{D}^{(i)}$, $\mathcal{D}^{(i)'}$). Form in $\mathcal{H} = \mathcal{H}^{(1)} \bar{\otimes} \mathcal{H}^{(2)}$ the tensor product $U(G) = (U^{(1)} \otimes U^{(2)})(G)$ (given by $G \ni g \to U^{(1)}(g) U^{(2)}(g)$ acting in $\mathcal{H}$).

The product space $\mathcal{D} = \mathcal{D}^{(1)} \hat{\otimes} \mathcal{D}^{(2)}$ is in general only a proper subset of the set of differentiable vectors of $U(G)$ (this latter set is of little interest, since $U(G)$ is not irreducible, except in very special cases). From the alternative set $\{p_{[n]}\}$ of norms on $\mathcal{D}$ given earlier, it is seen that $\mathcal{D}$ is actually the space of differentiable vectors of the representation $U(G \times G)$ of the product group $G \times G$, given by $G \times G \ni (g_1, g_2) \to U^{(1)}(g_1) \otimes U^{(2)}(g_2)$, since

$$\Delta = \Delta^{(1)} + \Delta^{(2)} = \sum_1^n J_k^{(1)^2} + \sum_1^n J_k^{(2)^2}$$

is evidently the Nelson operator of $U(G \times G)$.

$G \cong$ diagonal $(G \times G)$ is a subgroup of $G \times G$, and the reduction of $U(G)$ in UIR's of G is nothing but the reduction of $U(G \times G)$ (which is irreducible if $U^{(i)}(G)$ are) in UIR's of the subgroup G.

In this reduction, the transformation from the "$\mathcal{H}^{(1)} \bar{\otimes} \mathcal{H}^{(2)}$-basis" (product basis of $U^{(1)}(G) \otimes U^{(2)}(G)$, which in the case of a continuous basis will be a basis belonging to $\mathcal{D}^{(1)'} \otimes \mathcal{D}^{(2)'}$) to a basis (in general in $\mathcal{D}'$) realizing the reduction of U(G) into UIR's is given by (generalized) Clebsch-Gordan coefficients (CG-coef.):

Assume $\{E'(\lambda, i_\lambda)\} \subset \mathcal{D}'$ is a reduction basis (e.g., λ labels the eigenvalues of Casimirs of G). If $\varphi^{(i)} \in \mathcal{D}^{(i)}$, we have the expansion

$$\varphi^{(1)} \otimes \varphi^{(2)} \sim \int \sum_{i_\lambda} (E'(\lambda, i_\lambda), \varphi^{(1)} \otimes \varphi^{(2)}) E'(\lambda, i_\lambda) d\mu(\lambda).$$

If we have a discrete basis $\{e_n^{(i)}\}$ in $\mathcal{D}^{(i)}$,

$$\varphi^{(i)} = \sum \varphi_n^{(i)} e_n^{(i)},$$

we find

$$(E'(\lambda, i_\lambda), \varphi^{(1)} \otimes \varphi^{(2)}) = \sum_{m,n} C(\lambda, i_\lambda; (1)m, (2)n) \varphi_m^{(1)} \varphi_n^{(2)},$$

where the CG-coef. $C(\ldots) \equiv (E'(\lambda, i_\lambda), e_m^{(1)} \otimes e_n^{(2)})$ is in this case a "discrete-continuous" basis matrix element, a numerical function of (λ, i_λ) and (m,n).

In the general case the whole (m,n)-set of CG-coefficients will evidently be represented by an element in $\mathcal{B}(\mathcal{D}^{(1)}, \mathcal{D}^{(2)})$, indexed by (λ, i_λ).

Since the tensor product reduction problem is a special case of the subgroup reduction problem discussed in the last part of Section III, we should also expect to get "redundant" CG-coefficient, connecting both to non-U IR's and redundant UIR's of G, contained in U'(G) (= restriction of U'(G $\times$ G).

An example giving redundant UIR's is the reduction of a product $D_{j_1}^{+} \otimes D_{j_2}^{-}$ of a positive and a negative discrete series representation of SU(1,1), where UIR's of SU(1,1) belonging to the supplementary series ("exceptional interval") appear. This case, and also the other product representation cases for SU(1,1), have been treated—in a formal way—in articles by Holman and Biedenharn.[10] The framework given here provides a rigorous motivation for their method of deriving the CG-series and CG-coefficients by solving the difference eigenvalue equation of the Casimir operator of SU(1,1) in the discrete product basis

$$\{|j_1,m_1\rangle \otimes |j_2,m_2\rangle\}.$$

Conclusions

The main purpose of these lectures has been to advocate a certain standard solution of the problem of handling "improper, continuous bases" when using the "infinitesimal" method to study UIR's of Lie groups. The richer structure of $\mathcal{D} \subset \mathcal{H} \subset \mathcal{D}'$ (or a similar triplet) compared to the purely Hilbert space aspects stressed in the induction method of Mackey, with its central, very useful and beautiful induction-reduction theorem, seems to accommodate in a more natural way certain concepts arising in the reduction, as transformation coefficients between different bases, especially CG-coefficients in the case of product representations. Also, the richer IR structure in the "reduction" of U'(H) in $\mathcal{D}'$ raises new problems—intimately connected with the largely untouched problems of classification and "decomposition" of non-U, non-I representations of non-compact groups—which it should be interesting to study. These problems have also possibly a direct physical interest.

References

1. Yu. M. Berezanskij, Expansions in Eigenfunctions of Self-Adjoint Operators, Translations of Mathematical Monographs, Am. Math. Soc., Providence, Rhode Island (1968), Ch. V.
2. K. Maurin, General Eigenfunction Expansions and Unitary Representations of Topological Groups, Polish Scient. Publ., Warsaw (1968), Ch. II.
3. E. Nelson, W. F. Stinespring, Am. J. Math. 81, 547 (1959).
4. E. Nelson, Ann. Math. 70, 572 (1959).
5. A. A. Kirillov, Dokl. Akad. Nauk. 130, 966 (1960).
6. Iu. M. Shirokov, JETP 33, 1196 (1957).
7. J. S. Lomont and H. E. Moses, Journ. Math. Phys. 3, 405 (1962).
8. G. Lindblad and B. Nagel, Ann. Inst. H. Poincaré 13, 27 (1970).
9. H. Joos, Fortschritte der Physik 10, 65 (1962).
10. W. J. Holman and L. C. Biedenharn, Ann. Phys. 39, 1 (1966); 47, 205 (1968).

MACKEY-WIGNER AND COVARIANT GROUP REPRESENTATIONS*

U. Niederer
and
L. O'Raifeartaigh
Dublin Institute for Advanced Studies
Dublin, Ireland

1. Introduction

The invariance of a quantum theory under a transformation g is expressed by the invariance of the transition probabilities, i.e.,

$$|(f_1(g), f_2(g)| = |(f_1, f_2)|, \quad f \in \mathcal{H} \rightarrow f(g) \in \mathcal{H}, \tag{1.1}$$

where $\mathcal{H}$ is the Hilbert space of the theory (and for simplicity, we assume that there are no superselection rules). According to Wigner's theorem,[1] Eq. (1.1) implies the existence of a unitary or anti-unitary operator U(g) such that

$$f(g) = U(g)e^{i\delta(g,f)}f, \qquad f \in \mathcal{H}, \tag{1.2}$$

where $\delta(g, f)$ is a real number. If g is an element of a group G, then (1.2) in turn implies

$$U(gg') = U(g)U(g')e^{i\delta(g,g')}, \tag{1.3}$$

where $\delta(g', g)$ is real, i.e., that $\mathcal{H}$ carries a unitary or anti-unitary ray representation of the group. Whether or not the ray factor $e^{i\delta(g',g)}$ can be eliminated depends on the group structure.[2]

If the group G is continuous, then physical continuity demands the continuity of the physical states or rays, i.e.,

$$U(g')f \rightarrow U(g)e^{i\epsilon(g,f)}f, \quad g' \rightarrow g, \quad f \in \mathcal{H}, \tag{1.4}$$

where $\epsilon(g, f)$ is a real number, and from the linearity (or anti-linearity) of U(g), Eq. (1.4) implies

*Presented at NATO Summer School in Mathematical Physics, Istanbul, 1970.

$$U(g') \to U(g)e^{i\epsilon(g)}, \quad g' \to g, \tag{1.5}$$

i.e., that $\epsilon(g, f) = \epsilon(g)$ is actually independent of g. Note that $\delta(g', g) = 0$ in (1.3) implies $\epsilon(g) = 0$ in (1.5).

The particular group which we shall be interested in is the covering group $\mathcal{P}_+^\uparrow$ of the connected Poincaré group, as this group is the invariance group of any relativistic quantum theory. For the group $\mathcal{P}_+^\uparrow$, U(g) is unitary (because the group is connected) and the phases $\delta(g, g')$ can be eliminated.[4] Thus the Hilbert space of any relativistic quantum theory must carry a continuous unitary representation (CUR) of $\mathcal{P}_+^\uparrow$. For this reason the continuous unitary irreducible representations (CURs) of $\mathcal{P}_+^\uparrow$ have received considerable attention in the literature. They have been classified and each has appeared in various manifestations.

What we shall be concerned with in these notes is the connection between four such manifestations. All four of them take the form of transformations on functions spaces, namely,

(a) $m(g) \xrightarrow{g'} m(gg')$ on Mackey functions $m(g)$, $g \in \mathcal{P}_+^\uparrow$.

(b) $\omega(p) \xrightarrow{g} U(\Lambda(p)\Lambda\Lambda^{-1}(p'))\omega(p')e^{ip\cdot a}$ on Wigner functions $\omega(p)$, where $g = (a, \Lambda)$, $p_0 = \sqrt{\vec{p}^2 + m^2}$, $p' = \Lambda^{-1}p$.

(c) $\psi(p) \xrightarrow{g} D(\Lambda)\psi(p')e^{ip\cdot a}$ on covariant functions $\psi(p)$.

(d) $\phi(x) \xrightarrow{g} D(\Lambda)\phi(\Lambda^{-1}(x-a))$, on fields in configuration space.

The last three manifestations are familiar to physicists. However, all three of them incorporate a degree of arbitrariness. In case (b) the arbitrariness appears in the choice of representative $\Lambda(p)$ while for (c) and (d) it appears in the choice of the embedding representation $D(\Lambda)$. The manifestation (a) contains no such arbitrariness, and it links (b), (c) and (d) to the general theory of induced representations.[5] For these reasons we not only include (a) but take it as our starting point. Finally we should mention that, although our primary interest is $\mathcal{P}_+^\uparrow$, we find that the results are simpler and appear in better perspective if we consider general semi-direct products of the form $G = A \times S$, where A is abelian, although we shall always have $\mathcal{P}$ in mind and shall return to it at the appropriate point.

2. Mackey's Theory for General Groups

Let G be a locally compact separable group, H a closed subgroup and U(h), $h \in H$ a CUIR of H on a Hilbert space $\mathcal{H}_H$. Then a CUR of G is induced by the pair {H, U(h)} as follows:

Let $d\mu(c)$, $c \in C$ be the invariant† measure on the right coset space $C = G/H$ and let $m(g)$ be the vector-valued functions over G with values in $\mathcal{H}_H$ which satisfy the two conditions

i) $m(hg) = U(h)m(g)$ (subsidiary condition),

ii) $N^2(m) = \int d\mu(c)(m(g), m(g)) < \infty$ (square-integrability condition),

where the norm in the integrand is with respect to $\mathcal{H}_H$ and depends only on c on account of i). Then the induced representation is obtained by letting G act transitively on $m(g)$,

$$m(g) \xrightarrow{g'} (T_{g'}m)(g) = m(gg'). \tag{2.1}$$

From the definition it is clear that an induced representation is a generalization of the regular representation, to which it reduces for $\{H, U(h)\} = \{1, 1\}$. The group property follows from

$$(T_{g''}(T_{g'}m))(g) = (T_{g'}m)(gg'') = m(gg''g') = (T_{g''g'}m)(g), \tag{2.2}$$

and the unitarity follows from i) and ii) and the invariance of the measure $d\mu(c)$.

Condition i) means that although $m(g)$ is defined over G, it is essentially a function over C only, and for some purposes it is convenient to exhibit this property explicitly. This is done by means of a Mackey-Wigner transformation, as follows: For each $g \in G$, let $\gamma(g)$ be a representative element of the coset c to which g belongs, having the coset property

$$\gamma(gh) = \gamma(g) = \gamma(c) \tag{2.3}$$

and, if possible, chosen so that $\gamma(c)$ is continuous in c. Natural choices of $\gamma(c)$ may be suggested by the group structure or by outside (e.g., physical) considerations, and we shall give examples for $\mathcal{P}_+^\uparrow$. By definition $\gamma(g)g^{-1} \in H$. Then the Mackey-Wigner transformation is

$$\omega(g) = U(\gamma(g)g^{-1})m(g). \tag{2.4}$$

It is easy to verify that the Wigner functions $\omega(g)$ are functions over $C = G/H$, i.e., they satisfy

$$\omega(hg) = \omega(g) = \omega(c). \tag{2.5}$$

†If the measure is only quasi-invariant, the formalism must be slightly modified.[5]

This statement replaces the subsidiary condition. Since U(h) is unitary, the norm for the Wigner functions is

$$N^2(m) = N^2(\omega) = \int d\mu(c)(\omega(c), \omega(c)), \tag{2.6}$$

and, inserting (2.4) into (2.1), we find that the transformation law is

$$\omega(c) \xrightarrow{g} (T_g\omega)(c) = U(\gamma(c)g\gamma^{-1}(cg))\,\omega(cg). \tag{2.7}$$

Thus the Wigner functions eliminate the subsidiary condition and reduce the range of the arguments from G to G/H, but at the expense of introducing the arbitrary representatives $\gamma(c)$ and the 'generalized Wigner rotations' $U(\gamma(c)g\gamma^{-1}(cg))$.

3. Semi-Direct Products $G = A \times S$

We now specialize to semi-direct products $G = A \times S$, where the invariant subgroup A is abelian, and the group transformation is

$$(a, s)(a', s') = (a + \Lambda(s)a', ss'), \tag{3.1}$$

$\Lambda(s)$ being the linear transformation induced in A by S. The choice $\{H, U(h)\}$ which we wish to consider for this case is

$$\{H, U(h)\} = \{A \times \Sigma, V(a)V(\sigma)\} \tag{3.2}$$

where $V(a)$, $V(\sigma)$ are any CUIRs of A and Σ, and Σ is defined to be the little group of $V(a)$, i.e., the maximal closed subgroup of S leaving $V(a)$ invariant, i.e., satisfying

$$V(a) = V(\Lambda(\sigma)a), \qquad \sigma \in \Sigma \subset S \tag{3.3}$$

up to unitary equivalence. $V(a)$ must be chosen first, of course, and for abelian A, is one-dimensional. Note that for all $s \in S$

$$V_s(a) = V(\Lambda(s)a)$$

is still a CUIR of A. However, for $s \notin \Sigma$, $V_s(a)$ is not unitarily equivalent to $V(a)$. The set of all unitarily inequivalent $V_s(a)$ for all $s \in S$ is called the orbit of $V(a)$ and since

$$V_{\sigma s}(a) = V(\Lambda(\sigma)\Lambda(s)a) = V(\Lambda(s)a) = V_s(a), \tag{3.4}$$

we see that the elements of an orbit are in 1-1 correspondence with the cosets $S/\Sigma = (A\times S)/(A\times\Sigma) = G/H = C$. The CURs of G induced by $\{H, U(h)\}$ are equivalent for all $V_s(a)$ in the same orbit, but they are inequivalent for different orbits and for inequivalent choices of $V(\sigma)$.

Since $S/\Sigma = G/H$, a natural choice of representatives is

$$\gamma(c) = (o, s(c)), \tag{3.5}$$

where $s(c)$ is a set of representatives in S/Σ. Then, since

$$\begin{aligned}\gamma(c)g\gamma^{-1}(cg) &= (o, s(c)(a, s)(o, s(cs))^{-1} = \\ &= \Big(\Lambda(s(c))a, s(c)s\, s^{-1}(cs)\Big),\end{aligned} \tag{3.6}$$

the transformation law for the Wigner functions reduces to

$$\omega(c) \xrightarrow{(a,s)} (T_{(a,s)}\omega)(c) = V_c(a)V(s(c)s\, s^{-1}(cs))\omega(cs), \tag{3.7}$$

where $V_c(a) \equiv V_{s(c)}(a)$. This is just the transformation (b) of Section 1 and establishes the relationship between (a) and (b).

In the above, the abelian nature of A has not played an essential role, and the results actually go through with little modification for nonabelian A. However, the point of restricting ourselves to abelian A and for making the choice (3.2) for $\{H, U(h)\}$ is that under very mild mathematical conditions (which are certainly satisfied for $\mathcal{P}_+^\uparrow$), the CURs of $G = A\times S$ induced by $\{H, U(h)\}$ are <u>irreducible[3),4a),5) and exhaustive.</u>[3),5)]

A perhaps interesting question is what would happen if we induced a CUR of $A\times S$ with $\{S, U(s)\}$. The coset space in this case would be $A = A\times S/S$ and a natural choice of representative for the coset $(o, s)(a, 1)$ would be $(a, 1)$. These representatives would have the transformation law

$$(b, 1)(a, s) = (b+a, s) = (0, s)(\Lambda^{-1}(s)(a+b), 1) \quad \text{or}$$

$$b \xrightarrow{(a,s)} b^1 = \Lambda^{-1}(s)(a+b). \tag{3.8}$$

The Wigner functions would be, by definition,

$$\omega(b) = \omega(g) = U\Big((\Lambda^{-1}(s)a,\, 1)(a,\, s)^{-1}\Big)m(g) \;=\; U^{-1}(s)m(g),$$
$$g = (a,\, s), \qquad b = \Lambda^{-1}(s)a, \tag{3.9}$$

and would have the transformation law

$$\omega(b) \xrightarrow{(a,s)} (T_{(a,s)}\omega)(b) = U\Big((b,\, 1)(a,\, s)(b^1,\, 1)^{-1}\Big)\omega(b^1) = U(s)\omega(b^1), \tag{3.10}$$

with norm

$$N^2(\omega) = \int d\mu(a)\;(\omega(a),\, \omega(a)). \tag{3.11}$$

For $G = A \times S = \mathcal{P}_+^\uparrow$, $x = -a$, $\omega(x) = \phi(x)$, and $D(\Lambda) = U(s)$, the transformation (3.10) is formally the same as the transformation of fields (d) in Section 1. However, whereas (d) is supposed to carry an irreducible representation of $\mathcal{P}_+^\uparrow$, the representation (3.10) is highly reducible, and has the four-dimensional norm

$$N^2(\omega) = \int d^4x\; \phi^+(x)\phi(x). \tag{3.12}$$

The exact connection between the irreducible components of (3.10) and the $\phi(x)$ manifestation of the irreducible representations induced by $\{A\times\Sigma,\ V(a)V(\sigma)\}$ will be discussed in the section on locality in configuration space.

4. Covariance in Orbit Space

Let $G = A\times S$ as in the previous section and let $D(s)$ be an arbitrary continuous representation (CR) of S, which on restriction to Σ, is unitary and contains $V(\sigma)$, i.e.,

$$D(s) \;:\; D(\sigma) = V(\sigma) \oplus V_\perp(\sigma), \tag{4.1}$$

where the direct sum is formal and may, with appropriate modifications, be replaced by a direct integral or a mixture. The reducible Wigner functions

$$W(c) = \omega(c) \oplus \omega_\perp(c), \tag{4.2}$$

then have the transformation property

$$W(c) \xrightarrow{(a,s)} (T_{(a,s)}W)(c) = V_c(a)D(s(c)s\, s^{-1}(cs))W(cs), \tag{4.3}$$

and since D(s), unlike U(σ), is a representation of S, the D on the right hand side of (4.3) factors and we obtain

$$W(c) \xrightarrow{(a,s)} (T_{(a,s)}W)(c) = V_c(a)D(c)D(s)D^{-1}(cs)W(cs),$$
$$D(c) \equiv D(s(c)). \tag{4.4}$$

It follows that if we define the covariant orbital functions

$$\psi(c) = D^{-1}(c)W(c), \tag{4.5}$$

they have the covariant transformation law

$$\psi(c) \xrightarrow{(a,s)} (T_{(a,s)}\psi)(c) = V_c(a)D(s)\psi(cs). \tag{4.6}$$

Note that the complementary Wigner functions $\omega_\perp(c)$ may be zero, in which case W(c) is essentially irreducible. The norm for W(c) is

$$N^2(W) = \int d\mu(c)\langle W(c), W(c)\rangle = \int d\mu(c)\Big\{(\omega(c), \omega(c)) +$$
$$+ (\omega_\perp(c), \omega_\perp(c))_\perp\Big\}, \tag{4.7}$$

and hence the norm for the covariant function $\psi(c)$ is

$$N^2(\psi) = N^2(W) = \int d\mu(c)\langle\psi(c), D^\dagger(c)D(c)\psi(c)\rangle. \tag{4.8}$$

where here and henceforth the dagger is with respect to the inner-product $\langle\ ,\ \rangle$. For $\mathcal{P}_+^\uparrow$, the transformation (4.6) is identical to the transformation (c) of Section 1 and so Eqs. (4.1)-(4.6) establish the relationship between (b) and (c).

Note that each CUIR $U(\sigma) \subset D(\sigma)$ induces a CUR of G. Hence any Σ-invariant subsidiary condition in D(s)-space,

$$QW(c) = 0, \tag{4.9}$$

will be G-invariant. For the covariant functions $\psi(c)$, Eq. (4.9) takes the form

$$Q(c)\psi(c) = 0, \tag{4.10}$$

where

$$Q(c) = D^{-1}(c)QD(c), \tag{4.11}$$

and, using the fact that

$$D(c)D(s) = D(\sigma)D(cs), \qquad \sigma \in \Sigma, \tag{4.12}$$

and the Σ-invariance of Q, we see that $Q(c)$ is G-covariant in the sense that

$$D^{-1}(s)Q(c)D(s) = Q(cs). \tag{4.13}$$

The most interesting operator Q, is, of course,

$$P = P_{\perp}, \tag{4.14}$$

where $P_{\perp}$ is the projection operator on the Wigner functions $\omega_{\perp}(c)$, because (4.9) then implies

$$\omega_{\perp}(c) = 0,$$

which means that the $W(c)$ is essentially irreducible. Familiar examples of (4.10) for $Q = P_{\perp}$ are the Dirac and Lorentz conditions

$$(\gamma^{\mu}p_{\mu} - m)\psi(p) = 0, \qquad p^{\mu}\psi_{\mu}(p) = 0, \tag{4.15}$$

for spin $\frac{1}{2}$ and spin 1 wave-functions respectively.

If the CR $D(s)$ of S is pseudo-unitary with respect to a metric η, then, since $D(\sigma)$ is unitary, η can be normalized to 1 on $\omega(c)$, and then

$$P_{\perp} \supseteq \tfrac{1}{2}(1 - \eta). \tag{4.16}$$

Hence for $Q = P_{\perp}$, (4.9) implies

$$W(c) = \eta W(c). \tag{4.17}$$

The converse implication holds if and only if $Q = P_{\perp} = \frac{1}{2}(1 - \eta)$. From the definition of $\psi(c)$, Eq. (4.17) is equivalent to

$$D^\dagger(c)D(c)\psi(c) = D^\dagger(c)W(c) = D^\dagger(c)\eta\, W(c) =$$
$$= D^\dagger(c)\eta\, D(c)\psi(c) = \eta\psi(c). \quad (4.18)$$

Thus we obtain (4.18) as a consequence of the subsidiary condition (4.9) for $Q = P_\perp$, and the converse holds if, and only if, $P_\perp = \frac{1}{2}(1-\eta)$. Equation (4.18) allows us to write the norm (4.8) in the 'covariant' form

$$N^2(\psi) = \int d\mu(c)\langle\psi(c),\, D^\dagger(c)D(c)\psi(c)\rangle = \int d\mu(c)\langle\psi(c),\, \eta\psi(c)\rangle, \quad (4.19)$$

where 'covariant' here means that the integrand $\langle\psi(c),\, \eta\psi(c)\rangle$ is S-invariant.

5. Relationship of (c) to (a) and (d)

So far we have related the transformations (a) to (b) and then (b) to (c) of Section 1. We now wish to relate (a) to (c) directly, and to relate (c) to (d). To obtain the relation between (a) and (c), we simply combine the Mackey-Wigner and Wigner covariant transformations (2.4) and (4.5) respectively, and, since

$$\gamma(g)g^{-1} = (0, s(c))(a, s)^{-1} = (-\Lambda(\sigma)a, \sigma), \quad \sigma = s(c)s^{-1} \in \Sigma, \quad (5.1)$$

in this case, we obtain

$$\begin{aligned} \psi(c) = \psi(g) &= D^{-1}(c)U(-\Lambda(\sigma)a, \sigma)M(g) \\ &= D^{-1}(c)V^{-1}(\Lambda(\sigma)a)D(\sigma)M(g) \\ &= V^{-1}(a)D^{-1}(c)D(\sigma)M(g) \\ &= V^{-1}(a)D^{-1}(s)M(g), \end{aligned} \quad (5.2)$$

where, in an obvious notation,

$$M(g) = m(g) + m_\perp(g), \qquad D(\sigma) = V(\sigma) + V_\perp(\sigma). \quad (5.3)$$

One can verify directly that $\psi(c)$ as defined in (5.2) has the coset property

$$\psi(gh) = \psi(g) = \psi(c), \tag{5.4}$$

and the transformation property (4.6). The norm

$$\int d\mu(c) \langle \psi(c), D^\dagger(s) D(d) \psi(c) \rangle,$$

is easily seen to be equivalent to (4.8). Note that the representative $\gamma(g)$ has dropped out of (5.2). In this respect the transformation (5.2) is simpler than the two transformations which were combined to obtain it.

Finally, we wish to relate (c) to (d). To do this we define the 'Fourier transform'

$$\phi(a) = \int d\mu(c)\, V_c^{-1}(a)\psi(c), \qquad a \in A. \tag{5.5}$$

We then obtain the transformation law

$$\begin{aligned} \phi(b) \xrightarrow{(a,s)} (T_{(a,s)}\phi)(b) &= \int d\mu(c) V_c^{-1}(b)(T_{(a,s)}\psi)(c) \\ &= \int d\mu(c) V_c^{-1}(b) V_c(a) D(s) \psi(cs) \\ &= D(s) \int d\mu(c) V_c^{-1}(b-a)\psi(cs) \\ &= D(s) \int d\mu(c) V_{cs^{-1}}^{-1}(b-a)\psi(c) \\ &= D(s) \int d\mu(c) V_c^{-1}(\Lambda^{-1}(s)(b-a))\psi(c) \\ &= D(s)\, \phi(\Lambda^{-1}(s)(b-a)), \end{aligned} \tag{5.6}$$

which is the same as (d). The expression (4.8) for the norm may not, however, be local when expressed in terms of $\phi(b)$, a point which we shall be discussing in more detail later for $\mathcal{P}_+^\uparrow$.

The relationships between $m(g)$, $\omega(c)$, $\psi(c)$ and $\phi(a)$ are summarized in Table 1.

Mackey Functions

$m(hg) = U(h)m(g)$

$m(g) \xrightarrow{g'} m(gg')$

$\omega(c) = \omega(g) = U(\gamma(g)g^{-1})m(g)$ $\longrightarrow$

Wigner Functions

General G: $\omega(c) \xrightarrow{g} U(\gamma(c)g\gamma^{-1}(cg))\omega(cg)$

$G = A \times S$: $\omega(c) \xrightarrow{(a,s)} V(s(c)ss^{-1}(cs))\omega(cs)V_C(a)$

$\psi(c) = V^{-1}(a)D^{-1}(s)M(g)$ $\qquad$ $\psi(c) = D^{-1}(c)W(c)$

Covariant Functions

$G = A \times S$: $\psi(c) \xrightarrow{(a,s)} D(s)\psi(cs)V_C(a)$

$$\phi(b) = \int d\mu(c)V_C^{-1}(b)\psi(c), \quad b \in A$$

$\downarrow$

Functions over A (Configuration Space)

$G = A \times S$: $\phi(b) \xrightarrow{(a,s)} D(s)\phi(\Lambda^{-1}(s)(b-a))$

Table 1. A CUR of G is induced by $\{H, U(h)\}$. $C = G/H$. For $G = A \times S$, $\{H, U(h)\} = \{A \times \Sigma, V(a)V(\sigma)\}$, where Σ is the little group of $V(a)$. In general lower case letters denote elements of the set with the same capital letter. $V_C(a) = V(\Lambda(s)a)$, $s \in c$. $W(c) = \omega(c) + \omega_{\perp}(c)$, $M(g) = m(g) + m_{\perp}(g)$, where $D(\sigma) = V(\sigma) + V_{\perp}(\sigma)$. $\gamma(g) = \gamma(c)$ is a representative of the coset $c \in C$.

6. The Poincaré Group $\mathcal{P}_+^\uparrow$

The Poincaré group $\mathcal{P}_+^\uparrow$ is the semi-direct product $A_4 \times SL(2,C)$, where A_4 is the translation group in four dimensions and the action of $SL(2,C)$ on A_4 is the Lorentz action

$$a_\mu \xrightarrow{s} \Lambda_\mu^{\ \nu}(s)a_\nu = a_\mu^1 , \qquad a^1 = s^\dagger as, \quad a = a^\mu \tau_\mu = a\cdot\tau \tag{6.1}$$

where a_μ are the parameters of A_4, $s \in SL(2, C)$, τ_i, $i=1,2,3$ are the Pauli matrices and τ_0 is the unit 2×2 matrix. Note that the relationship between s and $\Lambda(s)$ is two to one.

The induction is with $\{A_4 \times \Sigma,\ V(a)\,V(\sigma)\}$ where $V(a)$ is any CUIR of A_4, and Σ is its little group. The CUIRs of A_4 are all of the form

$$V(a) = e^{ik\cdot a} \tag{6.2}$$

where k is any real 4-vector, and for convenience we take the Minkowski inner product. Then

$$V_s(a) = e^{ik\cdot\Lambda(s)a} = e^{i\Lambda^{-1}(s)k\cdot a} = e^{ip\cdot a}, \tag{6.3}$$

and it follows that the orbit of k is characterized by the values of $p = \Lambda^{-1}(s)k$, s $SL(2, C)$. From the known action of $\Lambda(s)$ on Minkowkian space, we then see that the orbits are

$$\begin{array}{llll} p^2 > 0,\ p_0 \gtrless 0 & \text{timelike} & \Sigma = SU(2) & p^2 = p^2 - p_1^2 - p_2^2 - p_3^2 \\ p^2 = 0,\ p_0 \gtrless 0 & \text{lightlike} & \Sigma = E(2) & \\ p^2 < 0, & \text{spacelike} & \Sigma = SU(1,1) & \\ p = 0, & \text{trivial} & \Sigma = SL(2,C) & \end{array} \tag{6.4}$$

From the definition of the little group, namely,

$$V_\sigma(a) = e^{ik\cdot\Lambda(\sigma)a} = e^{i\Lambda^{-1}(\sigma)k\cdot a} = e^{ik\cdot a}, \tag{6.5}$$

we see that the little groups Σ for the various orbits are as shown. Note that the coset variable c can be identified with the variable p (momentum) on each orbit. The general transformations (3.7), (4.6) (5.6) then reduce to the transformations (b), (c) and (d) of Section 1.

From now on we shall consider only the timelike orbits. For these orbits the CUIRs of $\mathcal{P}_+^\uparrow$ are characterized by $p^2 = m^2$, the sign of p_0, and the <u>spin</u> j, which is the non-negative half-integer characterizing the CUIR of the little group $\Sigma = SU(2)$. The points in the orbit are parametrized by the space momenta $\vec{p}(p_1 p_2 p_3)$, and $d\mu(p) = d^3p\,|\,(p_0)$. Henceforth, we shall normalize m^2 to unity.

The arbitrary quantity in the transformation (b) of Section 1 is the representative Lorentz transformation $\Lambda(p)$, or, to be more precise, $s(p)$. For $p^2 = 1$, $p_0 > 0$, we can choose $k = (1000)$ and then, from (6.1), $s(p)$ satisfies the equation

$$s^\dagger(p) 1 s(p) = s^\dagger(p) s(p) = p \cdot \tau , \tag{6.6}$$

which defines it up to an arbitrary rotation $s(p) \to us(p)$, $u\ u = 1$. Three standard choices of $s(p)$ are

$$\left.\begin{array}{ll} \text{i)} & s(p) = s^{-1}(\theta\phi) s(\lambda) s(\theta\phi) \\ \text{ii)} & s(p) = \qquad\qquad s(\lambda) s(\theta\phi) \end{array}\right\} \text{ where}$$

$$s(\lambda) = \begin{pmatrix} e^{\lambda/2} & 0 \\ 0 & e^{-\lambda/2} \end{pmatrix}, \quad s(\theta\phi) = \begin{pmatrix} \cos\theta/2 & \sin\theta/2\, e^{-i\psi} \\ -\sin\theta/2\, e^{i\psi} & \cos\theta/2 \end{pmatrix}$$

$$p_0 = \cosh\lambda, \quad \vec{p} = \sinh\lambda(\cos\theta,\ \sin\theta\cos\psi,\ \sin\theta\sin\psi)$$

$$\text{iii)} \quad s(p) = s(\mu) s(z) \qquad \text{where}$$

$$s(\mu) = \begin{pmatrix} e^{\mu/2} & 0 \\ 0 & e^{-\mu/2} \end{pmatrix}, \quad s(z) = \begin{pmatrix} 1 & z \\ 0 & 1 \end{pmatrix},$$

$$e^{\mu} \begin{pmatrix} 1 \\ z \end{pmatrix} = \begin{pmatrix} p_0 + p_3 \\ p_1 - ip_2 \end{pmatrix}$$

Choice i) is called the canonical choice[8)] and means that $\Lambda(p) = \Lambda(s(p))$ is a pure boost, i.e., is a Lorentz transformation in the 2-flat spanned by k and p. Choice iii) is used mainly because it is easy to manipulate and is useful in the 'infinite momentum[10)] limit' $p_3 \to \infty$. Both of these choices generalize immediately to other orbits. Choice ii) corresponds to a pure boost in the 3-direction followed by a rotation. It is made to obtain states of definite helicity.[9)]

For the transformations (c) and (d) of Section 1, the arbitrary quantity is the choice of embedding representation $D(\Lambda)$ (or, to be more precise, the embedding representation $D(s)$, $s \in SL(2, C)$). From the point of view of group theory, the choice of a unitary representation $D(s)$ would seem to be the simplest. However, since $SL(2, C)$ is non-compact, all the non-trivial unitary representations are infinite-dimensional, and not only would it be undesirable to introduce an infinite dimensional representation of $SL(2, C)$ to contain the finite dimensional representation $D^j(\sigma)$ of $SU(2)$, but it can be shown that from the physical point of view infinite dimensional representations of $SL(2, C)$ lead to serious difficulties with locality.[11)] For this reason, one restricts oneself to the non-unitary finite-dimensional representations.

The irreducible finite-dimensional representations of $SL(2, C)$, called $D^{(m,n)}(s)$ or $D(m, n)$, are unitary irreducible representations of the compact form $SU(2) \otimes SU(2)$, the labels m and n being just the spin labels for the invariant $SU(2)$ subgroups. The $SU(2)$ subgroup of $SL(2, C)$ is diagonal in $SU(2) \otimes SU(2)$, and with respect to it, $D(m, n)$ has the decompositions

$$D^{(m,n)}(\sigma) = \sum_{j=(m-n)}^{j=m+n} \oplus \; U^j(\sigma), \tag{6.7}$$

where j is integer-spaced. The integer-spacing of j means that $D(m, n)$ contains only integral spins, or only half-integral spins, i.e., that $(-1)^{2j} = (-1)^{2(m+n)}$. The representations $D(m, n)$ have the complex-conjugation property,

$$D^*(m, n) = CD(n, m)C^{-1}, \quad C^\dagger C = 1, \quad C^2 = (-1)^{2j} \tag{6.8a}$$

where in the conventional basis C is $\exp(i J_2)$. If the representation D is pseudo-unitary (in which case it must be reducible unless $m=n$), i.e., if

$$D^\dagger \eta D = \eta, \quad \eta = \eta^\dagger, \quad \eta^2 = 1, \tag{6.8b}$$

then the metric η plays also the role of a linear parity (space-reflexion) operator. The representations $D(m, o)$ and $D(o, m)$ are simply the complexifications of the representation $U^m(\sigma)$ of $SU(2)$, and hence are irreducible with respect to $SU(2)$. It follows that for

$D(s) = D(m, o)$ or $D(o, m)$ no subsidiary conditions are necessary. For the parity invariant combination $D(m, 1) \oplus D(o, m)$ there is one simple subsidiary condition.[4c] Other choices of

$$D(s) = \sum_{m,n} \oplus D(m, n)$$

are also popular, and two well-known classes are described in References 12 and 13.

For finite-dimensional $D(s)$, the invariant subsidiary operator $Q(p)$ of Section 4 can be made more explicit. First we note that since Q is an operator in the finite-dimensional space of $D(s)$, it is a finite tensor component with respect to $SL(2, C)$. On the other hand, Q is $SU(2)$-invariant. Hence from (6.7) we have, in an obvious notation,

$$Q = Q_0^{(0,0)} + Q_0^{(\frac{1}{2}\,\frac{1}{2})} + \ldots + Q_0^{(n,n)} \tag{6.9}$$

for some finite n, or in Lorentz tensor notation

$$Q = T + T_0 + T_{00} + \ldots + \underbrace{T_{00\ldots0}}_{2n} . \tag{6.10}$$

But since $k = (1\,0\,0\,0)$, Q can then be written as

$$Q = T + T_\mu k^\mu + T_{\mu\nu} k^\mu k^\nu + \ldots + T_{\mu\nu\ldots\sigma} k^\mu k^\nu \ldots k^\sigma , \tag{6.11}$$

and hence

$$Q(p) = D^{-1}(p) Q D(p) = T + T_\mu p^\mu + \ldots + T_{\mu\nu\ldots\sigma} p^\mu p^\nu \ldots p^\sigma . \tag{6.12}$$

Equation (6.12) exhibits the Poincaré invariance of $Q(p)$ explicitly, and shows that $Q(p)$ is a polynomial in p_μ.

7. Locality in Configuration Space

For $\mathcal{P}_+^\uparrow$, the configuration space functions $\phi(x)$ of (5.5) reduce to

$$\phi(x) = \int d\mu(p) e^{-ip\cdot x} \psi(\vec{p}), \quad x \in A_4 , \tag{7.1}$$

and we wish to investigate the conditions under which $\phi(x)$ is local, i.e., under which

1) $\phi(x)$ satisfies only local orbital conditions,
2) $\phi(x)$ satisfies only local subsidiary conditions, and
3) $\phi(x)$ has a local norm.

Here, by local orbital and subsidiary conditions, we mean conditions which are polynomial in $\partial_\mu = (\partial_0, \vec{\partial})$, and by a local norm we mean one which can be expressed as an integral of a finite number of space-time derivatives of $\phi^\dagger(x)$ and $\phi(x)$. We examine these conditions in turn.

1) Local orbital condition. This condition is already violated on a single timelike orbit, since the condition $p_0 > 0$, $p_0 < 0$ is non-local. To circumvent this difficulty we relax the irreducibility condition slightly and allow representations which are direct sums of irreducible representations on $p_0 > 0$ and $p_0 < 0$, i.e., we consider the functions

$$\begin{bmatrix} \phi_+(x) \\ \\ \phi_-(x) \end{bmatrix} = \begin{bmatrix} \int d\mu(p) e^{-ip\cdot x} u(\vec{p}) \\ \\ \int d\mu(p) e^{ip\cdot x} v(\vec{p}) \end{bmatrix}, \quad p_0 = +\sqrt{\vec{p}^2 + 1}, \tag{7.2}$$

where $u(\vec{p})$ and $v(-\vec{p})$ are the irreducible $\psi(\vec{p})$-functions for the orbits $p^2 = 1$, $p_0 > 0$, and $p^2 = 1$, $p_0 < 0$, respectively, and each orbit carries the same representation of $SL(2, C)$. Note that

$$u(\vec{p}) = D(\vec{p}) W_+(p), \qquad v(\vec{p}) = D(\vec{p}) W_-(-\vec{p}) \tag{7.3}$$

because $(\pm 1, 000) \rightarrow (\pm|p_0|, \pm\vec{p})$ under the same Lorentz transformation. The relaxation (7.2) does not, in itself, remove the non-locality of the orbital condition, since $\phi_\pm(x)$ still satisfy non-local orbital conditions $p_0 \gtrless 0$. However, the introduction of $\phi_\pm(\vec{x}, 0)$ allows us to make a (non-local) transformation from the (non-local) fields $\phi_\pm(\vec{x}, 0)$ to the fields $\phi(\vec{x}, 0)$ and $\dot{\phi}(\vec{x}, 0)$ where

$$\phi(\vec{x}t) = \phi_+(\vec{x}, t) + \phi_-(\vec{x}, t) = \int d\mu(p) \left\{ e^{-ip\cdot x} u(\vec{p}) + e^{ip\cdot x} v(\vec{p}) \right\}, \tag{7.4}$$

and these fields are local in the sense that the only orbital condition satisfied by $\phi(\vec{x}, t)$ is

$$(\Box + m^2)\phi(\vec{x},t) = (\partial_t^2 - \vec{\partial}^2 + m^2)\phi(\vec{x},t) = 0. \quad (7.5)$$

Note that at any fixed time t, the representation of $\mathcal{P}_+^\uparrow$ is carried not by $\phi(\vec{x},t)$ alone but by $\phi(\vec{x},t)$ and $\dot{\phi}(\vec{x},t)$, and that in the $\phi(\vec{x},t)$, $\dot{\phi}(\vec{x},t)$ basis the representation is not diagonal. Note also that the plus sign in (7.4) denotes ordinary vector sum in SL(2,C)-space and not direct sum.

2) Local subsidiary conditions. The subsidiary conditions for $u(\vec{p})$ and $v(\vec{p})$ are

$$Q_+(p)u(\vec{p}) = Q_-(p)v(\vec{p}) = 0, \quad Q_\pm(p) = D^{-1}(\vec{p})Q_\pm D(\vec{p}). \quad (7.6)$$

From Section 6 the $Q_\pm(p)$ are polynomials in p_μ, and hence there is a local subsidiary condition, namely,

$$Q_+(i\partial)\phi(x) = Q_-(-i\partial)\phi(x) = 0, \quad (7.7)$$

if, and only if,

$$Q_\pm(p) = Q_\mp(-p). \quad (7.8)$$

Let us now introduce the linear operator R defined by

$$R = R^\dagger = (-1)^{2m}, \qquad D(s) = \sum_{m,n} \oplus\, D^{(m,n)}(s). \quad (7.9)$$

R commutes with D(s) and has the property of commuting and anti-commuting with even and odd rank tensors, respectively. It follows from (6.12) that

$$RQ_\pm(p)R^{-1} = Q_\pm(-p), \quad (7.10)$$

and hence that a necessary and sufficient condition for (7.8) is

$$RQ_\pm R^{-1} = Q_\mp . \quad (7.11)$$

This equation defines Q_- in terms of Q_+ (or vice versa), and since R is SL(2,C)-invariant, it incorporates the demand that we have equivalent spin conditions on the two orbits. The result is that by choosing $Q_\pm$ so as to satisfy (7.11) the subsidiary conditions can always be made local.

3) Local norm. From (7.2) we see that the norm for $\phi(\vec{x},t)$ is

$$N^2(\phi) = \int d\mu(p)\left\{\langle u(\vec{p}), D^\dagger(\vec{p})D(\vec{p})\,u(\vec{p})\rangle + \langle v(\vec{p}), D^\dagger(\vec{p})D(\vec{p})\,v(\vec{p})\rangle\right\} \tag{7.12}$$

Substituting (7.5) into this expression, with $d\mu(p) = d^3p/p_0$, $p_0 > 0$, and $\langle a, b\rangle = a^\dagger b$, we obtain

$$\begin{aligned} N^2(\phi) = {} & \frac{1}{8(2\pi)^6}\int d^3(xyp)\,p_0^{-1}\left\{p_0^2\phi^\dagger(x)\Delta_+(\vec{p})\phi(y) + \dot{\phi}^\dagger(x)\Delta_+(\vec{p})\dot{\phi}(y)\right\} \\ & \left\{e^{i\vec{p}\cdot(\vec{x}-\vec{y})} + e^{-i\vec{p}\cdot(\vec{x}-\vec{y})}\right\} \\ & + \frac{i}{8(2\pi)^6}\int d^3(xyp)\left\{\phi^\dagger(x)\Delta_-(\vec{p})\dot{\phi}(y) - \dot{\phi}^\dagger(x)\Delta_-(\vec{p})\phi(y)\right\} \\ & \left\{e^{i\vec{p}\cdot(\vec{x}-\vec{y})} - e^{-i\vec{p}\cdot(\vec{x}-\vec{y})}\right\}, \quad p_0 > 0, \end{aligned} \tag{7.13}$$

where

$$\Delta_\pm(\vec{p}) = D^\dagger(\vec{p})D(\vec{p}) \pm D^\dagger(-\vec{p})D(-\vec{p}).$$

But by definition $s^\dagger(p)s(p) = p\cdot\tau$, $p_0 > 0$, and hence

$$D^\dagger(\vec{p})D(\vec{p}) = D^\dagger(s(\vec{p}))D(s(\vec{p})) = D(s^\dagger(p)s(p)) = D(p\cdot\tau), \quad p_0 > 0, \tag{7.14}$$

from which we see that each irreducible component $D(m,n)$ of $D^\dagger(\vec{p})D(\vec{p})$ is a polynomial in p_μ of order $2(m+n)$, and that $D^\dagger(\vec{p})D(\vec{p})$ is even and odd in p_μ for definite parity $(-1)^{2(m+n)} = \pm 1$, respectively. We can then construct Table 2 which appears on the following page. In the table we have utilized the fact that since $p_0^2 = \vec{p}^2 + 1$, p_0^2 can be eliminated, leaving terms of zero or first order in the non-local operator $p_0 > 0$. Comparing this table with Eq. (7.13), and taking the non-locality of p_0 into account, we see at once that we obtain a local expression if, and only if, the spin is half-integral, and then

Spin	Parity of SL(2, C)-Representation	$D^{\dagger}(\vec{p})D(\vec{p})$	$\Delta_{+}(\vec{p})$	$\Delta_{-}(\vec{p})$
Integral	$(-1)^{2j} = (-1)^{2(m+n)} = 1$	even in $(p_0, \vec{p})$	even in $\vec{p}$	$p_0 \times$ odd in $\vec{p}$
Half-Integral	$(-1)^{2j} = (-1)^{2(m+n)} = -1$	odd in $(p_0, \vec{p})$	$p_0 \times$ even in $\vec{p}$	odd in $\vec{p}$

TABLE 2

$$N^2(\phi) = \frac{1}{4(2\pi)^3}\int d^3x\Big\{\phi^\dagger(x)(\vec{\partial}^2+1)\delta_+(i\vec{\partial})\phi(x) + \dot{\phi}^\dagger(x)\delta_+(i\vec{\partial})\dot{\phi}(x) - i\phi^\dagger(x)\delta_-(i\vec{\partial})\dot{\phi}(x) + i\dot{\phi}(x)\delta_-(i\vec{\partial})\phi(x)\Big\}, \tag{7.15}$$

where

$$\Delta_+(\vec{p}) = p\,\delta_+(\vec{p}), \quad \Delta_-(\vec{p}) = \delta_-(\vec{p}), \quad \delta_\pm(\vec{p}) \text{ polynomial in } \vec{p}.$$

Thus a necessary and sufficient condition for a local norm is that the spin be half-integral.

Note that the above considerations apply quite independently of the existence of a pseudo-unitary metric η. If η exists and $Q = P_\perp$, the formalism simplifies somewhat. From the definitions of η in (6.8) and R in (7.9), we see that

$$\eta R = (-1)^{2(m+n)} R\eta = (-1)^{2j} R\eta, \tag{7.16}$$

and hence, from (7.11) that if we normalize η to 1 on $u(\vec{p})$ it takes the value $(-1)^{2j}$ on $v(\vec{p})$. Hence, making the appropriate modifications of (4.19) we have

$$N^2(\phi) = \int d\mu(p)\Big\{\langle u(\vec{p}), \eta u(\vec{p})\rangle + (-1)^{2j}\langle v(\vec{p}), \eta v(\vec{p})\rangle\Big\}, \tag{7.17}$$

and if we insert (7.4) into this expression we obtain

$$N^2(\phi) = \frac{1}{4(2\pi)^6}\int d^3(xyp)\,p_0^{-1}\Big\{p_0^2\phi^\dagger(x)\eta\phi(y) + \dot{\phi}^\dagger(x)y\dot{\phi}(y)\Big\}\Big\{e^{i\vec{p}\cdot(\vec{x}-\vec{y})} + (-1)^{2j}e^{-i\vec{p}\cdot(\vec{x}-\vec{y})}\Big\}$$
$$+ \frac{i}{4(2\pi)^6}\int d^3(xyp)\Big\{\phi^\dagger(x)\eta\dot{\phi}(y) - \dot{\phi}^\dagger(x)\eta\phi(y)\Big\}\Big\{e^{i\vec{p}\cdot(\vec{x}-\vec{y})} - (-1)^{2j}e^{-i\vec{p}\cdot(\vec{x}-\vec{y})}\Big\}. \tag{7.18}$$

We see that once again we obtain a local expression if, and only if, the spin is half-integral, and then the local expression is

$$N^2(\phi) = \frac{i}{2(2\pi)^3}\int d^3x\left\{\phi^\dagger(x)\eta\dot{\phi}(x) - \dot{\phi}^\dagger(x)\eta\phi(x)\right\}. \tag{7.19}$$

If the subsidiary conditions for $\phi(x)$ contain at least one Dirac equation

$$i\dot{\phi}(x) = (-i\vec{\alpha}\cdot\vec{\partial} + \beta)\phi(x), \tag{7.20}$$

as in References 12) and 13), Eq. (7.19) reduces still further to

$$N^2(\phi) = \frac{1}{(2\pi)^3}\int d^3x\,\phi^\dagger(x)\beta\eta\phi(x). \tag{7.21}$$

8. Physical Implications

Before going on to discuss the physical implications of the last section, it is useful to consider for a moment the representations (3.8)-(3.12) obtained by inducing with the subgroup S of $A \times S$. Since for $\mathcal{P}_+^\uparrow$, $S = SL(2, C)$, and we are interested in pseudo-unitary rather than unitary representations of $SL(2, C)$, we modify the original Mackey formalism by inserting the metric η in condition ii) of Section 2. We then find that for $\mathcal{P}_+^\uparrow$, the representations (3.8)-(3.12) can be written as

$$\phi(x) \xrightarrow{(a,s)} (T_{(a,s)}\phi)(x) = D(s)\phi(\Lambda^{-1}(s)(x-a)),$$

$$N_S^2(\phi) = \int d^4x\langle\phi(x), \eta\phi(x)\rangle. \tag{8.1}$$

The representation (8.1) is highly reducible, but a large reduction is effected by imposing the Klein-Gordon condition

$$(\Box + m^2)\,\phi(x) = 0. \tag{8.2}$$

The Klein-Gordon condition implies that

$$\partial^\mu j_\mu(x) = 0, \quad j_\mu(x) = \langle\phi(x), \eta\partial_\mu\phi(x)\rangle - \langle\partial_\mu\phi(x), \eta\phi(x)\rangle, \tag{8.3}$$

which means that it comes equipped with a local group-invariant bilinear, namely,

$$\tilde{N}^2(\phi) = \int d^3x \; j_0(x). \tag{8.4}$$

For half-integral spin $\tilde{N}(\phi)$ is the norm (7.17), but for integral spin, $\tilde{N}^2(\phi)$ is not positive definite[14] and hence $\tilde{N}(\phi)$ is not a norm. (Note that the indefiniteness of $\tilde{N}^2(\phi)$ does not contradict the definiteness of $\tilde{N}_S^2(\phi)$ in (8.1), because the solutions of (8.2) do not lie in the space of $\tilde{N}_S^2(\phi)$.)

We turn now to the physical significance of the results we have obtained concerning locality. First, we note that the usual minimal requirements for a relativistic theory are

i) Poincaré invariance,
ii) Positive probability, i.e., positive Poincaré invariant norm,
iii) Positive energy,
iv) Locality in the sense of Section 7,

and then we consider these requirements for first and second quantized theories separately.

First quantized theories. For these theories the fields $\phi(x)$ which we have discussed are elements of the underlying Hilbert space $\mathcal{H}$, and so they should satisfy all four requirements. But that is impossible, since, as we have seen, iii) and iv) are incompatible for fields of any spin and ii) and iv) are incompatible for fields of integer spin. The result is that there is no first quantized relativistic theory which satisfies all four requirements.

Second quantized theories. For second quantized theories the fields are not elements of the underlying Hilbert space $\mathcal{H}$, but operators on it. The conditions i)-iv) can then be shared out between $\phi(x)$ and $\mathcal{H}$ to remove the incompatibility as follows: Conditions i), ii) and iii) are demanded for $\mathcal{H}$, and are satisfied by using the non-local, positive energy, Wigner functions $\omega_+(\vec{p})$ (or the corresponding non-local, positive energy, covariant functions $u(\vec{p})$). Conditions i) and iv) are demanded for $\phi(x)$ and are satisfied by using quantized $\phi(x)$ which transform as in Section 1(d), for which no norm is necessary and for which the quantity $\tilde{N}^2(\phi)$ in (8.4) is interpreted as the charge.

References

1. E. Wigner, Group Theory (Academic Press, New York, 1959), p. 233. V. Bargmann, J. Math. Phys. 5, 862 (1964) and references contained therein. Some further references are given in G. Rasche and L. O'Raifeartaigh, Ann. Phys. 25, 155 (1963).

2. V. Bargmann, Ann. Math. 59, 1 (1954). D. Simms, Lecture Notes in Mathematics (Springer, Heidelberg (1968). K. Parthasarathy, ibid. (1969).
3. E. Wigner, Ann. Math. 40, 139 (1939). Y. Shirkov, JETP, 6, 919 (1958). H. Joos, Fort. Phys. 10, 65 (1962).
4. For previous work on this and related subjects see:
   a) A. Wightman, Nuovo Cimento Supplemento 14, 81 (1959);
   b) D. Pursey, Ann. Phys. 32, 159 (1965); A. McKerrell, Ann. Phys. 40, 237 (1966) (mass zero case);
   c) S. Weinberg, Phys. Rev. 133, B1318 (1964); 181, 1893 (1969);
   d) G. Parravicini, A. Sparzani, Nuovo Cim. 66A, 579 (1970); and references contained therein.
5. G. Mackey, Induced Representations of Groups and Quantum Mechanics (Benjamin, New York, 1963).
6. S. Schweber, Relativistic Quantum Field Theory (Row-Peterson, New York, 1961). R. Streater, A. Wightman, PCT Spin and Statistics and All That (Benjamin, New York, 1964).
7. S. Gasiorowicz, Elementary Particle Physics (Wiley, New York, 1967), Chapter 4.
8. L. Foldy, Phys. Rev. 102, 568 (1956).
9. M. Jacob, G. Wick, Ann. Phys. 7, 404 (1959).
10. H. Leutwyler, Acta Phys. Austriaca, Suppl. V, 320 (1968).
11. I. Grodsky, R. Streater, Phys. Rev. Letters 20, 695 (1968). R. Streater, Nobel Symposium 8. Elementary Particle Theory. Groups and Analyticity (Almqvist and Wiskell, Stockholm, 1968), p. 133. Oksak, I. Todorov, Princeton Institute for Advanced Study, 1970 (preprint).
12. W. Rarita, J. Schwinger, Phys. Rev. 60, 61 (1941).
13. V. Bargmann, E. Wigner, Proc. Nat. Acad. Sci. (U.S.A.) 34, 211 (1946).
14. W. Pauli, Phys. Rev. 58, 716 (1940).

Acknowledgements

The above presentation originated in some lectures given by one of the authors (O'R) at the Royal Irish Academy Summer School in Group Theory and at the Battelle Rencontres in Mathematics and Physics in 1969. The authors would like to thank many of the participants at these meetings for discussions and comments. In particular, they should like to thank Professor V. Bargmann, Professor C. C. Moore, Dr. J. T. Lewis and Dr. R. Musto.

DERIVATION OF THE LORENTZ AND GALILEI GROUPS FROM ROTATIONAL INVARIANCE†

Vittorio Gorini‡
Institut für theoretische Physik I
der Universität Marburg
Germany

1. Introduction

The purpose of this seminar is to prove a theorem which states that the only homogeneous kinematical groups which are compatible with the condition that inertial reference frames at rest and with the same space-time origin should be connected by space rotations (isotropy of space) are the Lorentz and Galilei groups, apart from the rotation group itself, which appears as a limiting case of the Lorentz group for vanishing invariant velocity.

We shall assume the relativity group to operate linearly on space-time, which can be seen to follow from a special formulation of the principle of space-time homogeneity, together with a weak continuity requirement.[1)]

It might seem that our result is not an unexpected one and, indeed, it has to be pointed out that, since the advent of relativity, there have appeared in the literature several derivations of the Lorentz transformations based on the customarily required symmetry properties of the space-time manifold, without the assumption of existence of an invariant velocity.[2)] However, to the extent of our knowledge, the existing derivations are often lacking rigour in the exploitation of the basic principles (such as space-time homogeneity and isotropy of space), and all are burdened by the use of more or less explicitly stated additional technical assumptions. Among the latter, not to mention the reciprocity principle, a significant role is played by the (often implicit) assumption that the kinematical group is a Lie group, or essentially equivalent conditions.[3)] On the other hand, we shall show that the result can be achieved without any

---

†Presented at the NATO Summer School in Mathematical Physics, Istanbul, 1970.

‡On leave of absence from Istituto di Fisica dell'Universitá, Milano, Italy. A. V. Humboldt fellow.

additional condition of this type.[4] This is remarkable, also in view of some recent proposals to change the space-time topology and, consequently, the topology of the kinematical group, on the basis of the algebraic and geometric structure of the group itself, once this is already known.[5]

The theorem which we give here is an improved version of a result which was proved in the last section of Reference 1.b). There, however, a rather strong continuity condition was used, which has later revealed itself to be unnecessary.

A thorough investigation of the possible kinematical groups has been recently carried out by Bacry and Lévy-Leblond.[6] The approach of these authors is a very general one in that they treat space-time translations as not necessarily commuting and give a purely group theoretical definition of space-time, as a homogeneous space of the kinematical group. On the other hand, the remarks made above apply to their paper, since they work with Lie algebras rather than with groups. Also, in order to limit the number of solutions, they require time reversal and parity to be automorphisms of the kinematical groups, which seems questionable. It has to be noted altogether that, upon restriction to homogeneous transformations, their result is as ours, apart from the additional appearance of the Carroll group[7] which, by its very definition as the limiting case of the Lorentz group for the transformation of space-like intervals with large $|\Delta\vec{r}|/\Delta t$ at low velocities, does not satisfy our conditions.

Our result is crucially dependent on the "richness" of the group of space rotations and hence on the dimensionality of space-time. For example, in a two-dimensional space-time, where the rotation group is reduced to the identity element, the equivalent statement of our theorem is no more true and many solutions exist other than the Galilei and the Lorentz ones, such as, f.i., the Parker group (containing the Lorentz group),[8] and the Galilei groups respectively with integer and rational values of the velocity.

We shall use the standard notation GL(4,R) for the group of all 4×4 non-singular matrices with real entries, under the usual operation of matrix multiplication. Latin indices run from 1 to 3 and Greek indices from 1 to 4. The summation convention is adopted. A matrix $G \in GL(4,R)$ is identified with the family of its matrix elements, $G = \{G_{\mu,\nu}\}_{\mu,\nu=1,2,3,4} = \{G_{\mu\nu}\}$.

2. The Theorem

It is convenient first to state the theorem and then to comment on its physical interpretation.

Theorem. Let $\mathcal{H}$ denote the following subset (which is trivially a subgroup) of GL(4,R):

$$\mathcal{H} = \{H \mid H \in GL(4,R);\ H_{i4} = 0,\ \ i = 1,2,3\}.$$

Let Φ denote the family of subgroups of GL(4,R) which is defined by

$$\mathcal{L} \in \Phi \text{ iff } \mathcal{L} \cap \mathcal{H} = \mathcal{C} = \Big\{C \mid C \in GL(4,R);\ C_{i4} = C_{4i} = 0,$$

$$i = 1,2,3;\ C_{44} = 1;\ C_{i\ell}C_{j\ell} = \delta_{ij};\ \det\{C_{i\ell}\} = 1\Big\}.$$

Then Φ can be indexed as $\Phi = \{\mathcal{G}(\lambda)\}_{\lambda \in [0,+\infty]}$, where

a) if $0 < \lambda < +\infty$,

$$\mathcal{G}(\lambda) = \Big\{L \mid L \in GL(4,R);\ g_{\mu\nu}(\lambda)L_{\mu\rho}L_{\nu\sigma} = g_{\rho\sigma}(\lambda);$$

$$g_{\mu\nu} = 0 \text{ if } \mu \neq \nu;\ g_{11} = g_{22} = g_{33} = -\lambda;\ g_{44} = 1;$$

$$\det L = 1;\ L_{44} \geqq 1\Big\};$$

b) $$\mathcal{G}(0) = \Big\{G \mid G \in GL(4,R);\ G_{4i} = 0,\ i = 1, 2, 3;$$

$$G_{44} = 1;\ G_{i\ell}G_{j\ell} = \delta_{ij},\ \det\{G_{ik}\} = 1\Big\};$$

c) $\mathcal{G}(+\infty) = \mathcal{C}$.

Comments. Since we have assumed linearity and we restrict ourselves to homogeneous transformations, we shall write the relation among the space-time coordinates of an event, as seen in two different inertial frames S and S', as

$$x'_\mu = L_{\mu\nu}x_\nu,\ x_4 = t,\ \{L_{\mu\nu}\} \in GL(4,R). \tag{1}$$

The condition for two frames S and S' to be at rest with respect to each other can be expressed by the requirement that any object which is at rest in S should be at rest in S' as well, which means that any world line which is parallel to the x_4-axis should be transformed by (1) into a world line which is parallel to the x'_4-axis. Obviously, for this to be true it is necessary and sufficient that

$$L_{i4} = 0,\ i = 1, 2, 3. \tag{2}$$

Therefore, denoting by $\mathcal{L}$ the kinematical group and by $\mathcal{L}_R$ its "rest" subgroup, we have

$$\mathcal{L}_R = \mathcal{L} \cap \mathcal{H}. \tag{3}$$

We assume

$$\mathcal{L}_R = \mathcal{C}. \tag{4}$$

Physically, this axiom states a) that in each frame clocks at different space points have been synchronized ($L_{4i} = 0$ for $L \in \mathcal{L}_R$, $i = 1, 2, 3$),[9] b) that time has a unidirectional flow and that all frames use the same time standard ($L_{44} = 1$ for $L \in \mathcal{L}_R$),[10] c) that with respect to each frame space is euclidean and isotropic and that the localization of events in space is given in terms of orthogonal, say, right-handed triads, all frames adopting the same standard of length ($L_{i\ell} L_{ij} = \delta_{\ell j}$, $\det\{L_{ij}\} = 1$ for $L \in \mathcal{L}_R$).[10] By relations (3) and (4), our kinematical group $\mathcal{L}$ is required to satisfy

$$\mathcal{L} \cap \mathcal{H} = \mathcal{C}. \tag{5}$$

We prove by our theorem that the only subgroups of GL(4,R) which satisfy (5) are, apart from $\mathcal{C}$ itself ($\mathcal{C} = \mathcal{G}(+\infty)$), the (proper orthochronous) Galilei group $\mathcal{G}(0)$ and the (proper orthochronous) Lorentz groups $\mathcal{G}(\lambda)$'s, corresponding to all possible values $c = \lambda^{-\frac{1}{2}}$ of the invariant velocity.

In order to prove the theorem we shall first state some lemmas.

<u>Lemma 1</u>. Let $G \in GL(4,R)$. Then G can be written at least in one way as a product

$$G = A\,X\,B, \tag{6}$$

where $A, B \in \mathcal{C}$ and

$$X_{21} = X_{31} = X_{24} = X_{34} = 0, \tag{7.a}$$

$$\Delta(X) = X_{11}X_{44} - X_{41}X_{14} > 0. \tag{7.b}$$

<u>Proof</u>. Write G as

$$G = \begin{bmatrix} \{G_{ik}\} & \{G_{i4}\} \\ \{G_{4k}\} & G_{44} \end{bmatrix} \tag{8}$$

Let $C, C' \in \mathcal{C}$. Then

$$C\,G\,C' = \begin{bmatrix} \{C_{ij}\;G_{j\ell}\;C'_{\ell k}\} & \{C_{i\ell}\;G_{\ell 4}\} \\ \{G_{4\ell}\;C'_{\ell i}\} & G_{44} \end{bmatrix} \tag{9}$$

Choose first C such that $C_{2\ell}G_{\ell 4} = C_{3\ell}G_{\ell 4} = 0$ and then C' such that $C_{2j}G_{j\ell}C'_{\ell 1} = C_{3j}G_{j\ell}C'_{\ell 1} = 0$. With these choices, $C\,G\,C'$ attains the form

$$C\,G\,C' = Y = \begin{bmatrix} Y_{11} & Y_{12} & Y_{13} & Y_{14} \\ 0 & Y_{21} & Y_{22} & 0 \\ 0 & Y_{31} & Y_{32} & 0 \\ Y_{41} & Y_{42} & Y_{43} & Y_{44} \end{bmatrix}, \quad Y_{44} = G_{44}. \tag{10}$$

Denote by P a rotation of π in the (x_1, x_2) plane:

$$P = \begin{bmatrix} -1 & 0 & 0 & 0 \\ 0 & -1 & 0 & 0 \\ 0 & 0 & 1 & 0 \\ 0 & 0 & 0 & 1 \end{bmatrix}. \tag{11}$$

Since $G \in GL(4,R)$, we have $\det Y \neq 0$, hence

$$\Delta(Y) = Y_{11}Y_{44} - Y_{41}Y_{14} \neq 0.$$

Then we can take $X = P^n Y$, $A = C^{-1}P^n$, $B = C'^{-1}$ with $n = 0$ or $n = 1$ according to whether $\Delta(Y) > 0$ or $\Delta(Y) < 0$.

Lemma 2. Let $\mathcal{L} \in \Phi$ and if $L \in \mathcal{L}$, let

$$L = A\,X\,B \tag{12}$$

be a decomposition of L as in Lemma 1. Then

$$X_{11} \neq 0 \quad \text{and} \quad L_{44} = X_{44} \neq 0. \tag{13}$$

Proof. One easily verifies that $X_{11} = 0$ (respectively, $X_{44} = 0$) implies $(XPX^{-1})_{i4} = 0$ (respectively, $(X^{-1}PX)_{i4} = 0$) and $(XPX^{-1})_{44} = -1$

(respectively $(X^{-1}PX)_{44} = -1$) and this contradicts (5) thus proving the lemma.[11)]

Now define $\mathcal{K}$ to be the subset of $\mathcal{L}$ formed by the matrices satisfying (7.a) (it is the set of velocity transformations along the x_1-axis) and let $\mathcal{K}_+$ denote the subset of $\mathcal{K}$ formed by the matrices satisfying (7.b). It is easy to control that both $\mathcal{K}$ and $\mathcal{K}_+$ are subgroups of $\mathcal{L}$ and that

$$\mathcal{K} = \mathcal{K}_+ \cup P\mathcal{K}_+ \qquad \mathcal{K}_+ \cap P\mathcal{K}_+ = \emptyset. \tag{14}$$

Next, consider the following subset of the real line R:

$$\vartheta = \{v \mid v \in R;\ v = -X_{14}/X_{11},\ X \in \mathcal{K}\}. \tag{15}$$

ϑ is the set of allowed velocities of S' relative to S and it is indeed a subset of R because of (13) (no infinite velocities!). By (14) we have

$$\vartheta = \{v \mid v \in R;\ v = -X_{14}/X_{11},\ X \in \mathcal{K}_+\}. \tag{16}$$

For every $v \in \vartheta$ define

$$\mathcal{K}_+(v) = \{X \mid X \in \mathcal{K}_+;\ -X_{14}/X_{11} = v\} \tag{17}$$

and note that

$$\mathcal{K}_+ = \bigcup_{v \in \vartheta} \mathcal{K}_+(v) \tag{18}$$

Further, define

$$\mathcal{C}_1 = \{C \mid C \in \mathcal{C};\ C_{11} = 1\}. \tag{19}$$

$\mathcal{C}_1$ is the subgroup of $\mathcal{C}$ formed by the matrices of the form

$$A(\alpha) = \begin{bmatrix} 1 & 0 & 0 & 0 \\ 0 & \cos\alpha & \sin\alpha & 0 \\ 0 & -\sin\alpha & \cos\alpha & 0 \\ 0 & 0 & 0 & 1 \end{bmatrix}, \quad \alpha \text{ real}. \tag{20}$$

Lemma 3. Let $v \in \vartheta$. Then

$$\mathcal{K}_+(v) = \{K \mid K = N(v)C;\ \ C \in \mathcal{C}_1\}, \tag{21}$$

where

$$N(v) = \begin{bmatrix} a(v) & 0 & 0 & -va(v) \\ 0 & e(v) & 0 & 0 \\ 0 & 0 & f(v) & 0 \\ c(v) & 0 & 0 & d(v) \end{bmatrix} = M(a(v), c(v), d(v), e(v), f(v)). \tag{22}$$

In Formula (22), the matrix elements $a(v)$, $c(v)$, and $d(v)$ are uniquely determined by v. As to the elements $e(v)$ and $f(v)$, they are uniquely determined by v, provided $e(v)$ is required to be positive. Further, depending on the value of v, $f(b)$ is either equal to $e(v)$ or to $-e(v)$.

Proof. Let X and Y be two arbitrary elements of $\mathcal{K}_+(v)$. One checks that $(XY^{-1})_{i4} = 0$, $i = 1, 2, 3$, so that, by (5), $XY^{-1} = C \in \mathcal{C}$. We have $X_{i1} = C_{ik} Y_{k1} = C_{i1} Y_{11}$. Since $X_{21} = X_{31} = 0$ and, by Lemma 2, $Y_{11} \neq 0$, we get $C_{21} = C_{31} = 0$ from which there follows that $C_{11} = \pm 1$. The circumstance $C_{11} = -1$ can be ruled out because it would imply $\Delta(X) = -\Delta(Y)$. Hence $C_{11} = 1$, namely $C = A(\gamma)$ for some γ. Now let K be any element of $\mathcal{K}_+(v)$. Then $KA(\alpha) \in \mathcal{K}_+(v)$ for every α. Setting $X = KA(\alpha)$, $Y = K$, we get $KA(\alpha)K^{-1} = A(\beta)$ for some β. In other words, for every $K \in \mathcal{K}_+(v)$ and for every α there exists a β such that

$$A(\beta)K = KA(\alpha). \tag{23}$$

Since $\det K \neq 0$, this equation admits of solutions only if $\beta = \alpha$ or $\beta = -\alpha$ (mod 2π).[12] As shown in Reference 12, if $\beta = \alpha$ (mod 2π), the general solution is of the form $K = M(a, c, d, e, e)A(\vartheta)$, where ϑ is arbitrary and e can be chosen to be positive. On the other hand, if $\beta = -\alpha$ (mod 2π), the general solution is of the form $K = M(a', c', d', e', -e')A(\vartheta')$, where ϑ' is arbitrary and e' can be chosen to be positive. Now let

$$K_1, K_2 \in \mathcal{K}_+(v) : K_{1,2} = M(a_{1,2}, c_{1,2}, d_{1,2}, e_{1,2}, f_{1,2} = \pm e_{1,2}).$$

There is a β such that $K_1 = A(\beta)K_2$ whence, first of all,

$$a_1 = a_2 = a, \quad c_1 = c_2 = c \text{ and } d_1 = d_2 = d.$$

Further,

$$(ad - bc)(\pm e_1^2) = \det K_1 = \det K_2 = (ad - bc)(\pm e_2^2).$$

Therefore, either $f_1 = e_1$, $f_2 = e_2$ or $f_1 = -e_1$, $f_2 = -e_2$ and since e_1 and e_2 have been chosen to be positive, $e_1 = e_2 = e$ in both cases.

The proof of the lemma is thus completed.

Now let $\mathfrak{N}$ denote the following subset of $\mathcal{K}_+$:

$$\mathfrak{N} = \{ X | X \in \mathcal{K}_+;\ X_{23} = 0;\ X_{22} > 0 \}. \tag{24}$$

$\mathfrak{N}$ is clearly a subgroup of $\mathcal{K}_+$ and it follows from Lemmas 1 and 3 that

$$\mathcal{L} = \{ L | L = AZB;\ A, B \in \mathcal{C};\ Z \in \mathfrak{N} \}. \tag{25}$$

By Lemma 3, $\mathfrak{N} \cap \mathcal{K}_+(v)$ contains exactly one element for every $v \in \vartheta$, and this is $N(v)$. In this way, we can define a (one-to-one) mapping N of ϑ onto $\mathfrak{N}$ as $N : v \to N(v)$.

We note that, in Formula (22),

$$a(v) \neq 0, \tag{26.a}$$

$$d(v) \neq 0, \tag{26.b}$$

$$e(v) > 0, \tag{26.c}$$

$$f^2(v) = e^2(v), \tag{26.d}$$

$$d(v) + vc(v) \neq 0. \tag{26.e}$$

Relations (26.a) and (26.b) follow from Lemma 2 (see (13)); relation (26.c) from the definition; relation (26.d) from Lemma 3 and relation (26.e) from the condition $\det N(v) \neq 0$.

Proof of the theorem. Note first that $\mathcal{C}_1 = \mathcal{K}_+ \cap \mathcal{C} = \mathcal{K}_+(0)$. Hence $0 \in \vartheta$ and we distinguish two cases, according to whether $\vartheta = \{0\}$ or $\vartheta \supset \{0\}$.

If $\vartheta = \{0\}$, we have $\mathfrak{N} = \{N(0)\} = \{1\}$, hence, by (25),

$$\mathcal{L} = \mathcal{C}. \tag{27}$$

If $\vartheta \supset \{0\}$, fix $v \in \vartheta$, $v \neq 0$. We have

$$N^{-1}(v) = \begin{bmatrix} \frac{d(v)}{a(v)(d(v)+vc(v))} & 0 & 0 & \frac{v}{d(v)+vc(v)} \\ 0 & e^{-1}(v) & 0 & 0 \\ 0 & 0 & f^{-1}(v) & 0 \\ \frac{-c(v)}{a(v)(d(v)+vc(v))} & 0 & 0 & \frac{1}{d(v)+vc(v)} \end{bmatrix} \tag{28}$$

Let $D(-\epsilon)$ denote a rotation of $-\epsilon$ in the (x_1, x_2) plane:

$$D(-\epsilon) = \begin{bmatrix} \cos\epsilon & -\sin\epsilon & 0 & 0 \\ \sin\epsilon & \cos\epsilon & 0 & 0 \\ 0 & 0 & 1 & 0 \\ 0 & 0 & 0 & 1 \end{bmatrix}, \quad 0 \leqq \epsilon \leqq 2\pi. \tag{29}$$

The matrix

$$L(\epsilon, v) = N^{-1}(v) \cdot D(-\epsilon) \cdot N(v) =$$

$$\begin{bmatrix} \frac{d\cos\epsilon + vc}{d+v\cdot c} & \frac{-de\sin\epsilon}{a(d+vc)} & 0 & \frac{dv(1-\cos\epsilon)}{d+vc} \\ \frac{a\sin\epsilon}{e} & \cos\epsilon & 0 & \frac{-va\sin\epsilon}{e} \\ 0 & 0 & 1 & 0 \\ \frac{c(1-\cos\epsilon)}{d+vc} & \frac{ce\sin\epsilon}{a(d+vc)} & 0 & \frac{d+vc\cos\epsilon}{d+vc} \end{bmatrix} \tag{30}$$

belongs to $\mathcal{L}$, being a product of three elements of $\mathcal{L}$. We immediately notice that

$$\frac{d+vc\cos\epsilon}{d+vc} > 0, \qquad \epsilon \in [0, 2\pi). \tag{31}$$

Indeed, the left hand side of (31) is a continuous function of ϵ which equals 1 for $\epsilon = 0$ and which, by Lemma 2 (see (13)), never becomes 0. Equation (31) gives the inequality

$$\left|\frac{v\,c(v)}{d(v)}\right| < 1 \quad \text{for every } v \in \vartheta . \tag{32}$$

We now proceed to factorize $L(\epsilon, v)$ according to (25). Define the matrices

$$A(\epsilon, v) = \begin{bmatrix} -\cos\gamma(\epsilon, v) & -\sin\gamma(\epsilon, v) & 0 & 0 \\ \sin\gamma(\epsilon, v) & -\cos\gamma(\epsilon, v) & 0 & 0 \\ 0 & 0 & 1 & 0 \\ 0 & 0 & 0 & 1 \end{bmatrix} \tag{33.a}$$

and

$$B(\epsilon, v) = \begin{bmatrix} \cos\gamma(\epsilon, v) & \sin\gamma(\epsilon, v) & 0 & 0 \\ -\sin\gamma(\epsilon, v) & \cos\gamma(\epsilon, v) & 0 & 0 \\ 0 & 0 & 1 & 0 \\ 0 & 0 & 0 & 1 \end{bmatrix}, \tag{33.b}$$

where

$$\cos\gamma(\epsilon, v) = \frac{de\,(1-\cos\epsilon)}{[d^2e^2(1-\cos\epsilon)^2 + a^2(d+vc)^2\sin^2\epsilon]^{\frac{1}{2}}} \tag{34.a}$$

and

$$\sin\gamma(\epsilon, v) = \frac{a(d+vc)\sin\epsilon}{[d^2e^2(1-\cos\epsilon)^2 + a^2(d+vc)^2\sin^2\epsilon]^{\frac{1}{2}}} \tag{34.b}$$

Note that $\cos\gamma(\epsilon, v)$ and $\sin\gamma(\epsilon, v)$ are defined in the limit $\epsilon \to +0$:

$$\cos\gamma(+0, v) = 0$$

and

$$\sin\gamma(+0, v) = \frac{a(d+vc)}{|a(d+vc)|} = 1.$$

We get

$$X(\epsilon, v) = A^{-1}(\epsilon, v)L(\epsilon, v)B^{-1}(\epsilon, v) =$$

$$= \begin{bmatrix} \frac{d+vc\cos\epsilon}{d+vc} & \varphi_1(a,c,d,e,v,\epsilon) & 0 & \varphi_2(a,c,d,e,v,\epsilon) \\ 0 & 1 & 0 & 0 \\ 0 & 0 & 1 & 0 \\ \varphi_3(a,c,d,e,v,\epsilon) & \varphi_4(a,c,d,e,v,\epsilon) & 0 & \frac{d+vc\cos\epsilon}{d+vc} \end{bmatrix}, \quad (35)$$

where

$$\varphi_1(a,c,d,e,v,\epsilon) =$$

$$= \frac{\sin\epsilon\{de^2(1-\cos\epsilon)[d^2e^2(1-\cos\epsilon)+a^2(d+vc)^2\cos\epsilon]-}{ae(d+vc)[d^2e^2(1-\cos\epsilon)^2+}$$

$$\frac{-a^2(d+vc)[a^2(d+vc)^2\sin^2\epsilon - de^2(1-\cos\epsilon)(d\cdot\cos\epsilon+vc)]\}}{+a^2(d+vc)^2\sin^2\epsilon)]}, \quad (35.a)$$

$$\varphi_2(a,c,d,e,v,\epsilon) = \frac{-v[d^2e^2(1-\cos\epsilon)^2+a^2(d+vc)^2\sin^2\epsilon]^{\frac{1}{2}}}{e(d+vc)}, \quad (35.b)$$

$$\varphi_3(a,c,d,e,v,\epsilon) = \frac{ce[2d(1-\cos\epsilon)+vc\sin^2\epsilon]}{(d+vc)[d^2e^2(1-\cos\epsilon)^2+a^2(d+vc)^2\sin^2\epsilon]^{\frac{1}{2}}} \quad (35.c)$$

and

$$\varphi_4(a,c,d,e,v,\epsilon) = \frac{c\cdot\sin\epsilon(1-\cos\epsilon)[e^2d-a^2(d+vc)]}{a(d+vc)[d^2e^2(1-\cos\epsilon)^2+a^2(d+vc)^2\sin^2\epsilon]^{\frac{1}{2}}}. \quad (35.d)$$

From (35) one sees immediately that $X(\epsilon, v) \in \mathcal{K}$. Further, by direct calculation one gets $\Delta(X(\epsilon, v)) = 1$ so that, actually, $X(\epsilon, v) \in \mathcal{K}_+$. Finally, $X_{23}(\epsilon, v) = 0$ and $X_{22}(\epsilon, v) = 1$ whence $X(\epsilon, v) \in \mathfrak{N}$, and we have

$$X(\epsilon, v) = N(w(\epsilon, v)), \quad (36)$$

where

$$w(\epsilon, v) = \frac{v[d^2e^2(1-\cos\epsilon)^2 + a^2(d+vc)^2\sin^2\epsilon]^{\frac{1}{2}}}{e(d+vc\cos\epsilon)} . \tag{37}$$

By Lemma 3 (see (22)) we must have

$$N_{12}(w(e, v)) = \varphi_1(a, c, d, e, v, \epsilon) = 0 \tag{38.a}$$

and

$$N_{42}(w(\epsilon, v)) = \varphi_4(a, c, d, e, v, \epsilon) = 0, \tag{38.b}$$

for every $\epsilon \in [0, 2\pi)$. One easily verifies that the necessary and sufficient condition for (38.a) and (38.b) to be satisfied is that

$$d(v)e^2(v) = a^2(v)[d(v) + vc(v)]. \tag{39}$$

Using (39) we obtain

$$d^2e^2(1-\cos\epsilon)^2 + a^2(d+vc)^2\sin^2\epsilon =$$

$$= de^2[2d(1-\cos\epsilon) + vc\sin^2\epsilon] \geqq 0. \tag{40}$$

Comparing (22) with (35) we have

$$N(w(\epsilon, v)) = \begin{bmatrix} a(w(\epsilon, v)) & 0 & 0 & -w(\epsilon, v)\, a(w(\epsilon, v)) \\ 0 & e(w(\epsilon, v)) & 0 & 0 \\ 0 & 0 & f(w(\epsilon, v)) & 0 \\ c(w(\epsilon, v)) & 0 & 0 & d(w(\epsilon, v)) \end{bmatrix} \tag{41}$$

where

$$a(w(\epsilon, v)) = d(w(\epsilon, v)) = \frac{d(v) + vc(v)\cos\epsilon}{d(v) + vc(v)}, \tag{42.a}$$

$$e(w(\epsilon, v)) = f(w(\epsilon, v)) = 1 \tag{42.b}$$

and, with the help of (40),

$$c(w(\epsilon, v)) = \frac{c(v)\{d(v)[2d(v)\cdot(1-\cos\epsilon)+vc(v)\sin^2\epsilon]\}^{\frac{1}{2}}}{d(v)(d(v)+vc(v))} \tag{42.c}$$

and

$$w(\epsilon, v) = \frac{v\{d(v)[2d(v)(1-\cos\epsilon)+vc(v)\sin^2\epsilon]\}^{\frac{1}{2}}}{d(v)+vc(v)\cdot\cos\epsilon} . \tag{42.d}$$

From (42.a,c,d) we get, for every $\epsilon \in [0, 2\pi)$,

$$1 + w^2(\epsilon, v)\,\frac{c(v)}{v\,d(v)} > 0, \tag{43}$$

$$a(w(\epsilon, v)) = \left[1 + \frac{c(v)}{v\,d(v)}\cdot w^2(\epsilon, v)\right]^{-\frac{1}{2}} \tag{44}$$

and

$$c(w(\epsilon, v)) = w(\epsilon, v))\cdot a(w(\epsilon, v))\,\frac{c(v)}{v\,d(v)}, \tag{45}$$

so that we can write

$N(w(\epsilon, v)) =$

$$\begin{bmatrix} \left[1+\frac{c(v)}{vd(v)}w^2(\epsilon,v)\right]^{-\frac{1}{2}} & 0 & 0 & -w(\epsilon,v)\left[1+\frac{c(v)}{vd(v)}w^2(\epsilon,v)\right]^{-\frac{1}{2}} \\ 0 & 1 & 0 & \\ 0 & 0 & 1 & \\ \frac{c(v)}{vd(v)}w(\epsilon,v)\left[1+\frac{c(v)}{vd(v)}w^2(\epsilon,v)\right]^{-\frac{1}{2}} & 0 & 0 & \left[1+\frac{c(v)}{vd(v)}w^2(\epsilon,v)\right]^{-\frac{1}{2}} \end{bmatrix}, \tag{46}$$

where $w(\epsilon, v)$ is given by (42.d).

By (31), $d(v)+vc(v)\cos\epsilon$ is different from 0 for every $\epsilon \in [0, 2\pi)$ and therefore $w(\epsilon, v)$ is a continuous function of ϵ which equals 0 iff $\epsilon = 0$ (compare (40)) and which has a definite sign for every $\epsilon \in (0, 2\pi)$. Then $w([0, 2\pi), v) = w([0, 2\pi], v)$ is a closed non-zero interval and we can write $w([0, 2\pi), v) = [0, w_0(v)]$ or $w([0, 2\pi), v) = [-w_0(v), 0]$ according to whether $w(\epsilon, v) \geqq 0$ or

$w(\epsilon, v) \leqq 0$. In the first case we have

$$w_o(v) = \sup_{\epsilon \in [0, 2\pi)} w(\epsilon, v);$$

in the second case,

$$w_o(v) = -\inf_{\epsilon \in [0, 2\pi)} w(\epsilon, v) .$$

From (46) we get immediately that $N^{-1}(w(\epsilon, v)) = N(-w(\epsilon, v))$. Therefor $-w(\epsilon, v) \in \vartheta$ and $\{N(-w(\epsilon, v)), N(w(\epsilon, v))\}_{\epsilon \in [0, 2\pi)} = N(\Gamma(v))$ where $\Gamma(v) \subseteq \vartheta$ is a closed symmetric non-zero interval, $\Gamma(v) = [-w_o(v), w_o(v)]$, $w_o(v) > 0$. We show that the ratio $c(v)/vd(v)$ has the same value $-\lambda$ for every $v \neq 0$ (and we can define $c(v)/vd(v) = -\lambda$ also for $v = 0$). Indeed, let $v' \in \vartheta$, $v' \neq 0$, $v' \neq v$. $\Gamma(v')$ and $\Gamma(v)$ are both symmetric intervals, hence either $\Gamma(v) \supseteq \Gamma(v')$ or $\Gamma(v') \supseteq \Gamma(v)$ and there exist $\epsilon, \epsilon' \in (0, 2\pi)$ such that $w(\epsilon, v) = w(\epsilon', v') \neq 0$. Therefore, $a(w(\epsilon, v)) = a(w, (\epsilon', v'))$ and from (44) we get

$$\frac{c(v)}{vd(v)} = \frac{c(v')}{v'd(v')} = -\lambda. \tag{47}$$

Then, for a given $v \in \vartheta$, $v \neq 0$, and for $w \in \Gamma(v)$, we have

$$N(w) = \begin{bmatrix} [1-\lambda w^2]^{-\frac{1}{2}} & 0 & 0 & -w[1-\lambda w^2]^{-\frac{1}{2}} \\ 0 & 1 & 0 & 0 \\ 0 & 0 & 1 & 0 \\ -\lambda w[1-\lambda w^2]^{-\frac{1}{2}} & 0 & 0 & [1-\lambda w^2]^{-\frac{1}{2}} \end{bmatrix}, \tag{48}$$

Suppose $\lambda < 0$. Choose a positive integer m such that $|\lambda|^{-\frac{1}{2}} \cdot tg(\pi/m) \in \Gamma(v)$. We have

$$[N(|\lambda|^{-\frac{1}{2}} \cdot tg(\pi/m))]^m = \begin{bmatrix} -1 & 0 & 0 & 0 \\ 0 & 1 & 0 & 0 \\ 0 & 0 & 1 & 0 \\ 0 & 0 & 0 & -1 \end{bmatrix} \tag{49}$$

which contradicts (5). Therefore, the circumstance $\lambda < 0$ must be ruled out.[13] We shall then distinguish two cases, according to whether $\lambda = 0$ or $\lambda > 0$.

a) $\lambda = 0$. In this case (48) writes

$$N(w) = \begin{bmatrix} 1 & 0 & 0 & -w \\ 0 & 1 & 0 & 0 \\ 0 & 0 & 1 & 0 \\ 0 & 0 & 0 & 1 \end{bmatrix}, \qquad (50)$$

for $w \in \Gamma(v) = [-w_o(v), w_o(v)]$, $w_o(v) > 0$. Equation (50) is the matrix of a special Galilei transformation along the x_1-axis. Hence $N(\Gamma(v))$ generates the one-dimensional Galilei group

$$\mathcal{G}_1(0), \quad \vartheta = (-\infty, +\infty)$$

and $N(v)$ is, of course, included in $\mathcal{G}_1(0)$ (choose a positive integer m such that $v/m \in \Gamma(v)$; then $N(v) = [N(v/m)]^m$). Thus we get from (25) that $\mathcal{L}$ is the (proper orthochronous homogeneous) Galilei group (in four dimensions):

$$\mathcal{L} = \mathcal{G}(0) = \Big\{ G \mid G \in GL(4,R);\ G_{4i} = 0,\ i = 1, 2, 3; \qquad (51)$$
$$G_{44} = 1;\ G_{i\ell}G_{j\ell} = \delta_{ij},\ \det\{G_{ik}\} = 1 \Big\}.$$

b) $\lambda > 0$. In this case, from Eqs. (43) and (47), we have that $|w| < \lambda^{-\frac{1}{2}}$, hence

$$\Gamma(v) = [-w_o(v), w_o(v)] \subset (-\lambda^{-\frac{1}{2}}, \lambda^{-\frac{1}{2}}).$$

The group generated by $N(\Gamma(v))$ is the group

$$\mathcal{G}_1(\lambda) = N[(-\lambda^{\frac{1}{2}}, \lambda^{-\frac{1}{2}})]$$

of special Lorentz transformations along the x_1-axis, corresponding to an invariant velocity $c = \lambda^{-\frac{1}{2}}$. We show that $\mathcal{G}_1(\lambda) = \mathfrak{n}$ by proving that $N(v)$ is included in $\mathcal{G}_1(\lambda)$. To this purpose, we must show that

$$\lambda v^2 < 1. \qquad (52)$$

From (47) we get

$$\left| \frac{vc(v)}{d(v)} \right| = \lambda v^2$$

which, combined with (32), gives (52). Then, using (25) and recalling the standard Wigner decomposition of the matrices of the Lorentz group,[14] we get that

$$\mathcal{L} = \mathcal{G}(\lambda) = \Big\{ L \,|\, L \in GL(4,R);\ g_{\mu\nu}(\lambda)\cdot L_{\mu\rho} L_{\nu\sigma} = g_{\rho\sigma}(\lambda);$$

$$g_{\mu\nu} = 0 \quad \text{if} \quad \mu \neq \nu;\ g_{11} = g_{22} = g_{33} = -\lambda;\ g_{44} = 1;$$

$$\det L = 1;\ L_{44} \geqq 1 \Big\}, \tag{53}$$

namely, $\mathcal{L}$ is the (proper orthochoronous) Lorentz group (in four dimensions) corresponding to an invariant velocity $c = \lambda^{-\frac{1}{2}}$.

The proof of our theorem is thus completed.

A last remark. It is perhaps interesting to note that the rotation group $\mathcal{C} = \mathcal{G}(+\infty)$ appears as a limit of the Lorentz group $\mathcal{G}(\lambda)$ as the invariant velocity $\lambda^{-\frac{1}{2}}$ tends to zero. This is a simple example of a limiting process, similar to group contraction, by which, however, the group dimensionality is lowered (the Lorentz group has dimension six while the rotation group has three). Although the result can also be obtained mathematically in a trivial way in terms of the usual group contraction by performing the limit $\mathcal{G}(\lambda) \rightarrow \mathcal{G}(0)$ and then by mapping $\mathcal{G}(0)$ homomorphically onto $\mathcal{G}(+\infty)$, it is clear that from a physical point of view the two procedures are of a different nature. Therefore, it might be worthwhile to inquire whether limiting processes on Lie groups which lower the dimension can be sensibly defined.[15] Physically, these might correspond, f.i., to a switching off of the interaction or to a transition to a non-relativistic limit on dynamical groups.[16),17]

Footnotes and References

1. a) V. Berzi and V. Gorini, J. Math. Phys. 10, 1518 (1969).
   b) V. Berzi and V. Gorini, Space-Time, Reference Frames and Relativistic Invariance: a Topological Approach, Milan University Preprint IFUM 108/FT, 1970. Submitted for publication.
2. A list of references can be found in Reference 1.a.

3. Compare, f.i., V. Lalan, Bull. Soc. Math. France 65, 83 (1937). Apart from our observation, Lalan's paper is rather a remarkable one in that it carries out a very careful analysis of the structure of the one-dimensional Lie subgroups of GL(2,R), with the choice of a velocity parameter as a coordinate over the group manifolds.
4. I am grateful to Professor H. D. Doebner for a discussion in which he stressed the importance of this point.
5. E. C. Zeeman, J. Math. Phys. 5, 490 (1964) and Topology 6, 161 (1967).
6. H. Bacry and J.-M. Lévy-Leblond, J. Math. Phys. 9, 1605 (1968).
7. J.-M. Lévy-Leblond, Ann. Inst. Henri Poincaré 3, 1 (1965). See also N. D. Sen Gupta, Nuovo Cimento 44, 512 (1966) and Indian J. Phys. 42, 528 (1968). The Carrol group might be termed the non-relativistic group for low energy tachyons.
8. L. Parker, Phys. Rev. 188, 2287 (1969).
9. Since we are ignoring the constancy of light velocity, it is advisable that the synchronization of clocks be carried out by mechanical means as, f.i., through bullets fired by standard guns at rest in the given frame. See, for example, J. Aharoni, The Special Theory of Relativity, Oxford University Press, London (1965), Chapter 1.
10. It is easy to devise conceptual experiments by which the standards of length and time in two different frames can be made the same. For example, we can make sure that the observers in frames S and S' use the same time standard, if both assume as unit of time the mean life of a given unstable particle measured at rest in the laboratory of each of the two observers.
11. Note that both XPX^{-1} and $X^{-1}PX$ are elements of $\mathcal{L}$, since $\mathcal{L}$ is a group and L, A, B and P are elements of $\mathcal{L}$.
12. V. Gorini and A. Zecca, J. Math. Phys. 11, 2226 (1970).
13. As it clearly appears from Eq. (48) together with (25), values of λ less than zero would correspond to kinematical groups isomorphic to the four-dimensional rotation group. However, these include "rest" transformations which contain time reversal (see (49)) and we exclude them by our assumption (5). As regards this point, compare assumption (3) of Reference 6 and the assumption of Section III of Reference 3.
14. E. P. Wigner, Ann. Math. 40, 149 (1939). R. Takahashi, Bull. Soc. Math. France 91, 306 (1963).

15. This point was raised in discussions with Dr. O. Melsheimer.
16. H. D. Doebner and O. Melsheimer, J. Math. Phys. 9, 1638 (1968) and 11, 1463 (1970).
17. J. J. Aghassi, P. Roman and R. M. Santilli, Phys. Rev. D 1, 2753 (1970), J. Math. Phys. 11, 2297 (1970) and Boston University preprint BU-EP-1 70, March 1970.

GENERALIZED EIGENVECTORS AND GROUP REPRESENTATIONS—THE CONNECTION BETWEEN REPRESENTATIONS OF SO(4,1) AND THE POINCARÉ GROUP†

A. Böhm
University of Texas
Austin, Texas

Introduction

Rarely has a mathematical structure been so eagerly accepted by physicists as the Rigged Hilbert Space $\Phi \subset \mathcal{H} \subset \Phi^{\times}$. The main reason for this is probably its ability to elevate the Dirac formalism of quantum mechanic that is "scarcely to be surpassed in brevity and elegance" (v. Neumann) into an equally beautiful and mathematically rigorous theory. Furthermore, the Rigged Hilbert Space is an important means of investigating such mathematical structures as representations of non-compact groups which have become very important in theoretical physics. The employment of the R. H. Sp. provides the possibility for working with algebraic (infinitesimal) methods in the representation theory of non-compact groups, which are so familiar to physicists from compact group representations.

The aim of the present lectures is to display the mathematical beauty and strength of the R. H. Sp. method in non-compact group representations with a physically important example. The main theme is the connection between representations of SO(4,1) and the Poincaré group. As an introduction, the irreducible unitary representations of SO(4,1) will be derived in the first lecture. The second lecture gives some relevant properties of SO(4,1) and Poincaré group representations. In the third lecture the contraction of a principle series representation of SO(4,1) into representations of the Poincaré group is described. The last lecture gives a construction of representations of the Lie algebra of SO(4,1) in terms of generators of unitary Poincaré group representations.

The basic concepts of R. H. Sp. and group representations have been described in the lectures of Nagel.[1] For further material

---

†Presented at the NATO Summer School in Mathematical Physics, Istanbul, 1970.
Supported in part by AEC, NSF and NATO.

on the R. H. Spaces, one will have to refer to the literature[2),3),4)] which will be introduced only briefly here.

I will also have to assume that the basic properties of the representations of the Poincaré group $\mathcal{P}$,[5)] the Lorentz group SO(3,1)[6)] and of the groups SO(n),[7)] n=2, 3, 4, 5, are known, and can report only briefly some of their properties here in order to establish the notation.

The irreducible representation spaces of SO(4) are characterized by two numbers: $\mathcal{H}(k_0, n)$ where $k_0 = \pm$ half-integer or $\pm$ integer and $n = 1, 2, \ldots$. The reduction of $\mathcal{H}(k_0, n)$ with respect to SO(3) is given by

$$\mathcal{H}(k_0, n) = \sum_{j=|k_0|}^{|k_0|+n-1} \mathcal{M}^j .$$

The basis in $\mathcal{H}(k_0, n)$ is denoted by

$$|j_3, j; (k_0, n)\rangle; \quad -j \leqq j_3 \leqq +j.$$

For the calculation, we shall use a more convenient notation and call

$$j_3 = m_{21}, \quad j = m_{31}, \quad k_0 = m_{41}, \quad |k_0| + n = m_{42} + 1.$$

The eigenvalue of the Casiminoperator of SO(4) is:

$$\Delta_{SO(4)} = \vec{V}^2 + \vec{M}^2 = k_0^2 + (|k_0| + n)^2 - 1 = m_{41}^2 + m_{42}^2 + 2m_{42}$$

where M_i are the generators of SO(3) and V_i, M_i, $i=1,2,3$, the generators of SO(4).

The irreducible representation spaces of SO(3,1) are characterized by two numbers $\mathcal{H}(k_0, \nu)$, where k_0 has the same spectrum as above, $|k_0|$ is again the lowest angular momentum j that occurs and ν is any real number $\geqq 0$ for the principal series representation, which we will only use here.[6)] (For the supplementary series representations $k_0 = 0$ and $0 \leqq i\nu \leqq 1$.)

I. Unitary Irreducible Representations of SO(4,1)

We denote the Hermitian generators of SO(4,1) by

$$L_{\alpha\beta} \quad \alpha, \beta = 0, 1, 2, 3, 4 \quad \text{with} \quad L_{\alpha\beta} = -L_{\beta\alpha} . \tag{1}$$

They fulfill the commutation relations (c. r.):

$$\left[L_{\alpha\beta}, L_{\gamma\delta}\right] = -i\left(g_{\alpha\gamma} L_{\beta\delta} + g_{\beta\delta} L_{\alpha\gamma} - g_{\alpha\delta} L_{\beta\gamma} - g_{\beta\gamma} L_{\alpha\delta}\right)$$
$$g_{oo} = +1, \; g_{aa} = -1, \quad a = 1, 2, 3, 4. \tag{2}$$

For physical application it is convenient to use the notation

$$V_\mu = L_{4\mu}, \; L_{\mu\nu} \quad \nu, \mu = 0, 1, 2, 3$$
$$N_i = L_{oi} \quad M_i = \tfrac{1}{2} \epsilon_{ijk} L_{jk} \tag{3}$$

where $L_{\mu\nu}$ generate the homogeneous Lorentz group SO(3,1):

$$\left[L_{\mu\nu}, L_{\rho\sigma}\right] = -i\left(g_{\mu\rho} L_{\nu\sigma} + g_{\nu\sigma} L_{\mu\rho} - g_{\mu\sigma} L_{\nu\rho} - g_{\nu\rho} L_{\mu\sigma}\right) \tag{4}$$

and V_μ is a Lorentz-vector operator,

$$\left[L_{\mu\nu}, V_\rho\right] = i\left(g_{\nu\rho} V_\mu - g_{\mu\rho} V_\nu\right) \tag{5}$$

with the additional property

$$\left[V_\mu, V_\nu\right] = iL_{\mu\nu} \, . \tag{6}$$

SO(4,1) has two independent Casimiroperators, for which one conveniently and commonly chooses

$$Q = -\tfrac{1}{2} L_{\alpha\beta} L^{\alpha\beta} = V_\mu V^\mu - \tfrac{1}{2} L_{\mu\nu} L^{\mu\nu} = V_\mu V^\mu + \vec{N}^2 - \vec{M}^2 \tag{7}$$

$$Q_{(4)} = -W_\alpha W^\alpha, \qquad W_\alpha = \tfrac{1}{8} \epsilon^{\alpha\beta\gamma\delta\epsilon} L_{\beta\gamma} L_{\delta\epsilon} \tag{8}$$

with (3) $Q_{(4)}$ can also be written:

$$Q_{(4)} = (\vec{M} \cdot \vec{N})^2 - W_\mu W^\mu, \quad W_\mu = \tfrac{1}{2} \epsilon_{\mu\nu\rho\sigma} V^\nu L^{\rho\sigma} \, . \tag{9}$$

Before we start with the evaluation of the unitary irreducible representations (u. i.r.) of SO(4,1) we have to make some general remarks concerning representations of non-compact groups and their Lie algebras.

The way we shall find all the u. i. r. of SO(4,1) will be to find irreducible representations of the Lie algebra L(SO(4,1)) and then show that these representations of the Lie algebra are connected with u. i. r. of SO(4,1); (in fact, we shall not even go over to the representations of the group, so that our task is to find all the representations of L(SO(4,1)) which are connected with u. i. r. of SO(4,1)). So we will not have to find all the representations of the c. r. (2) but only those which fulfill the following additional conditions (Ladder representations):

1) There is a dense subspace D of the representation space $\mathcal{H}$ which is invariant with respect to all $L_{\alpha\beta}$ [8),9)] and all its (finite but arbitrary large) products and $L_{\alpha\beta}$ is a symmetric operator on D.

2) Every irreducible representation space $\mathcal{H}(k_0, n)$ of SO(4) lies completely in D.

3) Every irreducible representation space $\mathcal{H}(k_0, n)$ occurs at most once.

4) The operator

$$\Delta_{SO(4,1)} = \sum L_{\alpha\beta} L_{\alpha\beta} = V_o^2 + \vec{V}^2 + \vec{N}^2 + \vec{M}^2$$

(Nelson operator)

is essentially self-adjoint (e.s.a.) on D.

These conditions do not constitute a minimal requirement but are interrelated with each other.

The conditions 1) and 4) assure that the representation of the Lie algebra integrates to a unitary representation of the group (Nelson Theorem).[10)] (I.e., if $L_{\alpha\beta}$ is a representation of the Lie algebra fulfilling 1) and 4), then

$$e^{i\epsilon^{\alpha\beta} L_{\alpha\beta}}$$

gives a unitary representation of the group element $g(\epsilon^{\alpha\beta})$.)

Condition 3) is always fulfilled for irreducible unitary representations of SO(4,1).[11)] So 3) is in fact a consequence of 1) and 4).

Condition 2) allows only for representations of SO(4,1) in which the maximal compact subgroup is diagonal.

In the following evaluation of the representations we will require only conditions 1), 2), 3). We shall follow the following procedure: We start from an irreducible finite dimensional representation space $\mathcal{H}(k_o, n)$ of $SO(4)_{V_i M_i}$ and apply to its basis vectors $|j_3, j; k_o, n\rangle$ the SO(4)-vector-operator (V_o, N_i) and reach by this neighboring representations $\mathcal{H}(k_o \pm 1, n \pm 1)$. This we do a finite but arbitrary large number of times. In this way we obtain the space

$$\Psi = \sum_{\substack{\text{algebraic} \\ k_o, n}} \mathcal{H}(k_o, n) , \tag{10}$$

the algebraic direct sum of the $\mathcal{H}(k_o,n)$. Then we require on Ψ: $L^{\dagger}_{\alpha\beta} = L_{\alpha\beta}$ and have therewith obtained in Ψ a space D which fulfills the conditions 1), 2), 3). (There are, however, other possible choices for D, e.g., the later introduced space Φ). The representation of L(SO(4,1)) is algebraically irreducible on Ψ (i.e., there is no invariant subspace, closed or non-closed, of Ψ). The completion of Ψ with respect to the scalar product gives the irreducible representation space $\mathcal{H}$ of L(SO(4,1)), and the representation so obtained is usually called a (singleton) ladder representation. We then show that the representations obtained in this way all fulfill condition 4), i.e., are integrable.[12)]

The unitary irreducible representations of SO(4,1) have been found by Thomas, Newton and Dixmier.[13)] We will derive them here by a method that is not only very simple but also permits an immediate generalization to all the groups SO(n,1).[14)]

We define the quantities $A_{\alpha\beta}$ on D by

$$\begin{aligned} L_{\alpha\beta} &= i\sqrt{g_{\alpha\alpha}}\,\sqrt{g_{\beta\beta}}\,A_{\alpha\beta} \qquad \alpha, \beta = 0, 1, 2, 3, 4 \\ g_{oo} &= +1, \quad g_{aa} = -1 \qquad a = 1, 2, 3, 4 \end{aligned} \tag{11}$$

then it follows from the c.r. of $L_{\alpha\beta}$ (2) that $A_{\alpha\beta}$ fulfill the following c.r.

$$\left[A_{\alpha\beta}, A_{\gamma\delta}\right] = -\delta_{\alpha\gamma} A_{\beta\delta} - \delta_{\beta\delta} A_{\alpha\delta} + \delta_{\beta\gamma} A_{\alpha\delta} + \delta_{\alpha\delta} A_{\beta\gamma} \tag{12}$$

and from the fact that $L_{\alpha\beta}$ is symmetric, it follows that

$$\begin{aligned} &\text{(a)} \quad A^{+}_{ab} = -A_{ab} \qquad a,b = 1, 2, 3, 4 \\ &\text{(b)} \quad A^{+}_{oa} = +A_{oa}. \end{aligned} \tag{13}$$

The relations (12) are the c.r. of SO(5). Would we have instead of (13b)

$$A_{oa}^{+} = -A_{oa} \tag{13b^1}$$

our task would have been to find all the skew-symmetric representations of the Lie algebra L(SO(5)) which is equivalent to finding all the unitary representations of the group SO(5), because of the compactness. All linear representations of (12) which fulfill (13a) (i.e., all linear representations of L(SO(5)) which contain only skew-symmetric representations of L(SO(4))) will include the skew-symmetric representations of L(SO(5)) (unitary representations of SO(5)) and the skew-symmetric representations of L(SO(4,1)).

In order to derive the skew-symmetric irreducible representations of L(SO(4,1)) we will first briefly describe the well-known unitary representations of SO(5) and then see what has to be changed if the relation ($13b^1$) is replaced by (13b).

The unitary irreducible representation of SO(5) are characterized by two numbers m_{51}, m_{52}, integer or half-integer. The basis vectors in an irreducible representation space in which the following chain of subgroups is diagonal

$$SO(5) \supset SO(4)_{1,2,3,4} \supset SO(3)_{1,2,3} \supset SO(2)_{1,2} \tag{14}$$

are most conveniently labeled by the (modified[14]) Gelfand-Zetlin[7]) pattern

$$(m_{ij}) = \begin{bmatrix} m_{51} & m_{52} \\ m_{41} & m_{42} \\ m_{31} & \\ m_{21} & \end{bmatrix} \tag{15}$$

where (m_{41}, m_{42}), m_{31} and m_{21} are integers or half-integers characterizing the irreducible representations of SO(4), SO(3) and SO(2), respectively.

They fulfill the following relation

$$\begin{aligned}
&\text{(a)}\quad |m_{41}| \leqq m_{51} \leqq m_{42} \leqq m_{52}\\
&\text{(b)}\quad |m_{41}| \leqq m_{31} \leqq m_{42}\\
&\text{(c)}\quad |m_{21}| \leqq m_{31}\,.
\end{aligned}\tag{16}$$

The action of the generators $A_{\alpha\beta}$ on the basis vectors are given by:

$$A_{12}|(m_{ij})\rangle = im_{21}|(m_{ij})\rangle \tag{17a}$$

$$A_{23}|(m_{ij})\rangle = A(m_{21})\left|\begin{bmatrix} m_{51} & m_{52} \\ m_{41} & m_{42} \\ m_{31} & \\ (m_{21}+1) & \end{bmatrix}\right\rangle - A(m_{21}-1)\left|\begin{bmatrix} m_{51} & m_{52} \\ m_{41} & m_{42} \\ m_{31} & \\ (m_{21}-1) & \end{bmatrix}\right\rangle \tag{17b}$$

$$A_{34}|(m_{ij})\rangle = B(m_{31})\left|\begin{bmatrix} m_{51} & m_{52} \\ m_{41} & m_{42} \\ (m_{31}+1) & \\ m_{21} & \end{bmatrix}\right\rangle + iC|(m_{ij})\rangle - B(m_{31}-1)\left|\begin{bmatrix} m_{51} & m_{52} \\ m_{41} & m_{42} \\ (m_{31}-1) & \\ m_{21} & \end{bmatrix}\right\rangle \tag{17c}$$

$$A_{40}|(m_{ij})\rangle = D(m_{41})\left|\begin{bmatrix} m_{51} & m_{52} \\ (m_{41}+1) & m_{42} \\ m_{31} & \\ m_{21} & \end{bmatrix}\right\rangle + E(m_{42})\left|\begin{bmatrix} m_{51} & m_{52} \\ m_{41} & (m_{42}+1) \\ m_{31} & \\ m_{21} & \end{bmatrix}\right\rangle \tag{17d}$$

$$-D(m_{41}-1)\left|\begin{bmatrix} m_{51} & m_{52} \\ (m_{41}-1) & m_{42} \\ m_{31} & \\ m_{21} & \end{bmatrix}\right\rangle - E(m_{42}-1)\left|\begin{bmatrix} m_{51} & m_{52} \\ m_{41} & (m_{42}-1) \\ m_{31} & \\ m_{21} & \end{bmatrix}\right\rangle$$

and the matrix elements are:

$$D(m_{41}) =$$

$$\frac{1}{2}\sqrt{(m_{31}+\tfrac{1}{2})^2-(m_{41}+\tfrac{1}{2})^2}\sqrt{\frac{[(z_{51}+\frac{1}{2})^2-(m_{41}+\frac{1}{2})^2][(z_{52}+3/2)^2-(m_{41}+\frac{1}{2})^2]}{[(m_{42}+1)^2-m_{41}^2][(m_{42}+1)^2-(m_{41}+1)^2]}} \tag{18d}$$

$$E(m_{42}) =$$

$$\frac{1}{2}\sqrt{[(m_{31}+\tfrac{1}{2})^2-(m_{42}+3/2)^2]\frac{[(z_{51}+\frac{1}{2})^2-(m_{42}+3/2)^2][(z_{52}+3/2)^2-(m_{42}+3/2)^2]}{[m_{41}^2-(m_{42}+1)^2][m_{41}^2-(m_{42}+2)^2]}} \tag{18e}$$

$$B(m_{31}) = \sqrt{m_{21}^2-(m_{31}+1)^2}\sqrt{\frac{[m_{41}^2-(m_{31}+1)^2][(m_{42}+1)^2-(m_{31}+1)^2]}{(m_{31}+1)^2[4(m_{31}+1)^2-1]}} \tag{18b}$$

$$C = \frac{m_{21}\, m_{41}\, (m_{42}+1)}{m_{31}\, (m_{31}+1)} \tag{18c}$$

$$A(m_{21}) = \frac{1}{2}\sqrt{(m_{31}+\tfrac{1}{2})^2 - (m_{21}+\tfrac{1}{2})^2} \tag{18a}$$

where

$$z_{51} = m_{51}\,,\quad z_{52} = m_{52} \tag{19}$$

are integers or half-integers.

These numbers m_{ij} are connected with the numbers j_3, j and (k_o, n) that usually characterize the irreducible representation of SO(2), SO(3) and SO(4) respectively by:

$$j_3 = m_{21}\,,\quad j = m_{31}\,,\quad k_o = m_{41}\,,\quad |k_o| + n = (m_{42}+1) = \ell\,. \tag{20}$$

As the representations of compact groups are well-known, we do not want to give a derivation of (16), (17), (18), we just remark that they can be easily obtained in the following way: One starts with an irreducible representation of SO(4) which is given by (17a), (17b), (17c), (16b) and (16c) (the matrix elements of the remaining operators A_{ab} are obtained from these by a straightforward use of the c.r. of SO(4)). As the c.r. of SO(5) are equivalent to the c.r. of SO(4) (i.e., the relation (12) with α, β replaced by $a, b = 1, 2, 3, 4$) and the additional relations

$$\left[A_{i(i+1)}, A_{40}\right] = 0 \qquad i = 1, 2, 3$$

$$\left[A_{34}, [A_{40}, A_{34}]\right] = A_{40} \tag{21}$$

$$\left[A_{40}, [A_{40}, A_{34}]\right] = -A_{34}\,,$$

one uses the matrix elements of SO(4) given by (17a), (17b) and (17c) and the relations (21) to obtain the expression (17d) for A_{40} and a set of recurrence relations for the coefficient functions D and E. The solution of these recurrence relations gives the expressions for D and E in (18). The skew-symmetry ($13b^1$) then gives (19) and (16a).

For the skew-symmetric representations of L(SO(4,1)), all the above used requirements are valid except ($13b^1$); instead of ($13b^1$)

we have (13b). (In fact, the expressions (17) and (18) for arbitrary complex values of z_{51} and z_{52} give all the linear representations of L(SO(5)) or L(SO(4,1)) in which SO(4) is diagonal.) We obtain, therefore, the skew-symmetric representations of L(SO(4,1)) if we use (17), (18), (16b,c) and (13b). From (13b) we obtain the values of z_{51} and z_{52} that characterize the skew-symmetric "irreducible" ladder representation of L(SO(4,1)) and the spectrum of m_{41} and m_{42} in these irreducible representations (i.e., the analogue of (16a)).

From (18d) and (18e) we see that the matrix elements are symmetric with respect to the exchange of $z_{51} \longleftrightarrow z_{52}+1$, so that such an exchange will lead to equivalent representations. Therefore, we need to consider the consequence of a given relation on one of these two quantities z_{51} or $(z_{52}+1)$ only.

From the relation (16b) (and the fact that m_{41} and m_{42} are both integers or both half-integers), we conclude that there are both maximum and minimum values for m_{41} and m_{42} and that

$$m_{41}^{max} = m_{42}^{min}. \tag{22}$$

From (17d) we see that this can only be true if

$$D(m_{41}^{max}) = 0 \qquad E(m_{42}^{min} - 1) = 0 \tag{23}$$

and from this and (18e), (18f), we obtain the condition on z_{51} (or on $z_{52}+1$, but according to the above remarks we can choose z_{51}):

$$(z_{51} + \tfrac{1}{2})^2 = (m_{41}^{max} + \tfrac{1}{2})^2 = (m_{42}^{min} + \tfrac{1}{2})^2. \tag{24}$$

From this we see that z_{51} is integer or half-integer

$$z_{51} = m_{51}, \tag{25}$$

and if we take the two square roots of (24) we see that we can equivalently characterize the SO(4,1) irreducible representation by

$$m_{51} = m_{41}^{max} = m_{42}^{min} \tag{26}$$

or by

$$m_{51} = -m_{41}^{max} - 1 \tag{26^1}$$

(because $(z_{51}+\frac{1}{2})$ and $-(z_{51}+\frac{1}{2})$ give equivalent representations). We choose $m_{51} \geqq 0$ and obtain then from (26) a restriction on the spectrum of the SO(4) irreducible representation of SO(4,1) characterized by (m_{51}, z_{52}), where z_{52} is still completely arbitrary; this spectrum is given by:

$$|m_{41}| \leqq m_{51} \leqq m_{42}. \tag{27}$$

What we have actually found so far is that all linear irreducible representations of L(SO(4,1)) in which SO(4) is diagonal (and each irreducible representation of SO(4) is contained at most once) are characterized by a non-negative integer or half-integer m_{51}, and by an arbitrary complex number z_{52} and that the reduction of this irreducible representation with respect to SO(4) has to fulfill the restriction (27); <u>in specific cases further specific restrictions on the spectrum of SO(4) irreducible representations will occur</u>.

We now use the requirement of skew-symmetry of the matrix representation of A_{40}, (13b), to determine the allowed values of z_{52} for skew-symmetric representations of L(SO(4,1)). From (13b), i.e.,

$$\left\langle \begin{bmatrix} m_{51} & m_{52} \\ m_{41}+1 & m_{42} \\ m_{31} & \\ m_{21} & \end{bmatrix} \middle| A_{40} \middle| (m_{ij}) \right\rangle = \left\langle (m_{ij}) \middle| A_{40} \middle| \begin{bmatrix} m_{51} & m_{52} \\ m_{41}+1 & m_{42} \\ m_{31} & \\ m_{21} & \end{bmatrix} \right\rangle^*$$

follows with (17d) $D(m_{41}) = -D(m_{41})^*$ and similarly $E(m_{42}) = -E(m_{42})^*$. Consequently, the square roots in (18f) and (18e) must be purely imaginary, i.e., the expressions under the root have to be real and negative. The reality is fulfilled in one of the two following cases

$$z_{52} + 3/2 = iy_{52}, \quad y_{52} \text{ real} \tag{28a}$$

and

$$z_{52} = x_{52}, \qquad x_{52} \text{ real.} \tag{28b}$$

We see that in case (28a) $[(z_{52}+3/2)^2 - (m_{42}+3/2)^2]$ or $[(z_{52}+3/2)^2 - (m_{41}+\frac{1}{2})^2]$ in the expressions for $E(m_{42})$ and $D(m_{41})$ can never become zero; consequently in this case we cannot obtain any restrictions on the spectrum of (m_{41}, m_{42}) in addition to (27).

However, for case (28b) when x_{52} is integer or half-integer further restrictions on the spectrum of (m_{41}, m_{42}) will appear.

From the requirement of negativity it follows from (18d) and (18e) that y_{52} and x_{52} have to fulfill the following additional conditions:

$$(z_{52} + 3/2)^2 \leqq (m_{41} + \tfrac{1}{2})^2 \quad \text{for } m_{51} \neq 0 \tag{29}$$

$$(z_{52} + 3/2)^2 \leqq (m_{42} + 3/2)^2 . \tag{29^1}$$

For $m_{51} \neq 0$ it follows from (27) that (29^1) is always fulfilled if (29) is, so that for $m_{51} \neq 0$ we need only investigate the consequences of condition (29). For $m_{51} = 0$ it follows from (18d) and (27) that $D(m_{41}) \equiv 0$ and $m_{41} \equiv 0$; consequently, (29) need not be fulfilled and (29^1) becomes an independent condition which has to be fulfilled.

For the case (28a), conditions (29), (29^1) are satisfied for arbitrary real y_{52} ($y_{52} \neq 0$). Therewith we have obtained one series of skew-symmetric irreducible representations of L(SO(4,1)), which we call the principle continuous series. These irreducible representations are characterized by (r, ρ) with $r = m_{51} = z_{51}$ integer or half-integer and $\rho = y_{52} = -i(z_{52} + 3/2)$ arbitrary real, and their reduction with respect to SO(4) is given by (27). As not $(z_{52} + 3/2)$ but only $(z_{52} + 3/2)^2$ appears in all expressions for the representation, y_{52} and $-y_{52}$ determine equivalent representations, so that (r, ρ) with $\rho > 0$ already determines all the inequivalent principle continuous series representations. We have excluded here the case $\rho = y_{52} = i(z_{52} + 3/2) = 0$, because for the case that z_{52} is integer or half-integer, further restrictions on the spectrum of (m_{41}, m_{42}) might appear, as remarked above. (These representations are encountered later and are the $\pi_{m_{51}, (m_{52}+2) = \frac{1}{2}}$.)

In the following lectures we shall only make use of the principle continuous series representations;* however, for the sake of completeness, we will derive here all the representations of L(SO(4,1)) which fulfill the conditions 1.), 2.) and 3.) stated at the beginning and show that they all integrate to unitary representations of SO(4,1).

Of the representations that are obtained in case (28b), we first derive those for which the spectrum of (m_{41}, m_{42}) fulfills

---

*The results of the rest of this section except for pages 218 and 219 are not needed for the understanding of the following sections.

restrictions in addition to (27); these are the descrete series representations, for which x_{52} is integer or half-integer. We first derive the representations with $m_{51} \neq 0$; then, as mentioned before, we need only require that (29) is fulfilled.

For

$$x_{52} = m_{52} = \text{integer or half-integer} \qquad (28b^1)$$

follows from the matrix elements (18d) and (17d) that $|m_{41}|$ is (in addition to (27)) also restricted from below. Because there exists an m_{41}^{min}, as

$$\begin{aligned} D(m_{41}^{min} - 1) &= 0 \quad \text{for} \\ (m_{52} + 3/2)^2 &= (m_{41}^{min} - 1 + \tfrac{1}{2})^2 \end{aligned} \qquad (30)$$

and there exists an m_{41}^{max}, because

$$\begin{aligned} D(m_{41}^{max}) &= 0 \quad \text{for} \\ (m_{52} + 3/2)^2 &= (m_{41}^{max} + \tfrac{1}{2})^2 . \end{aligned} \qquad (31)$$

We first consider the representations with $m_{41}^{min} \neq 0$ and $m_{41}^{max} \neq 0$. As again only $(m_{52} + 3/2)^2$ determines the representation, we can choose either $(m_{52} + 3/2)$ or $-(m_{52} + 3/2)$ to characterize inequivalent representations; we make the choice such that $m_{52} + 2$ is positive. Then it follows from (30) that

$$m_{52} + 2 = m_{41}^{min} \quad \text{for} \quad m_{41} \text{ positive} \qquad (30a)$$

and from (31) that

$$m_{52} + 2 = -m_{41}^{max} \quad \text{for} \quad m_{41} \text{ negative.} \qquad (31a)$$

For (30a) the spectrum of m_{41} can never become negative and for (31a) the spectrum of m_{41} can never become positive. So the value of m_{51} and m_{52} is not sufficient to specify the irreducible representation of this kind, but we need in addition the sign (m_{41}). So we have two classes of representations of this kind:

$$\pi^{+}_{m_{51},(m_{52}+2)} \quad \text{for sign}(m_{41}) = +1$$

and

$$\pi^{-}_{m_{51},(m_{52}+2)} \quad \text{for sign}(m_{41}) = -1.$$

The spectrum of m_{41} and m_{42} is given by (30a), (31a) and (27) to be:

$$m_{52}+2 \leqq m_{41} \leqq m_{51} \leqq m_{42} \quad \text{for} \quad \pi^{+}_{m_{51},m_{52}+2} \tag{30b}$$

and

$$m_{52}+2 \leqq -m_{41} \leqq m_{51} \leqq m_{42} \quad \text{for} \quad \pi^{-}_{m_{51},m_{52}+2}. \tag{31b}$$

The possible values of m_{52} for these representations follow from (30b) and (31b):

$$m_{51} = \tfrac{1}{2}, 1, 3/2, 2 \ldots \qquad m_{52}+2 = m_{51}, m_{51}-1, \ldots 1 \text{ or } \tfrac{1}{2}. \tag{32}$$

We have still to consider the case when $m_{41}^{min} = 0$ and $m_{41}^{max} = 0$: For $m_{41}^{min} = 0$, we obtain from (18d) with $D(m_{41}^{min} - 1) = 0$:

$$(x_{52} + 3/2)^2 = 1/4 \tag{33}$$

and for $m_{41}^{max} = 0$ we obtain from (18a) with $D(m_{41}^{max}) = 0$ also

$$(x_{52} + 3/2)^2 = 1/4. \tag{33^1}$$

So we see from (33) and (33^1) that for the value

$$x_{52} + 2 = m_{52} + 2 = 0 \tag{34}$$

(or $x_{52} + 2 = 1$, which gives an equivalent irreducible representation), we have an irreducible representation of SO(4,1) for which $m_{41}^{max} = 0$ as well as $m_{41}^{min} = 0$, so that for the value (34) of $x_{52} = m_{52}$, the spectrum of (m_{41}, m_{42}) is (using (27):

$$m_{41} \equiv 0 \quad \text{and} \quad m_{51} \leqq m_{42}. \tag{35}$$

So we have another discrete series of representations in addition to (30b) and (31b), which we call $\pi_{m_{51},0}$, and which is characterized by:

$$x_{52} + 2 = 0 \qquad m_{51} = 1, 2, 3, \ldots \tag{35a}$$

whose spectrum of SO(4) representations is given by (35) (as $m_{41} = 0$, m_{51} cannot be half-integer and $m_{51} = 0$ we excluded). It is easily seen that for all above representations (29) is fulfilled.

We now consider the case $m_{51} = 0$, $x_{52} = m_{52} =$ integer or half-integer. Then (27) reads

$$m_{41} \equiv 0 \leqq m_{42}$$

and (27¹)

$$D(m_{41}) \equiv 0.$$

From (29¹) it follows because of (27¹) ($m_{42}^{min} = 0$) that

$$(m_{52} + 3/2)^2 \leqq (3/2)^2. \tag{36}$$

From (36) it follows that the only integer values m_{52} can have are

$$m_{52} + 3/2 = \pm 3/2, \ \pm 1/2 \tag{37}$$

For the two equivalent irreducible representations $m_{52} + 3/2 = \pm 3/2$, one sees from (18e) that $E(m_{42}^{min}) = E(0) = 0$. From this follows that $E(m_{42}) \equiv 0$ so that

$$m_{51} = 0 \qquad m_{52} + 3/2 = \pm 3/2 \tag{38}$$

is the trivial, one-dimensional representation. The two equivalent irreducible representations

$$m_{52} + 3/2 = \pm 1/2 \quad m_{51} = 0 \tag{39}$$

will be included into another continuous series.

We now derive the continuous series representations of case (28b):

$$z_{52} = x_{52} = \text{continuous.} \tag{28b²}$$

In this case the spectrum of SO(4) representations (m_{41}, m_{42}) is always given by (27).

We again consider first $m_{51} \neq 0$ and start with m_{51} = half-integer. Then it follows from (27) that (the smallest value of $((m_{41}+\frac{1}{2})^2) = 0$, so that (29) gives the condition

$$0 \leqq x_{52} + 3/2 \leqq 0. \tag{40}$$

So we obtain the series of irreducible representations:

$$m_{51} = 1/2,\ 3/2,\ \ldots \qquad x_{52} + 3/2 = 0. \tag{40^1}$$

These representations are a limiting case of the principle series representations $(r, \rho) = (m_{52}, y_{52} = 0)$ and we shall include them in the principle series.

For m_{51} = integer $\neq 0$ follows from (27) that (the smallest value of $(m_{41}+\frac{1}{2})^2) = \frac{1}{4}$ so that (29) gives the condition

$$-1/2 \leqq x_{52} + 3/2 \leqq +1/2. \tag{41}$$

Because the representations characterized by $(x_{52}+3/2)$ and $-(x_{52}+3/2)$ are equivalent, we obtain already all inequivalent irreps if we take instead of (41):

$$0 \leqq x_{52} + 3/2 \leqq 1/2. \tag{41^1}$$

The representations characterized by

$$m_{52} = 1, 2, 3 \ldots \quad x_{52} + 3/2 = 1/2$$

are already contained in the discrete series $\pi_{m_{51},0}$. So the supplementary continuous series representations are given by:

$$m_{51} = 1,\ 2,\ 3,\ \ldots \qquad 0 \leqq (x_{52} + 3/2) < 1/2. \tag{42}$$

The irreducible representations

$$m_{51} = 1,\ 2,\ 3,\ \ldots \qquad x_{52} + 3/2 = 0 \tag{43}$$

are limiting cases of the principle series representations

$$(r, \rho) = (m_{51}, y_{52} = 0).$$

We finally consider the case $m_{51} = 0$. Then it follows from (27) that (the smallest value of $(m_{42}+3/2)^2) = (3/2)^2$, so that (29^1)

$$-3/2 \leqq x_{52} + 3/2 \leqq 3/2. \tag{44}$$

Again for inequivalent representations we need only consider

$$0 \leqq x_{52} + 3/2 \leqq 3/2. \tag{44^1}$$

We exclude from this series the trivial representation (38) and obtain the other supplementary continuous series

$$m_{51} = 0 \qquad 0 \leqq x_{52} + 3/2 < 3/2. \tag{45}$$

We remark that the irreducible representation (39) is contained in this series.

Therewith we have completed the classification of all irreducible representations of L(SO(4,1)) that fulfill the condition 1.), 2.), 3.) of page 200. In Table I[14] we collect our results. We remark that we meet a certain arbitrariness in the listing of certain representations into the various series, because certain limiting cases belong to more than one series. The notation π for the discrete series is that of Dixmier;[13] the notation (r,ρ) of the principle continuous series is the one, which we will employ in the following; for the supplementary continuous series we have not introduced a new notation but just given the equation number where they appear in the text. In Table II we list all the matrix elements of the generators in the irreducible representation (m_{51}, z_{51}) of SO(4,1). For the SO(4)-basis vectors in the irreducible representation space $\mathcal{H}(m_{51}, z_{52})$, we use the notation $|j_3, j, k_0, \ell\rangle$; the connection between the two sets of labels is given by (20). For the matrix elements of M_i, V_3 and V_0, the expressions in Table II are just rewritten from expressions in (17) and (18) using the present notation (20) and the connections (11) and (3). The remaining matrix elements can then be calculated from these relevant matrix elements by using, e.g., the commutation relations $[V_0, V_3] = iN_3$ and $[N_3, M_\pm] = \pm N_\pm$.

The connection between our basis vectors $|j_3, j, k_0, \ell\rangle$ and the basis vectors in Reference 15 which we call $|j, m, n, \ell\rangle_S$ here is given by:

$$|j, m, n, \ell\rangle_S = (+i)^j (-i)^\ell (+i)^n |j_3, j, k_0, \ell\rangle$$

where $m = j_3$, $n = k_0$.

TABLE I

Representations	Conditions for m_{51} & z_{52}	SO (4)-Content
$(r,\rho)=(m_{51},y_{52})$	$m_{51}=0,\ \frac{1}{2},\ 1\ \ldots$ $z_{52}+\frac{3}{2}=iy_{52},\ 0<y_{52}$	$\lvert m_{41}\rvert \leqq m_{51} \leqq m_{42}$ 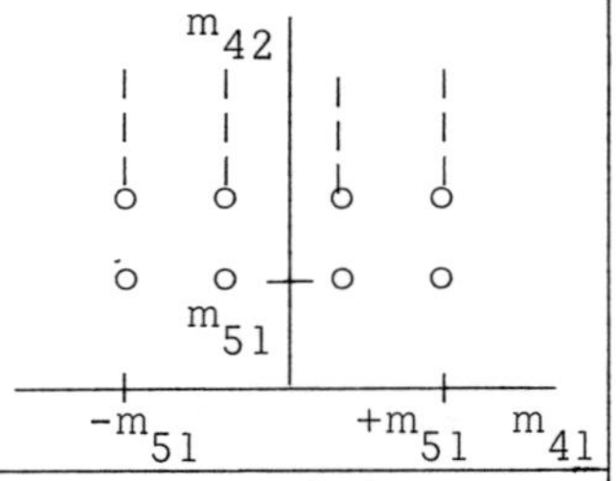
(42)	$m_{51}=1,\ 2,\ 3,\ \ldots$ $z_{52}=x_{52},\ 0\leqq x_{52}+\frac{3}{2}<\frac{1}{2}$	
(45)	$m_{51}=0$ $z_{52}=x_{52};\ 0\leqq x_{52}+\frac{3}{2}<\frac{3}{2}$	
$\pi_{m_{51},0}$	$m_{51}=1,\ 2,\ 3\ \ldots$ $z_{52}=m_{52},\ m_{52}+2=1$ or $m_{52}+2=0$	$m_{41}=0,\ m_{51}\leqq m_{42}$ 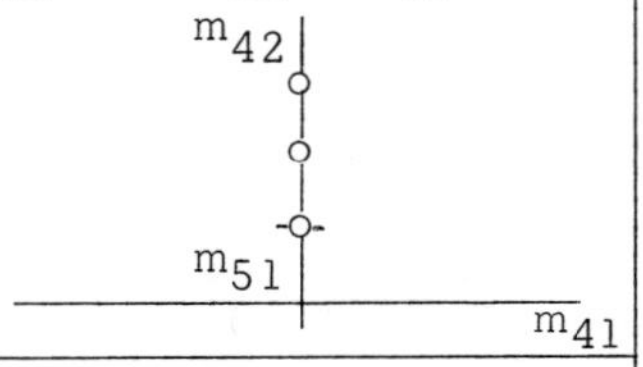
$\pi^{+}_{\ell,q}=$ $\pi^{+}(m_{51},m_{52}+2)$	$m_{51}=\frac{1}{2},\ 1,\ \frac{3}{2}\ \ldots$ $z_{52}=m_{52},\ m_{52}+2=$ $m_{51},\ m_{51}-1\ \ldots$ 1 or $\frac{1}{2}$	$m_{52}+2\leqq m_{41}\leqq m_{51}\leqq m_{42}$ 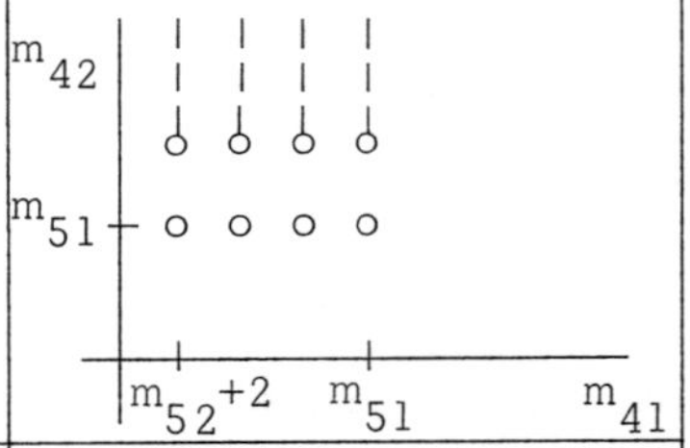
$\pi^{-}_{\ell,q}=$ $\pi^{-}(m_{51},m_{52}+2)$	The same as for $\pi^{+}(m_{51},\ m_{52})$	$m_{52}+2\leqq -m_{41}\leqq m_{51}\leqq m_{42}$ 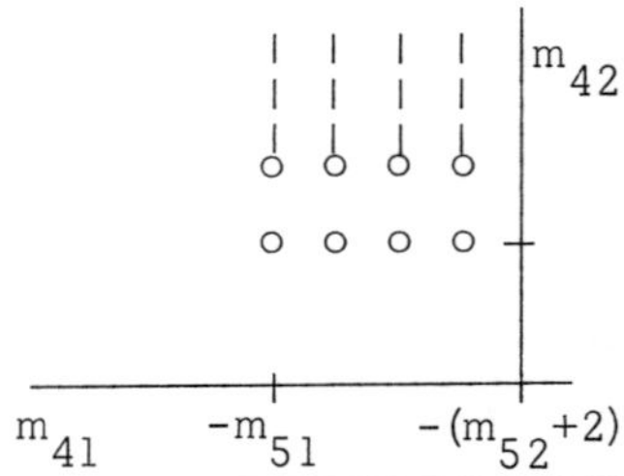

TABLE II

$$M_3 |j_3, j, k_o, \ell\rangle = j_3 |j_3, j, k_o, \ell\rangle$$

$$M_\pm |j_3, j, k_o, \ell\rangle = \sqrt{(j\mp j_3)(j\pm j_3+1)} \quad |j, j_3\pm 1, k_o \ell\rangle$$

with $M_\pm = M_1 \pm iM_2$

$$V_3 |j_3, j, k_o, \ell\rangle = -\alpha(j+1, k_o, \ell) \sqrt{(j+j_3+1)(j-j_3+1)} \quad |j_3, j+1, k_o, \ell\rangle$$

$$+ \frac{j_3 k_o \ell}{j(j+1)} \; |j, j_3, k_o, \ell\rangle$$

$$+ \alpha(j, k_o, \ell) \sqrt{(j+j_3)(j-j_3)} \quad |j_3, j-1, k_o, \ell\rangle$$

where

$$\alpha(j, k_o, \ell) = \sqrt{\frac{(k_o^2 - j^2)(\ell^2 - j^2)}{j^2(4j^2-1)}}$$

$$V_\pm |j_3, j, k_o, \ell\rangle = \pm\alpha(j+1, k_o, \ell)\sqrt{(j\pm j_3+1)(j\pm j_3+2)} \quad |j_3\pm 1, j+1, k_o, \ell\rangle$$

$$+ \frac{k_o \ell}{j(j+1)} \sqrt{(j\mp j_3)(j\pm j_3+1)} \quad |j_3\pm 1, j, k_o, \ell\rangle$$

$$\pm \; \alpha(j, k_o, \ell) \sqrt{(j\mp j_3)(j\mp j_3-1)} \quad |j_3\pm 1, j-1, k_o, \ell\rangle$$

with $V_\pm = V_1 \pm iV_2$

$$V_o |j_3, j, k_o, \ell\rangle = \frac{a(k_o, \ell)}{2} \sqrt{(\ell+j+1)(\ell-j)} \quad |j_3, j, k_o, \ell+1\rangle$$

$$+ \frac{b(k_o, \ell)}{2} \sqrt{(j-k_o)(j+k_o+1)} \quad |j_3, j, k_o+1, \ell\rangle$$

$$- \frac{c(k_o, \ell)}{2} \sqrt{(j+k_o)(j-k_o+1)} \quad |j_3, j, k_o-1, \ell\rangle$$

$$+ \frac{d(k_o, \ell)}{2} \sqrt{(\ell+j)(\ell-j-1)} \quad |j_3, j, k_o, \ell-1\rangle$$

TABLE II (Continued)

---

where

$$a(k_o,\ell) = -i\sqrt{\frac{[(m_{51}+\frac{1}{2})^2-(\ell+\frac{1}{2})^2][(z_{52}+\frac{3}{2})^2-(\ell+\frac{1}{2})^2]}{[k_o^2-\ell^2][k_o^2-(\ell+1)^2}}$$

$$b(k_o,\ell) = +i\sqrt{\frac{[(m_{51}+\frac{1}{2})^2-(k_0+\frac{1}{2})^2][(k_0+\frac{1}{2})^2-(z_{52}+\frac{3}{2})^2]}{[\ell^2-k_0^2][\ell^2-(k_0+1)^2]}}$$

$$d(k_o,\ell) = -a(k_o,\ell-1)$$

$$c(k_o,\ell) = b(k_o-1,\ell)$$

$$N_3|j_3,j,k_o,\ell\rangle =$$

$$= -\beta(j+1,j_3)\Big[a(k_o,\ell)\sqrt{(j+1)^2-k_o^2)(\ell+j+2)(\ell+j+1)}\ |j_3,j+1,k_o,\ell+1\rangle$$

$$-b(k_o,\ell)\sqrt{(\ell^2-(j+1)^2)(j+2+k_o)(j+1+k_o)}\ |j_3,j+1,k_o+1,\ell\rangle$$

$$c(k_o,\ell)\sqrt{(\ell^2-(j+1)^2)(j+2-k_o)(j+1-k_o)}\ |j_3,j+1,k_o-1,\ell\rangle$$

$$-d(k_o,\ell)\sqrt{((j+1)^2-k_o^2)(\ell-j-1)(\ell-j-2)}\ |j_3,j+1,k_o,\ell-1\rangle\Big]$$

$$-i\frac{j_3}{2j(j+1)}\Big[a(k_o,\ell)k_o\sqrt{(\ell+j+1)(\ell-j)}\ |j_3,j,k_o,\ell+1\rangle$$

$$+b(k_o,\ell)\cdot\ell\sqrt{(j-k_o)(j+k_o+1)}\ |j_3,j,k_o+1,\ell\rangle$$

$$+c(k_o,\ell)\ \ell\sqrt{(j+k_o)(j-k_o+1)}\ |j_3,j,k_o-1,\ell\rangle$$

$$-d(k_o,\ell)k_o\cdot\sqrt{(\ell+j)(\ell-j-1)}\ |j_3,k_o,\ell-1\rangle\Big]$$

TABLE II (Continued)

$$+\beta(j,j_3)\Big[a(k_o,\ell)\sqrt{(j^2-k_o^2)(\ell+1-j)(\ell-j)}\quad |j_3,j-1,k_o,\ell+1\rangle$$

$$+b(k_o,\ell)\sqrt{(\ell^2-j^2)(j-k_o-1)(j-k_o)}\quad |j_3,j-1,k_o+1,\ell\rangle$$

$$-c(k_o,\ell)\sqrt{(\ell^2-j^2)(j+k_o+1)(j+k_o)}\quad |j_3,j-1,k_o-1,\ell\rangle$$

$$-d(k_o,\ell)\sqrt{(j^2-k_o^2)(\ell+j)(\ell+j-1)}\quad |j_3,j-1,k_o,\ell-1\rangle\Big]$$

where

$$\beta(j,j_3) = \frac{1}{2j}\sqrt{\frac{(j-j_3)(j+j_3)}{(2j+1)(2j-1)}}$$

$$N\pm|j_3,j,k_o,\ell\rangle = \mp\gamma(j+1,\ \pm j_3+1)\times$$

$$\times\Big[a(k_o,\ell)\sqrt{((j+1)^2-k_o^2)(\ell+j+2)(\ell+j+1)}\quad |j_3\pm 1,j+1,k_o,\ell+1\rangle$$

$$-b(k_o,\ell)\sqrt{(\ell^2-(j+1)^2)(j+2+k_o)(j+1+k_o)}\quad |j_3\pm 1,j+1,k_o+1,\ell\rangle$$

$$+c(k_o,\ell)\sqrt{(\ell^2-(j+1)^2)(j+2-k_o)(j+1-k_o)}\quad |j_3\pm 1,j+1,k_o-1,\ell\rangle$$

$$-d(k_o,\ell)\sqrt{((j+1)^2-k_o^2)(\ell-j-1)(\ell-j-2)}\quad |j_3\pm 1,j+1,k_o,\ell-1\rangle\Big]$$

$$-i\sqrt{\frac{(j\mp j_3)(j\pm j_3+1)}{2j(j+1)}}\Big[a(k_o,\ell)k_o\sqrt{(\ell+j+1)(\ell-j)}\quad |j_3\pm 1,j,k_o,\ell+1\rangle$$

$$+b(k_o,\ell)\,\ell\sqrt{(j-k_o)(j+k_o+1)}\quad |j_3\pm 1,j,k_o+1,\ell\rangle$$

$$+c(k_o,\ell)\,\ell\sqrt{(j+k_o)(j-k_o+1)}\quad |j_3\pm 1,j,k_o-1,\ell\rangle$$

$$-d(k_o,\ell)k_o\sqrt{(\ell+j)(\ell-j-1)}\quad |j_3\pm 1,j,k_o,\ell-1\rangle\Big]$$

TABLE II (Continued)

$\mp\gamma(j,\mp j_3)\ a(k_o,\ell)\sqrt{(j^2-k_o^2)(\ell+1-j)(\ell-j)}\ \ \vert j_3\pm 1, j-1, k_o, \ell+1\rangle$
$+b(k_o,\ell)\sqrt{(\ell^2-j^2)(j-k_o-1)(j-k_o)}\ \ \vert j_3\pm 1, j-1, k_o+1, \ell\rangle$
$-c(k_o,\ell)\sqrt{(\ell^2-j^2)(j+k_o+1)(j+k_o)}\ \ \vert j_3\pm 1, j-1, k_o-1, \ell\rangle$
$-d(k_o,\ell)\sqrt{(j^2-k_o^2)(\ell+j)(\ell+j-1)}\ \ \vert j_3\pm 1, j-1, k_o, \ell-1\rangle\ \Big]$
with $N\pm = N_1 \pm iN_2$
where $$\gamma(j,j_3) = \frac{1}{2j}\sqrt{\frac{(j+j_3)(j+j_3-1)}{(2j+1)(2j-1)}}\ .$$

The calculation of the eigenvalues of the Casimir operators (7) and (8) in an irreducible representation characterized by (m_{51}, z_{52}) is straightforward though lengthy. From the matrix elements of the generators, given in Table II or (17) and (18), one obtains:

$$Q \overset{\text{def}}{=} \alpha^2 = -z_{51}(z_{51}+1) + 9/4 - (z_{52}+3/2)^2 \qquad (46)$$

$$Q_{(4)} = z_{51}(z_{51}+1)[\,1/4 - (z_{52}+3/2)^2]. \qquad (47)$$

For the principle series representations (r,ρ), this gives $(r = z_{51},\ \rho = -i(z_{52}+3/2))$:

$$Q = \alpha^2 = -r(r+1) + 9/4 + \rho^2 \qquad (46')$$

$$Q_{(4)} = r(r+1)(\rho^2+\tfrac{1}{4}) = r(r+1)\alpha^2 + (r-1)r(r+1)(r+2)\,. \qquad (47')$$

To show that condition 4) is fulfilled is very easy. We write the Nelson operator

$$\Delta_{SO(4,1)} = Q + 2\Delta_{SO(4)} = \alpha^2 + 2\Delta_{SO(4)} \tag{48}$$

where α^2 is the eigenvalue of Q, (46'). $\Delta_{SO(4)}$ is the central elliptic element in a unitary representation of the group SO(4) and therefore by a theorem of Nelson and Stinespring[33] essentially self-adjoint. As α^2 is a number, it follows that $\Delta_{SO(4,1)}$ is e.s.a. on Ш. We remark that we have not used any property of SO(4,1) here except that it is semi-simple, from which it follows that we can write the Nelson operator as a sum of the Casimir operator and the Casimir operator of the maximal compact subgroup. From this we see that the Nelson operator of a semi-simple Lie group is e.s.a. for ladder representations,[6] i.e., for any representation that fulfills conditions 1.) and 2.).

It is also easy to explicitly show that $\Delta_{SO(4)}$ is e.s.a. without making use of the Nelson Stinespring theorem, by making use of one of the many criteria for the e.s. adjointness of a symmetric operator Δ: Δ is e.s.a. if $(\Delta + 1)^{-1}$ is bounded with dense domain. As Ш is dense in $\mathcal{H}$ and the spectrum of

$$(\alpha^2 + 1 + \Delta_{SO(4)})^{-1}$$

is

$$\frac{1}{\alpha^2 + k_0^2 + (|k_0| + n)^2}$$

i.e., bounded, it follows that $\Delta_{SO(4,1)}$ is e.s.a.

II. Some Properties of the Representation Spaces of SO(4,1) and the Poincaré Group

SO(4,1) and the Poincaré group $\mathcal{P}$ have both the homogeneous Lorentz group SO(3,1) $\underset{\text{Local}}{\cong}$ SL(2,c) as a subgroup. To study the connection between SO(4,1) irreducible representations (irreps) and irreps of $\mathcal{P}$, we should first find out which place SO(3,1) has in both groups. This can be done easiest with the help of some further mathematical tools.[1),2),3)]

We consider the space

$$Ш = \sum_{\text{alg.}} \mathcal{H}(k_o, n) \tag{1}$$

where the sum runs over an arbitrary high but finite set of elements (k_o, n) in the irrep space (r, ρ) of SO(4,1), cf. (I.27). In Ψ we introduce a topology by the following countable number of compatible scalar products.[1),16)]

$$(\phi, \psi)_p = (\phi, (\Delta+1)^p \psi), \quad \psi, \phi \in \Psi$$

$$, \quad \Delta = \sum_{\substack{\alpha, \beta \\ \alpha > \beta}} L_{\alpha\beta}^2$$

where $(\cdot, \cdot)$ is the scalar product in the representation space $\mathcal{H}(r, \rho)$ of SO(4,1), with respect to which the generators $L_{\mu\nu}$, V_μ are symmetric. The completion of Ψ with respect to this topology we call Φ.

Φ has the following properties:[16),1)]

1) $\Phi \subset \mathcal{H}$ is dense,

2) $\mathcal{E}(SO(4,1))$ is an algebra of continuous (with respect to the topology of Φ, τ_Φ) operators on Φ.[9)]

3) Φ is nuclear.

From these properties it follows that the nuclear spectral theorem is applicable and that there exists a complete set of simultaneous generalized eigenvectors in $\Phi^\times$, the conjugate of Φ for every complete system of commuting operators essentially self-adjoint (e.s.a.) on Φ. We choose as this system of e.s.a. operators

$$Q_1 = -\frac{1}{2} L_{\mu\nu} L^{\mu\nu}, \quad Q_2 = \frac{1}{8} \epsilon^{\mu\nu\rho\sigma} L_{\mu\nu} L_{\rho\sigma}, \quad \vec{M}^2, \quad M_3$$

$$(Q_1 = \vec{N}^2 - \vec{M}^2 \qquad Q_2 = \vec{N} \cdot \vec{M}). \tag{3}$$

As the e.s.-adjointness of $\vec{M}^2$ and M_3 is apparent, we just remark on the e.s.-adjointness of the Casimir operators of SL(2,c): as $\mathcal{H}(r, \rho)$ also carries a unitary representation of $SO(3,1) \to U(SO(3,1))$ (which one obtains if one restricts oneself to SO(3,1) in the representation of $SO(4,1) \to U(SO(4,1))$ and Q_1 and Q_2 are in the center of $\mathcal{E}(SO(3,1))$, it follows that Q_1 and Q_2 are e.s.a. on the Garding domain D_G.[11)] And as[1)] $D_G \subset \Phi$, Q_1 and Q_2 are also e.s.a. on Φ.

The eigenvalues of Q_1 and Q_2 are denoted by:[6)]

$$Q_1 = -k_o^2 + \nu^2 + 1, \quad Q_2 = k_o \nu \tag{4}$$

where (k_o, ν) are numbers that characterize the unitary irreps of SO(3,1). For the principle series representations $\nu \geqq 0$ and $k_o = \pm$ integer or $\pm$ half-integer. $|k_o|$ is the smallest j that appears in the irrep (k_o, ν), and we will see later that its spectrum in an irrep. of SO(4,1) is the same as the spectrum of the k_o used for the denotion of the SO(4) representations.

As we are considering a principle series representation (r, ρ) of SO(4,1), we expect that it will contain only principle series representations of its subgroup SO(3,1). The task is now to find which irreps (k_o, ν) of SO(3,1) are contained in the irrep (r, ρ) of SO(4,1) and how many times does each irrep (k_o, ν) appear. This task was solved by Ström[18] and we give here only the result:

In an irrep (r, ρ) the spectrum of k_o is $k_o = \pm r, \pm(r-1), \pm\ldots\pm 1/2$ or 0 and the spectrum of ν is the positive real line $0 \leqq \nu < \infty$ with multiplicity two or:

$$\mathcal{H}(r,\rho) \underset{SO(3,1)}{\Longrightarrow} \sum_{\eta=1,2} \sum_{\substack{k_o=0,\pm 1\ldots \\ \pm 1/2, \pm 3/2\ldots}}^{\pm r} \int_{\nu=0}^{\infty} \oplus\, \mathcal{H}^{\eta}(k_o, \nu) \frac{k_o^2+\nu^2}{(2\pi^4)^2} d\nu \tag{5}$$

where η is the degeneracy index distinguishing between the two equivalent irreps (k_o, ν) that appear in (r, ρ).

From this we conclude that the operators (3) or equivalently the numbers k_o, ν, j, j_3 connected with their eigenvalues are not sufficient to characterize the basis of generalized eigenvectors in $\mathcal{H}(r,\rho)$. In addition to the operators (3) we have to take any operator that commutes with all the elements of (3) and the spectrum of which consists of two elements (corresponding to the two values for η in (5)). For this operator we choose the operator I[19] that has the following properties:

$$I L_{\mu\nu} I^{-1} = L_{\mu\nu} \qquad I V_{\mu} I^{-1} = -V_{\mu} . \tag{6}$$

I can be defined by its action on the SO(4) basis of $\mathcal{H}(r,\rho)$

$$I | j, j_3, k_o, n\rangle = (-1)^{j+n} | j, j_3, -k_o, n\rangle \tag{7}$$

It is a matter of straightforward calculation using the matrix elements of $L_{\mu\nu}$ and V_{μ} given by (I.17), (I.18) or Table II, Section I to check that from (7) follows (6).

The operators

$$Q_1, Q_2, \vec{M}^2, M_3, I \tag{8}$$

constitute then a complete system of commuting e.s.a. operators in $\Phi \subset \mathcal{H}(r,\rho)$ and we can apply the nuclear spectral theorem.

We remark that from the commutation relations follows only that the elements of (8) commute on Φ, whereas for the applicability of the spectral theorem we need that they are strongly commuting. (I.e., their spectral resolutions commute or the sum of the squares of each pair of elements of (8) is e.s.a. on Φ.) The strong commutativity of Q_1 and Q_2 follows, because $Q_1{}^2 + Q_2{}^2$ is again an element of the center of the group representation $U(SO(3,1)_{L_{\mu\nu}})$ and therefore e.s.a.[17] As $U(SO(3)_{M_i})/\Phi$ commutes with Q_1, Q_2 and $U(SO(3)_{M_i})$ is bounded, it follows that $U(SO(3)_{M_i})$ and Q_1 are strongly commuting. The compactness of SO(3) then leads immediately to the strong commutativity of $\vec{M}^2$, M_3 with Q_1 and with each other. Similar arguments for the strong commutativity apply to the other commuting systems, e.g., (14). The basis of generalized eigenvectors of (8) we call:

$$|j_3, j, k_o, \nu, \pm, (r,\rho)\rangle \in \Phi^x. \tag{9}$$

It has the property

$$\begin{aligned}
I \; |j_3, j, k_o, \nu, \pm, (r,\rho)\rangle &= \pm |j_3, j, k_o, \nu, \pm, (r,\rho)\rangle \\
M_3 \; |j_3, j, k_o, \nu, \pm, (r,\rho)\rangle &= j_3 |j_3, j, k_o, \nu, \pm, (r,\rho)\rangle \\
\vec{M}^2 \; |j_3, j, k_o, \nu, \pm, (r,\rho)\rangle &= j(j+1) |j_3, j, k_o, \nu, \pm, (r,\rho)\rangle \\
Q_1 \; |j_3, j, k_o, \nu, \pm, (r,\rho)\rangle &= -(k_o^2 - \nu^2 - 1) |j_3, j, k_o, \nu, \pm, (r,\rho)\rangle \\
Q_2 \; |j_3, j, k_o, \nu, \pm, (r,\rho)\rangle &= k_o \nu |j_3, j, k_o, \nu, \pm, (r,\rho)\rangle
\end{aligned} \tag{9'}$$

$$(\phi, \psi) = \sum_{\eta=\pm} \sum_{\substack{k_o=0,\pm1\ldots \\ (\text{or } =\pm\frac{1}{2}\pm\frac{3}{2}\ldots)}}^{r} \sum_{j=|k_o|}^{\infty} \sum_{j_3=-j}^{+j} \int_{\nu=0}^{\infty} \langle\phi|j_3, j, k_o, \nu, \eta\rangle \langle \eta, \nu, k_o, j, j_3|\psi\rangle \, d\nu \, \frac{k_o^2 + \nu^2}{(2\pi^4)^2} \,. \tag{9''}$$

If we compare the spectrum of k_0 given by (5) with the spectrum of k'_0 given by (I.27) for the reduction of (r,ρ) with respect to the SO(4) irreps (k'_0, n), we see that the spectrum of k_0 and k'_0 is the same in an irrep of SO(4,1).

We shall now briefly review some well known facts of the representations of the Poincaré group.[5),20)]

The generators of $\mathcal{P}$ are P_μ, $L_{\mu\nu}$ and the commutation relations:

$$[P_\mu, P_\nu] = 0 \tag{10a}$$

$$[P_\rho, L_{\mu\nu}] = i(g_{\mu\rho}P_\nu - g_{\nu\rho}P_\mu) \tag{10b}$$

$$[L_{\mu\nu}, L_{\rho\sigma}] = -i(g_{\mu\rho}L_{\nu\sigma} + g_{\nu\sigma}L_{\mu\rho} - g_{\mu\sigma}L_{\nu\rho} - g_{\nu\rho}L_{\mu\sigma}). \tag{10c}$$

We define the operators

$$M^2 = P_\mu P^\mu \qquad M = \sqrt{M^2}$$
$$w^\mu = \tfrac{1}{2}\epsilon^{\mu\nu\rho\sigma} P_\nu L_{\rho\sigma} = -i[P^\mu, Q_2], \tag{11}$$

$$\hat{w}^\mu = M^{-1} w^\mu, \tag{12}$$

$$\hat{W} = -\hat{w}_\mu \hat{w}^\mu = \tfrac{1}{2} L_{\mu\nu} L^{\mu\nu} - L_{\rho\mu} L^{\mu}{}_{\sigma} \cdot P^\rho P^\sigma M^{-2}. \tag{13}$$

M^2 and W commute with every $P_\mu, L_{\mu\nu}$.

The class of irreps we are interested in $(P_\mu P^\mu > 0)$ are then characterized by (m, s, ϵ) where

m^2 = eigenvalue of M^2
$s(s+1)$ = eigenvalue of $\hat{W}$
ϵ = sign eigenvalue of P_0

A complete system of commuting self-adjoint operators in the irrep space $\mathcal{H}(m,s,\epsilon)$ of the group $\mathcal{P}$: $(a\Lambda) \to U(a\Lambda)$ is given by

$$P_i, \quad S_3 = U(L^{-1}(p))\, W_3 U(L(p)). \tag{14}$$

That P_i is self-adjoint (e.s.a. on the Garding domain) is clear because it is a representative of a generator in the space of a global representation of the group $\mathcal{P}$. S_3 is self-adjoint, because it is bounded and symmetric.

The set of generalized eigenvectors of (14) is the canonical basis of the Poincaré group representation:

$$|p_i, s_3; (m, s, \epsilon)\rangle \tag{15}$$

which has under $U(\cdot, \Lambda)$ the following transformation properties

$$U(\Lambda)|p_i s_3, s, m, \epsilon\rangle = \sum_{s_3'} |(\Lambda p)_i s_3'\rangle D^s_{s_3' s_3}(R(\Lambda, p)) \tag{16}$$

where $R(\Lambda, p)$ is the Wigner rotation:

$$R(\Lambda, p) = L(\Lambda p)\Lambda L^{-1}(p) \tag{17}$$

and $L(p)$ is the rotation-free Lorentz transformation with the property

$$L(p)p = (\epsilon m, 0, 0, 0) \quad (\text{"boost"}). \tag{18}$$

$D^s(R)$ is the representation matrix of the rotation R in the irrep of $SO(3)$ with highest weight s.

Before proceeding we should like to make some remarks concerning the canonical basis vectors (15): they are much used and well accepted but still their mathematical properties seem to be widely unknown. For the nuclear spectral theorem, one needs a nuclear space Ψ dense in $\mathcal{H}(s, m, \epsilon)$ on which the system of commuting operators is e.s.a.; then there exist a complete set of generalized eigenvectors in the space Ψ^x conjugate to Ψ. It can be shown (Ch. VI, Sec. 6 of Reference 2) that for any unitary representation U_G in $\mathcal{H}$ of any Lie group G, one can construct a nuclear space $\Psi \subset \mathcal{H}$ dense, such that the elements of $\mathcal{E}(G)$ are represented by τ_Ψ-continuous operators (i.e., operators that are continuous with respect to the topology in Ψ), and the central symmetric elements and the generators are e.s.a. (Ψ is essentially the Garding domain endorsed with a suitable nuclear topology.) An extremely important topic is the question of how wide the space $\Psi \subset \mathcal{H}$ may be taken to be: the wider the space Ψ the narrower the space Ψ^x and consequently the more the information we have about the nature of the generalized eigenvectors. With the construction of the topology by the Nelson operator, as done for the case of $SO(4,1)$ in (2), one achieves the most ideal situation: τ_Φ is the coarsest topology (and therewith Φ the largest space) such that the enveloping algebra of $SO(4,1)$ is represented by continuous operators.[1)]

We therefore construct also for the representations of the Poincaré group a countably Hilbert space, $\hat{\Phi}$, by introducing into a dense invariant domain in $\mathcal{H}$, e.g., the Garding domain D_G, the topology τ_{Φ} by the countable number of scalar products

$$(\phi,\psi)_p = (\phi,(\Delta_{\mathcal{P}} + 1)^p\psi) \qquad \phi,\psi \in D_G$$

where

$$\Delta_{\mathcal{P}} = P_0^2 + \vec{P}^2 + \vec{N}^2 + \vec{M}^2$$

is the e.s.a. Nelson operator of $\mathcal{P}$. It has been shown recently by Nagel[1] that for the irreps $(m>0, s, \epsilon)$ $\hat{\Phi}$ is nuclear. So $\hat{\Phi} \subset \mathcal{H}(m,s,\epsilon) \subset \hat{\Phi}^x$ is a rigged Hilbert space and the nuclear spectral theorem is applicable. Therefore the canonical basis vectors $|p_i s_3\rangle$ are elements of the space $\hat{\Phi}^x$.

Another system of commuting operators in an irrep of the Poincaré group is given by

$$Q_1 = -1/2\, L_{\mu\nu}L^{\mu\nu}, \quad Q_2 = 1/8\, \epsilon^{\mu\nu\rho\sigma} L_{\mu\nu} L_{\rho\sigma} \quad \vec{M}^2, \; M_3. \tag{19}$$

The self-adjointness of these operators (or the e.s.a. on the Garding domain) follows from the same argument as in the case of SO(4,1) for the system (3). We can again introduce a basis of simultaneous generalized eigenvectors

$$|j_3, j, k_0, \nu; \; (s,m,\epsilon)\rangle \qquad \in \hat{\Phi}^x. \tag{20}$$

The reduction of the irreps $\mathcal{H}(m,s,\epsilon)$ of $\mathcal{P}$ with respect to SO(3,1) has been given by Joos and others.[20] The result is:

$$\mathcal{H}(s,m,\epsilon) \xrightarrow[SO(3,1)]{} \sum_{\substack{k_0 = \pm 0, \pm 1 \ldots \\ \text{or } \pm\frac{1}{2}, \pm\frac{3}{2} \ldots}}^{s} \int_{\nu=0}^{\infty} \oplus\, \mathcal{H}(k_0,\nu) \frac{k_0^2+\nu^2}{(2\pi^4)^2}\, d\nu \tag{21}$$

i.e., in an irrep (s,m,ϵ) the spectrum of $k_0 = \pm s, \pm(s-1) \ldots \pm\frac{1}{2}$ or 0 and the spectrum of ν is the positive real line $0 \leqq \nu < \infty$ with multiplicity one.

We notice the similarity between the reduction of (r,ρ) of $SO(4,1)$, (5), and (s,m,ϵ) of $\mathcal{P}$ with respect to $SO(3,1)$. The only distinction is that for (r,ρ) each generalized eigenspace $\Phi^x_{j_3 j}(k_0,\nu)$ of Q_1, Q_2 $\vec{M}^2, M^3$ is two-dimensional, or

$$\Phi^x(k_0,\nu) = \mathcal{H}^{\eta=+}(k_0,\nu) \oplus \mathcal{H}^{\eta=-}(k_0,\nu)$$

whereas for (s,m,ϵ) it is one-dimensional, or $\Phi^x(k_0,\nu) = \mathcal{H}(k_0,\nu)$. So for $\mathcal{H}(s,m,\epsilon)$ (19) is already a complete system and we need no further label for the generalized eigenvectors (20).

A similarity of the kind like the one between (5) and (21) we had actually expected from comparison with the case where the subgroup with respect to which one reduces is compact. (E.g., $SO(3,1) > SO(3)$ versus $E_3 > SO(3)$), the new feature is only the doubling of the spectrum of ν.)

III. Contraction of Irreducible Representations of SO(4,1) into Irreducible Representations of the Poincaré Group

We shall now investigate the connection between $\mathcal{P}$ and $SO(4,1)$ further and first study the contraction[21] of an $SO(4,1)$ irrep (r,ρ) to a representation of $\mathcal{P}$.

Comparing the commutation relations of (I.4), (I.5), (I.6) for $SO(4,1)$ and (II.10) of $\mathcal{P}$, we see that they only differ in the relations (II.10a) for P_μ and V_μ. If we define

$$P_\mu^{(\lambda)} = \lambda V_\mu \tag{1}$$

and use (I.5) (I.6), we obtain

$$\left[P_\mu^{(\lambda)}, P_\nu^{(\lambda)}\right] = \lambda^2 i L_{\mu\nu} \quad \text{and} \quad \left[P_\rho^{(\lambda)}, L_{\mu\nu}\right] = i\left(g_{\mu\rho} P_\nu^{(\lambda)} - g_{\nu\rho} P_\mu^{(\lambda)}\right) \tag{2}$$

and we see that for $\lambda \to 0$ these commutation relations go into the commutation relations of $\mathcal{P}$ In taking this limit we have, however, to make sure that $P_\mu^{(\lambda)} \not\to 0$ if we want to get an irrep of the kind (s,m,ϵ) and not an unfaithful representation of the Poincaré group. This can be achieved if we let the representation of V_μ grow, i.e., let $\alpha \to \infty$ (or $\rho \to \infty$) while $\lambda \to 0$. It should be remarked that this contraction through a sequence of representations is only possible if there exists such a sequence of representations. Had ρ^2 or α^2

an upper bound (as it is the case in the discrete series representations of SO(4,1)), i.e., would it not have been possible to go with $\alpha^2 \to +\infty$ then we could not have reached the representations of the Poincaré group with $m^2 > 0$.

For the Casimir operator (I.7), (I.8) and (I.46'), (I.47') of SO(4,1), we obtain:

$$\lambda^2 Q = P_\mu^{(\lambda)} P^{(\lambda)\mu} - \tfrac{1}{2}\lambda^2 L_{\mu\nu} L^{\mu\nu} = \lambda^2 \alpha^2 \tag{3}$$

$$\lambda^2 Q_{(4)} = \lambda^2 (\vec{M}\cdot\vec{N}) - w_\mu^{(\lambda)} w^{(\lambda)\mu} = r(r+1)\lambda^2\alpha^2 + \lambda^2 (r-1)r(r+1)(r+2)$$

with (4)

$$w_\mu^{(\lambda)} = \tfrac{1}{2}\epsilon_{\mu\nu\rho\sigma} P^{(\lambda)\nu} L^{\rho\sigma} .$$

And if we now perform the limiting process:

$$\lambda \to 0 \qquad \alpha \to \infty \qquad \text{such that} \qquad \lambda^2\alpha^2 \to m^2 \tag{5}$$

where m^2 is an arbitrarily chosen constant and call

$$P_\mu = \lim_{\substack{\lambda \to 0 \\ \alpha \to \infty}} P_\mu^{(\lambda)}; \qquad \text{where} \quad \alpha^2\lambda^2 \to m^2 \tag{6}$$

then we obtain from (3) and (4)

$$P_\mu P^\mu = m^2 \tag{7}$$

$$-w_\mu w^\mu = m^2 r(r+1), \; w_\mu = \lim w_\mu^{(\lambda)} = \tfrac{1}{2}\epsilon_{\mu\nu\rho\sigma} P^\nu L^{\rho\sigma} . \tag{8}$$

So we see that with $r = s$ we can obtain one of the two irrep $(s = r, m, \epsilon = \pm 1)$ or both of them. To investigate this question further we have to see what happens to the generalized basis vectors (9) in this limiting process. As the $SO(3,1)_L$ subgroup is unaffected in this contraction, (generalized) eigenvectors of Q_1, Q_2, $\vec{L}^2$, L_3 in (r,ρ) go into generalized eigenvectors of the same operators in (s, m, ϵ). It remains to check what happens to I in this contraction process. There we have to distinguish two cases between which we can choose:

1.) I changes the sign λ
2.) I does not change the sign λ.

In case 1) we have from (6) and (1)

$$IP_\mu I^{-1} = \lim \lambda V_\mu = P_\mu . \tag{9}$$

In case 2) we have

$$IP_\mu I^{-1} = -\lim \lambda V_\mu = -P_\mu . \tag{10}$$

So in case 2.) I transforms from an irrep of $\mathcal{P}$ with sign $p_o = \pm 1$ into an irrep of $\mathcal{P}$ with sign $p_o = \mp 1$, whereas in case 1.) I does not transform between inequivalent irrep spaces of $\mathcal{P}$.

In case 2) we shall introduce in $\Phi^x(k_o, \nu)$ of $\mathcal{H}(r,\rho)$ the new basis

$$|j_3, j, k_o, \nu, (r,\rho), 1\rangle =$$
$$= \frac{1}{\sqrt{2}}(|j_3, j, k_o, \nu, (r,\rho), +\rangle + |j_3, j, k_o, \nu, (r,\rho), -\rangle)$$
$$|j_3, j, k_o, \nu, (r,\rho), 2\rangle = \tag{11}$$
$$= \frac{1}{\sqrt{2}}(|j_3, j, k_o, \nu, (r,\rho) +\rangle - |j_3, j, k_o, \gamma, (r,\rho) -\rangle)$$

then

$$\lim |j_3, j, k_o, \nu, (r,\rho), 1\rangle = |j_3, j, k_o, \nu, (r=s, m, \epsilon=+)\rangle$$

and (12)

$$\lim |j_3, j, k_o, \nu, (r,\rho), 2\rangle = |j_3, j, k_o, \nu, (r=s, m, \epsilon=-)\rangle$$

and the irrep (r,ρ) of SO(4,1) contracts into the direct sum of the irreps $(r,m,+)$ and $(r,m,-)$ of $\mathcal{P}$:

$$(r,\rho) \rightarrow (s=r, m, +) \oplus (s=r, m, -); \tag{13}$$

here m is the arbitrarily chosen number in limiting process (5).

As it is conventional (though not at all compelling if one drops the requirement that time inversion is represented by an anti-unitary operator) to consider only positive energy representations, $\epsilon = +1$, of $\mathcal{P}$ we shall choose case 1).

In case 1), λ is in the space $\mathcal{H}(r,\rho)$ an operator with two eigenvalues. We therefore replace

$$\lambda \to \lambda\Lambda_o \quad \text{with } \lambda > 0 \tag{14}$$

and

$$I\Lambda^o I^{-1} = -\Lambda^o; \quad \text{spectrum } \Lambda^o = \{+1, -1\} \tag{15}$$

$$\Lambda^o V_\mu = V_\mu \Lambda^o \qquad [\Lambda^o, U(SO(3,1)_{L_{\mu\nu}}] = 0. \tag{16}$$

The irrep (r,ρ) of SO(4,1) contracts now into the direct sum of two equivalent irreps $(r,m,+)$ of $\mathcal{P}$.

$$(r,\rho) \to (s=r,m,+) \oplus (s=r,m,+) \tag{17}$$

and the generalized basis vectors (9) of $\mathcal{H}(r,\rho)$ go into generalized basis vectors of $\mathcal{H}(r,m,+) \oplus \mathcal{H}(r,m,+)$. Instead of using the generalized basis vectors (9) in which I is diagonal, we introduce a new basis in which Λ^o is diagonal:

$$|j_3, j, k_o, \nu, (r, \ ), \lambda_\pm\rangle = \frac{1}{\sqrt{2}} (|j_3, j, k_o, \nu, (r,\rho), +\rangle \pm |j_3, j, k_o, \nu, (r,\rho), -\rangle) \tag{18}$$

where

$$\Lambda^o |j_3, j, k_o, \nu, (r,\rho), \lambda_\pm\rangle = \pm |j_3, j, k_o, \nu (r,\rho) \lambda_\pm\rangle \tag{18^1}$$

where the new observable Λ^o is diagonal.

[There is one feature which—though not immediately connected with our subject of contraction—might be worth mentioning here: From (16) follows that there is no element in the enveloping algebra $\mathcal{E}(SO(4,1))$ that changes the eigenvalue of Λ^o. So the space Φ (and correspondingly the space Φ^x) decomposes into

$$\Phi = \Phi^{\lambda+} \oplus \Phi^{\lambda-},$$

where

$$\Lambda^o \phi^{\lambda\pm} = \pm\phi^{\lambda\pm} \quad \text{with} \quad \phi^{\lambda\pm} \in \Phi^{\lambda\pm},$$

and each of the two subspaces is a τ_Φ-closed subspace of Φ invariant under $\mathcal{E}(SO(4,1))$. So we have in Φ a reducible representation of $\mathcal{E}(SO(4,1))$, though the representation of the group SO(4,1) in $\mathcal{H}(r,\rho)$, and therewith also in Φ, is irreducible ($\tau_{\mathcal{H}}$-irreducible and τ_Φ-irreducible respectively). This seems to be a general feature of unitary irreducible representations of non-compact Lie groups.[22)]]

Summarizing our results, we have seen that in the contraction of SO(4,1) to the Poincaré group the irrep (r,ρ) of the principle series goes into the direct sum of two irreps of the Poincaré group with $s=r$ and $m=$arbitrary and we are free to arrange the contraction of the representation such that these two irreps are equivalent, (17), or inequivalent, (13). We can, of course, identify the two spaces $\mathcal{H}(r,\rho)$ and $\mathcal{H}(s{=}r,m,+) \oplus \mathcal{H}(s{=}r,m,+)$ and consider their contraction as a transition process between operators in one and the same Hilbert space. We know that the generalized basis vectors $|j_3,j,k_0,\nu,(r,\rho)\lambda_\pm\rangle$ are $\in \Phi^x$, the conjugate to the countably Hilbert space Φ. Of the vectors $|j_3,j,k_0,\nu,(s,m,+)\lambda_\pm\rangle$, we know that they are $\in \hat{\Phi}^x$ the conjugate of the nuclear countably Hilbert space $\hat{\Phi}$. We would, of course, suspect that the generalized basis for $\mathcal{P}$: $|j_3,j,k_0,\nu(s,m,+)\lambda_\pm\rangle$ and the generalized basis for SO(4,1): $|j_3,j,k_0,\nu(r,\rho)\lambda_\pm\rangle$ are elements of the same space. But as the connection here is clouded by the limiting process (5), we shall investigate this point after we have discussed the connection between $\mathcal{P}$ SO(4,1) and $\mathcal{P}$ which in a certain sense is the inversion of the contraction process.

IV. Construction of SO(4,1) Representations in Unitary Representations of the Poincaré Group

We start with an irrep space $\mathcal{H}(s,m,+)$ of $\mathcal{P}$. Then we said in Section II that we can always find a nuclear countably-Hilbert space $\hat{\Phi}_{(1)}$ (in Section II we called it $\hat{\Phi}$) such that

$$\hat{\Phi}_{(1)} \subset \mathcal{H}(s,m,+) \subset \hat{\Phi}^x_{(1)} \tag{1}$$

is a Rigged Hilbert space and $\mathcal{E}(\mathcal{P})$ is an algebra of continuous operators in $\hat{\Phi}_{(1)}$.[16)] The topology in $\hat{\Phi}_{(1)}$ was given by a countable number of scalar products:

$$(\phi,\psi)_p = (\phi,(\Delta_{\mathcal{P}}+1)^p\psi) \tag{2}$$

where $\Delta_{\mathcal{P}}$ is the e.s.a. Nelson operator of $\mathcal{P}$:

$$\Delta_{\mathcal{P}} = P_o^2 + \vec{P}^2 + \vec{N}^2 + \vec{M}^2. \tag{3}$$

ϕ, ψ in (2) can be taken as elements of a dense invariant subspace D of $\mathcal{H}(s,m,+)$, e.g., the Garding space D_G, and $\hat{\Phi}_{(1)}$ is then the completion of D with respect to the topology given by (2).

On $\hat{\Phi}_{(1)}$ we define

$$\begin{aligned} B_\mu &= P_\mu + \frac{\lambda}{2m} \{P^\rho, L_{\rho\mu}\}, \quad \text{with} \quad \lambda = \text{const.} > 0 \\ &= P_\mu - i\frac{\lambda}{2m}[P_\mu, Q_1]. \end{aligned} \tag{4}$$

As $\mathcal{E}(\mathcal{P})$ is an algebra of continuous operators in $\hat{\Phi}_{(1)}$, it is clear that B_μ are continuous operators in $\hat{\Phi}_{(1)}$ and B_μ is symmetric in $\mathcal{H}(s,m,+)$.

A straightforward computation using the c.r. (II.10) of $\mathcal{P}$ shows that[23)]

$$[B_\mu, B_\nu] = i\lambda^2 L_{\mu\nu} \tag{5}$$

$$[L_{\mu\nu}, B_\rho] = i(g_{\nu\rho}B_\mu - g_{\mu\rho}B_\nu). \tag{6}$$

So $(1/\lambda)B_\mu$ and $L_{\mu\nu}$ gives us a symmetric representation of the Lie algebra of SO(4,1) in the irrep space $\mathcal{H}(s,m,+)$ of $\mathcal{P}$.

The question arises, whether $(1/\lambda)B_\mu, L_{\mu\nu}$ generate a global representation of the group SO(4,1), i.e., whether the representation of the Lie algebra of SO(4,1) given by B_μ and $L_{\mu\nu}$ is integrable. Before we can investigate this question, we will have to study the property of this representation of the Lie algebra $\mathcal{L}$(SO(4,1)).

We shall first calculate the Casimiroperators of $\mathcal{L}(SO(4,1)_{B_\mu L_{\mu\nu}})$, (I.7), (I.8), (I.9):

$$\lambda^2 Q = B_\mu B^\mu - \frac{\lambda^2}{2} L_{\mu\nu} L^{\mu\nu} \tag{7}$$

and

$$\lambda^2 Q_{(4)} = \lambda^2 Q_2^2 - \omega_\mu \omega^\mu \tag{8}$$

where

$$\omega^{\mu} = \tfrac{1}{2}\epsilon^{\mu\nu\rho\sigma} B_{\nu} L_{\rho\sigma} = -i[B^{\mu}, Q_2]. \tag{9}$$

In the following calculation we express B_{μ} by P_{μ} and $L_{\mu\nu}$.[24)]
Inserting (4) into (9), one obtains

$$\omega^{\mu} = w^{\mu} + \frac{\lambda}{2m}\{P^{\mu}, Q_2\}. \tag{10}$$

From which one calculates

$$\omega_{\mu}\omega^{\mu} = w_{\mu}w^{\mu} + \frac{\lambda^2}{m^2}(m^2 Q_2^2 + \tfrac{1}{4} w_{\mu}w^{\mu})$$
$$w_{\mu}w^{\mu} = -m^2 s(s+1). \tag{11}$$

From (4) we obtain further

$$B_{\mu}B^{\mu} = m^2 + \frac{9}{4}\lambda^2 + \frac{\lambda^2}{m^2} P^{\rho}P^{\sigma}L_{\rho\mu}L^{\mu}{}_{\sigma}. \tag{12}$$

Inserting (12) into (7) and (11) into (8), we obtain

$$\lambda^2 Q = m^2 + \frac{9}{4}\lambda^2 - \lambda^2 s(s+1) \tag{13}$$

$$\lambda^2 Q_{(4)} = (m^2 + \tfrac{1}{4}\lambda^2) s(s+1). \tag{13}$$

So we see that in the irreducible representation space $\mathcal{H}(m,s,+)$ of $\mathcal{P}$ the Casimiroperators of $(SO(4,1)_{B_{\mu}L_{\mu\nu}})$ are constants. If we further use (13) to express $\lambda^2 Q_{(4)}$ by Q and s, we obtain:

$$\lambda^2 Q_{(4)} = \lambda^2 Q\, s(s+1) + \lambda^2\, s(s+1)(s-1)(s+2). \tag{15}$$

Because $m^2 > 0$, we have $Q > 9/4 - s(s+1)$ and if we compare this with (I.46')(I.47'), we see that the eigenvalues of the Casimir-operators Q and $Q_{(4)}$ in this representation of the Lie algebra $\mathcal{L}(SO(4,1)_{(1/\lambda)B_{\mu}L_{\mu\nu}})$ are the same as the eigenvalues of the Casimiroperators in the principle series representation $(r=s,\ \rho = \sqrt{m^2/\lambda^2})$ of the group SO(4,1). From this we conclude that if the representation of $\mathcal{L}(SO(4,1)_{(1/\lambda)B_{\mu}L_{\mu\nu}})$ is integrable, it must integrate to a principle series representation of SO(4,1). Of the principle series representation (r,ρ) of SO(4,1), however, we know the reduction with respect to the group $SO(3,1)_{L_{\mu\nu}}$, which is given by (II.5) and this is quite distinct from the reduction of the representation space $\mathcal{H}(s,m,+)$ with respect to the subgroup $SO(3,1)_{L_{\mu\nu}}$ given in (II.21). Therefore in $\mathcal{H}(s,m,+)$ cannot be given a principle series representation of SO(4,1) and thus we conclude that the representation of the Lie-algebra of SO(4,1) given by $(1/\lambda)B_\mu$ and $L_{\mu\nu}$ on $\mathcal{H}(s,m,+)$ is non-integrable and only partially integrable on the $SO(3,1)_{L_{\mu\nu}}$ subgroup.

Let us now take a second replica of the irrep space $\mathcal{H}(s,m,+)$ of $\mathcal{P}$ as above and consider the direct sum

$$\mathcal{H} = \mathcal{H}(s,m,+) \oplus \mathcal{H}(s,m,+). \tag{16}$$

To distinguish between these two equivalent irrep spaces, we introduce the operator Λ^o with eigenvalue +1 on the first and eigenvalue -1 on the second of these two spaces. By I we denote the operator which transforms between $\mathcal{H}^{\lambda+}(s,m,+)$ and $\mathcal{H}^{\lambda-}(s,m,+)$. Then clearly

$$I\,\Lambda^o\,I^{-1} = -\Lambda^o. \qquad \text{(II.15), (17)}$$

We write the operators on this direct sum in the form:

$$P_\mu = P_\mu^{\lambda+} + P_\mu^{\lambda-} = \begin{pmatrix} P_\mu & 0 \\ 0 & P_\mu \end{pmatrix}, \quad L_{\mu\nu} = \begin{pmatrix} L_{\mu\nu} & \\ & L_{\mu\nu} \end{pmatrix},$$

$$\Lambda^o = \begin{pmatrix} +1 & \\ & -1 \end{pmatrix}, \quad I = \begin{pmatrix} 0 & 1 \\ 1 & 0 \end{pmatrix} \tag{18}$$

We call $\hat{\Phi}_{(1)}$ the countably-Hilbert space constructed as described by the norms (2) in $\mathcal{H}^{\lambda+}(s,m,+)$ and $\hat{\Phi}_{(-1)}$ the countably-Hilbert space constructed in the same way in $\mathcal{H}^{\lambda-}(s,m,+)$:

$$\Lambda_o \Phi_{(\pm 1)} = \pm\Phi_{(\pm 1)} \tag{19}$$

We then take the direct sum of the linear spaces $\hat{\Phi}_{(1)}$ and $\hat{\Phi}_{(-1)}$:

$$\hat{\Phi} = \hat{\Phi}_{(1)} \dotplus \hat{\Phi}_{(-1)} \tag{20}$$

(i.e.,

$$\tilde{\Phi} = \{\phi = (\phi_1, \phi_{-1}) \mid \phi_i \in \hat{\Phi}_i\}, \tag{20^1}$$

$\tilde{\Phi}$ is nuclear because $\hat{\Phi}_{(+1)}$ and $\hat{\Phi}_{(-1)}$ are (Reference 2, Ch. I, 9).

Then we have in

$$\hat{\Phi} \subset \mathcal{H} \subset \hat{\Phi}^{x} \tag{21}$$

a rigged Hilbert space and P_μ and $L_{\mu\nu}$ of (18) are continuous operators in $\tilde{\Phi}$. In $\tilde{\Phi}$ we define

$$\begin{aligned} B_\mu &= \Lambda^o P_\mu + \frac{\Lambda^o \lambda}{2m} \{P^\rho, L_{\rho\mu}\} \\ &= \Lambda^o \left(P_\mu - i \frac{\lambda}{2m} [P_\mu, Q_1]\right) \end{aligned} \tag{4^1}$$

where $\lambda > 0$ is the same constant as before. As before, it follows that $(\Lambda^{o2} = 1)$:

$$[B_\mu, B_\nu] = i\lambda^2 L_{\mu\nu} \tag{5^1}$$

$$[L_{\mu\nu}, B_\rho] = i(g_{\nu\rho} B_\mu - g_{\mu\rho} B_\nu) \tag{6^1}$$

so that B_μ and $L_{\mu\nu}$ give a symmetric representation of $\mathcal{L}(SO(4,1))$ in $\mathcal{H}$ of continuous operators in $\hat{\Phi}$. It is clear that we will find again for the Casimir operators

$$\lambda^2 Q = m^2 + \frac{9}{4}\lambda^2 - \lambda^2 s(s+1) \tag{13^1}$$

$$\lambda^2 Q_{(4)} = (m^2 + \tfrac{1}{4}\lambda^2)\, s(s+1). \tag{14^1}$$

So we have obtained a representation of $\mathcal{L}(SO(4,1))$ with the same eigenvalues of the Casimiroperators as in the above case on a space twice as big. But (4^1) does not give the direct sum representation of two of the representations (4) because

$$B_\mu = \begin{pmatrix} P_\mu + \frac{\lambda}{2m}\{P^\rho, L_{\rho\mu}\} & 0 \\ 0 & -P_\mu - \frac{\lambda}{2m}\{P^\rho, L_{\rho\mu}\} \end{pmatrix} \tag{22}$$

The operator B_μ ($\mu = 0,1,2,3$) defined by (4^1) or (22) is an extension of the second kind in the sense of Naimark[25] of the operator $B_\mu/\Phi_{(1)}$ defined by (4).

The difference between the representation (4) of $\mathcal{L}(SO(4,1))$ in $\mathcal{H}(s,m,+)$ and (4^1) in $\mathcal{H}$ is that in $\mathcal{H}$ there is defined an operator I, (17), which does not transform out of $\mathcal{H}$, with the properties:

$$\begin{aligned} I B_\mu I^{-1} &= -B_\mu \\ I L_{\mu\nu} I^{-1} &= L_{\mu\nu} \end{aligned} \tag{23}$$

whereas such an operator cannot be defined in $\mathcal{H}(s,m,+)$. In fact, it follows from (4) that an operator that commutes with $L_{\mu\nu}$ and anti-commutes with B_μ must also anti-commute with P_μ and therefore transform out of $\mathcal{H}(s,m,+)$. From (II.7) we know that an operator with the property (23) exists in an irrep space of $\mathcal{L}(SO(4,1))$ in which SO(4) can be diagonalized—like, e.g., the irrep space $\mathcal{H}(r,\rho)$ of the principle series representation of SO(4,1). From this we see again that the representation of $\mathcal{L}(SO(4,1))$ on $\mathcal{H}(s,m,+)$ given by (4) cannot integrate to a representation of the group.

On the other hand the representation (4^1) of $\mathcal{L}(SO(4,1))$ in $\mathcal{H}$ has such an operator I; it also has the same spectrum of SO(3,1)-group representations as $\mathcal{H}(r,\rho)$. These properties are, of course, not enough to conclude that the representation (4^1) of $\mathcal{L}(SO(4,1))$ integrates to a representation of the group SO(4,1), but it still allows for this possibility.

A sufficient condition[10] for the integrability of the representation of $\mathcal{L}(SO(4,1))$ is that the Nelson operator is e.s.a. on a common invariant dense domain of $\mathcal{L}(SO(4,1))$.

We therefore introduce the Nelson operators of $\mathcal{L}(SO(4,1)_{(1/\lambda)B_\mu, L_{\mu\nu}})$ in $\mathcal{H}$ by:

$$\hat{\Delta}_{SO(4,1)} = \frac{1}{\lambda^2} B_0^2 + \frac{1}{\lambda^2}\vec{B}^2 + \vec{N}^2 + \vec{M}^2. \tag{24}$$

Because $Q = \alpha^2 = m^2/\lambda^2 + 9/4 - s(s+1)$, we can write (24)

$$\hat{\Delta}_{SO(4,1)} = \alpha^2 + 2\frac{1}{\lambda^2}\vec{B}^2 + 2\vec{M}^2 = \alpha^2 + 2\hat{\Delta}_{SO(4)} \tag{25}$$

where $\hat{\Delta}_{SO(4)}$ is the Nelson operator of $\mathcal{L}(SO(4)_{B_i M_i})$:

$$\hat{\Delta}_{SO(4)} = \frac{1}{\lambda^2}\vec{B}^2 + \vec{M}^2 \tag{26}$$

of which we do not yet know whether it is the Casimiroperator of a group representation of SO(4).

We can also express $\hat{\Delta}_{SO(4,1)}$ by the Nelsonoperator of $\mathcal{P}$ on $\mathcal{H}$:

$$\Delta_{\mathcal{P}} = P_o^2 + \vec{P}^2 + \vec{N}^2 + \vec{M}^2 \tag{28}$$

and obtain after some simple calculation

$$\hat{\Delta}_{SO(4,1)} = \Delta_{\mathcal{P}} + Q_1 + \frac{1}{\lambda^2}2\, B_o^2 - \frac{1}{\lambda^2}2\, P_o^2 - \frac{9}{4} + s(s+1). \tag{29}$$

As we have a global representation of $\mathcal{P}$, $\Delta_{\mathcal{P}}$ (and also Q_1 and P_o and P_o^2) is e.s.a. on D_G and consequently also on $\hat{\Phi}$.

In the same way we can define the Nelsonoperator of $\mathcal{L}(SO(4,1)_{(1/\lambda)B_\mu^{(i)}, L_{\mu\nu}^{(i)}})$ in $\mathcal{H}^{(\lambda_i)}(s,m,+)$:

$$\begin{aligned} \hat{\Delta}_{(i)} = \hat{\Delta}_{SO(4,1)/\hat{\Phi}_{(i)}} &= \frac{1}{\lambda^2} B_o^{(i)^2} + \frac{1}{\lambda^2}\vec{B}^{(i)^2} + \vec{N}^{(i)^2} + \vec{M}^{(i)^2} \\ &= \alpha^2 + 2\hat{\Delta}^{(i)}_{SO(4)} \end{aligned} \tag{30}$$

where

$$B_\mu^{(i)} = B_{\mu/\hat{\Phi}_{(i)}} \tag{31}$$

is defined by (4) (B_μ there is called $B_\mu^{(1)}$ here) and is the restriction of B_μ given by (4^1) to the subspace $\hat{\Phi}_{(i)}$, (i=+1, -1), and similarly

$$L_{\mu\nu}^{(i)} = L_{\mu\nu/\hat{\Phi}_{(i)}} \tag{32}$$

$$\hat{\Delta}^{(i)}_{SO(4)} = \hat{\Delta}_{SO(4)/\hat{\Phi}_{(i)}} \tag{33}$$

$\tilde{\Delta}_{SO(4,1)}$ is the Naimark extension[25)] of $\hat{\Delta}_{(1)}$.

Of $\hat{\Delta}_{(1)}$ we know the following properties: $(\hat{\Delta}_{(1)}+1)$ is symmetric and positive definite; this is a consequence of (30) (the r.h.s. of (30) is $\geqq 0$ as it is the sum of squares of symmetric operators); therefore, its deficiency indices[26)] are equal: (n,n). It is not e.s.a. because otherwise the representation of $\mathcal{L}(SO(4,1)_{B_\mu^{(1)} L_{\mu\nu}^{(1)}})$ in $\mathcal{H}(s,m,+)$ must be integrable[10)] to a unitary representation of the group SO(4,1), which, as we have seen above, cannot be the case; therefore, $n > 0$.

From this it follows that $\hat{\Delta}_{SO(4,1)} + 1 = \hat{\Delta}_{(1)} \oplus \hat{\Delta}_{(-1)} + 1$ is also a positive definite symmetric operator and that its deficiency indices are (2n,2n) (Lemma 6, Reference 25a). As a consequence of the Nelson theorem, it then follows that the representation of $\mathcal{L}(SO(4,1)_{B_\mu L_{\mu\nu}})$ given by (4^1) on $\hat{\Phi} = \hat{\Phi}_{(1)} \dotplus \hat{\Phi}_{(-1)} \subset \mathcal{H}$ is <u>not integrable to a unitary representation</u> of the group SO(4,1).

There remains the following possibility for $\mathcal{L}(SO(4,1)_{B_\mu L_{\mu\nu}})$ of (4^1) to be connected with a unitary representation of the principle series of the group SO(4,1). This is, that there exists an extension of $L_{\mu\nu}$ and B_μ of (4^1) to a larger invariant dense domain $\supset \hat{\Phi}$ on which the extension of $\hat{\Delta}_{SO(4,1)}$ is e.s.a. (The extended B_μ, B_μ^{ext} can, in general, no more be written as a function of P_μ and $L_{\mu\nu}$ as given in (4^1); only their restrictions $B_\mu = B_\mu^{ext}/\hat{\Phi}$ are equal to the r.h.s. of (4^1).) In the following we shall discuss this possibility. Before that we should make one further remark concerning the integrability of the representation of $\mathcal{L}(SO(4,1)_{B_\mu^{(i)} L_{\mu\nu}^{(i)}})$ in $\mathcal{H}(s,m,+)$. From (30) we see that from the non-integrability of the representation of $\mathcal{L}(SO(4,1)_{B_\mu^{(i)}, L_{\mu\nu}^{(i)}})$ follows that $\hat{\Delta}_{(1)}$ is not e.s.a. and therefore that already the representation of the compact subalgebra $\mathcal{L}(SO(4)_{B_i^{(i)} L_{ij}^{(i)}})$ is non-integrable which was to be expected because otherwise we would have had a dense set of analytic vectors in $\hat{\Phi}_{(i)}$.

We introduce into the representation space $\mathcal{H}$ of the Poincaré group a new topology by the following countable number of scalar products.

$$\langle \psi, \phi \rangle_p = \left(\psi, (\hat{\Delta}_{SO(4,1)} + 1)^p \phi \right), \quad p = 1, 2, \ldots \qquad \psi, \phi \in D_G \subset \hat{\Phi} \subset \mathcal{H}. \tag{34}$$

Equation (34) gives a suitable set of compatible norms because $\hat{\Delta}_{SO(4,1)}$ is symmetric and positive definite.

The completion of D_G with respect to the topology given by (34) (which we shall call $\tilde{\tau}$) we call $\tilde{\Phi}$.

Then one easily sees:

$$\hat{\Phi} \subset \tilde{\Phi}. \tag{35}$$

The proof of this statement (given in the appendix of this section) is simple; it makes use of the fact that $\hat{\Delta}_{SO(4,1)}+1$ is an element of $\mathcal{E}(\ )$ and employs a lemma of Nelson.[10]

It is obvious that an equivalent topology $\tilde{\tilde{\tau}}$ can be introduced by the countable number of scalar products.

$$[\psi, \phi]_p = \sum_{q=0}^{p} \langle \psi, \phi \rangle_q, \quad p=1,2,\dots. \tag{34^1}$$

For $\tilde{\Phi}$ we shall use (34) or (34^1), whichever will be more convenient.

One cannot prove that $\hat{\Phi} \supset \tilde{\Phi}$, because $P_0^2 \notin \mathcal{E}(SO(4,1)_{B_\mu, L_{\mu\nu}})$ on D_G.

We have now obtained the following chain of spaces

$$\hat{\Phi} \subset \tilde{\Phi} \subset \mathcal{H} \subset \tilde{\Phi}^x \subset \hat{\Phi}^x \tag{36}$$

and $B_\mu, L_{\mu\nu}$ and $\hat{\Delta}_{SO(4,1)}$ can be extended to $\tilde{\tau}$-continuous operators on $\tilde{\Phi}$[9] (closure by continuity) which we call $\tilde{B}_\mu$, $\tilde{L}_{\mu\nu}$, $\tilde{\Delta}_{SO(4,1)}$. This extension is unique and for the extended operators we have again (24):

$$\tilde{\Delta}_{SO(4,1)} = \frac{1}{\lambda^2}\tilde{B}_0^2 + \frac{1}{\lambda^2}\vec{\tilde{B}}^2 + \vec{\tilde{N}}^2 + \vec{\tilde{M}}^2 \tag{24^1}$$

(However, P_μ cannot be extended to such operators so that $\tilde{B}_\mu$ is not connected with generators of $\mathcal{P}$ by relations like (4^1).)

The difference between $\tilde{\Phi}$ and the space Φ in Section II is that in Φ, $\Delta_{SO(4,1)}$ is e.s.a. and Φ is nuclear. The nuclearity was proven[16] to be a consequence of the spectrum of $\Delta_{SO(4,1)}$ in the global representation (r,ρ) of SO(4,1). The nuclearity of $\tilde{\Phi}$ is, therefore, not guaranteed (however, if it should turn out that the extension of $\hat{\Delta}_{SO(4,1)}$ to $\tilde{\Phi}$ is e.s.a., then the nuclearity of $\tilde{\Phi}$ follows from the integrability).

In the same way as was done for $\mathcal{H}=\mathcal{H}^{\lambda+}(s,m,+) \oplus \mathcal{H}^{\lambda-}(s,m,+)$ we can introduce a new countably Hilbert space topology into each of the irreducible representation spaces of $\mathcal{P}$, $\mathcal{H}^\lambda(s,m,+)$:

$$\langle \psi_1, \phi_1 \rangle_p = \left(\psi_1, (\hat{\Delta}_{(1)} + 1)^p \phi_1\right) \tag{34_1}$$

$$\psi_1, \phi_1 \in D_{G(1)} \subset \hat{\Phi}_{(1)} \subset \mathcal{H}(s, m, +).$$

By completion with respect to this topology we then obtain $\tilde{\Phi}_{(1)}$ and

$$\Phi_{(1)} \subset \tilde{\Phi}_{(1)} \subset \mathcal{H}(s, m, +) \subset \tilde{\Phi}^x_{(1)} \subset \Phi^x_{(1)} \tag{36_1}$$

(and equivalently for $\tilde{\Phi}_{(-1)}$).

We can then extend $L_{\mu\nu}$, $B_\mu^{(1)}$ (B_μ of (4)) and $\hat{\Delta}_{(1)}$ to $\tilde{\tau}_{(1)}$-continuous operators in $\tilde{\Phi}_{(1)}$. As we know that we cannot have a unitary representation (r, ρ) of the group SO(4,1) in $\mathcal{H}(s, m, +)$, we conclude that the extension of $\hat{\Delta}_{(1)}$ to $\tilde{\Phi}_{(1)}$ cannot be e.s.a. Let us indicate the extension of the operators to $\tilde{\Phi}_{(1)}$ (and similarly $\tilde{\Phi}_{(-1)}$) by $\tilde{\ }$. Then we have:

$\hat{\Delta}_{(1)}$ has the deficiency indices (n, n), $\tilde{\Delta}_{(1)}$ has the deficiency indices (m, m), where $m \leq n$ but $m > 0$, and the same for $\tilde{\Delta}_{(-1)}$.

The direct sum $\tilde{\Delta}_{(1)} \oplus \tilde{\Delta}_{(-1)}$ with the domain $\tilde{\Phi}_{(1)} \dot{+} \tilde{\Phi}_{(-1)}$ in $\mathcal{H}^{\lambda+}(s, m, +) \oplus \mathcal{H}^{\lambda-}(s, m, +)$ has then the deficiency indices $(2m, 2m)$,[25a)] $n \geq m > 0$. The question then remains is there an extension of $\tilde{\Delta}_{(1)} \oplus \tilde{\Delta}_{(-1)}$ to an invariant dense domain of $\mathcal{H}$ (perhaps $\tilde{\Delta}_{SO(4,1)}$, the extension of $\hat{\Delta}_{SO(4,1)}$ to $\tilde{\Phi}$) which is e.s.a.?

To discuss this question we now consider in more detail the operator

$$S = \hat{\Delta}_{SO(4,1)} + 1 = \hat{\Delta}_{(1)} \oplus \hat{\Delta}_{(-1)} + 1. \tag{37}$$

This is a positive definite symmetric operator in $\mathcal{H}$ with deficiency indices $(2n, 2n)$, $n > 0$; it is a $\tau_{\hat{\Phi}}$-continuous operator with invariant domain $D(S) = \hat{\Phi} = \hat{\Phi}_{(1)} \dot{+} \hat{\Phi}_{(-1)} \subset \mathcal{H}$. Let us then consider the sequence of spaces

$$D(S) \subset \bigcap_{q \geq 0} \Phi_q = \tilde{\Phi} \subset \cdots \tilde{\Phi}_q \subset \cdots \tilde{\Phi}_2 \subset \tilde{\Phi}_1 \subset \mathcal{H} \tag{38}$$

where each Hilbert space $\tilde{\Phi}_q$ is obtained by completion of $\hat{\Phi}$ with respect to the one scalar product $[\psi, \phi]_q$ given by (34^1).[27)]

The theory of extensions of semi-bounded operator gives us then the following result:[28)]

There exists exactly one self-adjoint extension S_μ (among all the self-adjoint extensions) of S with the property

$$D(S) \subset D(S_\mu) \subset D[S] = D[S_\mu] \tag{39}$$

where $D[S] = \tilde{\Phi}_1$ and $D[S_\mu] = D_1$ is the completion of $\hat{\Phi}$ with respect to the scalar product $[\psi, \phi]_{\mu, p=1}$ where

$$[\psi, \phi]_{\mu, p} = \sum_{q=0}^{p} \langle \psi, \phi \rangle_{\mu, q} = \sum_{q=0}^{p} (\psi, S_\mu^q \phi) . \tag{34_μ^1}$$

This is the Friedrichs-extension with the following property:

$$m(S_\mu) = m(S), \quad \left(m(A) = \inf_{f \in D(A)} \frac{(f, Af)}{(f, f)} \right) \tag{40}$$

$$(f, Sg) = (f, S_\mu g) \quad \text{for} \quad f, g \in D[S] . \tag{41}$$

So S_μ is the closure of S in the completion of $D(S)$ with respect to the scalar product $[\phi, \psi]_1$.

In analogy to $\tilde{\tau}$, we can now equip the space $\hat{\Phi}$ with the topology $\tilde{\tau}_\mu$ given by the countable number of scalar products (34_μ^1). The completion of $\hat{\Phi}$ with respect to $\tilde{\tau}_\mu$ we call D and the completion with respect to the q-th norm we call D_q; then we have in analogy to (38):

$$D(S) \subset D(S_\mu) \subset \bigcap_{q \geqq 0} D_q = D \subset \cdots D_q \subset \cdots D_1 = \Phi_1 = D[S_\mu] \subset \mathcal{H} . \tag{38_μ}$$

The restrictions of S_μ to D, $S_{\mu/D}$, is e.s.a. and

$$\overline{S_{\mu/D}} = S_\mu . \tag{42}$$

($\bar{A}$ denotes the closure of an operator A in $\mathcal{H}$.) (The simple proof of (42) is given in Appendix B of this section.)

Therewith we have constructed on $D \supset D(S)$ an extension of the operator S which is e.s.a. and which leaves D invariant. If we could construct a <u>unique</u> extension of the generators $B_\mu, L_{\mu\nu}$ to D, which leave D invariant, then the representation of $\mathcal{L}(SO(4,1))$ given by this extension would be integrable.

There is already a unique extension of S and the generators, namely the extension $\tilde{S} = \tilde{\Delta}_{SO(4,1)} + 1$, $\tilde{B}_\mu, \tilde{L}_{\mu\nu}$ to the space $\tilde{\Phi}$. If we can show that $\tilde{S} = S_{\mu/D}$ or equivalently $\tilde{\Phi} = D$, then D is the space of differentiable vectors of a principle series representation of SO(4,1) with the generators $\tilde{B}_\mu$, $\tilde{L}_{\mu\nu}$. (It is, in fact, already enough to show that $\tilde{\Phi} \supset D$ because if already the restriction of $\tilde{S}$ to a subspace D is e.s.a. then $\tilde{S}$ itself is also e.s.a.) Unfortunately, we can formulate this only as a conjecture:

> The space $\tilde{\Phi}$ of (38) contains the space D of (38_μ). Consequently, the Friedrichs extension of the direct sum of Nelson operators $\hat{\Delta}_{(+1)}, \hat{\Delta}_{(-1)}$ of the <u>Lie algebra</u> (4) of SO(4,1) is the Nelson operator of a principle series representation of the <u>group</u> SO(4,1).

This conjecture, which immediately suggests itself, is not as obvious as it looks. The proof will have to make use of the fact that $\hat{\Phi} = \hat{\Phi}_{(1)} \dot{+} \hat{\Phi}_{(-1)}$.

We summarize the discussion of this section:

The representation of $\mathcal{L}(SO(4,1))$ defined by (4) on an irrep space $\mathcal{H}(s,m,+)$ is not integrable to a global representation of the group SO(4,1). As a consequence, also the direct sum representation (4^1) in

$$\mathcal{H}(s,m,+) \oplus \mathcal{H}(s,m,+)$$

(the Naimark extension of (4)) is not integrable. However, it appears that there is a unique extension of this direct sum representation that integrates to the principle series representation $(r,\rho) = (s, \sqrt{m^2/\lambda^2})$ of the group SO(4,1).

<u>Appendix to Section IV</u>

Appendix A. Proof of (35).

From (29) we see that $(\hat{\Delta}_{SO(4,1)} + 1)^p$ is an element of the enveloping algebra of $\mathcal{P}$ on $\hat{\Phi}$ of order $\leqq 4p$, because B_0^2 is of order 4 by (4^1). Then Lemma 6.3 of Reference 10 states that

$$\left(\psi, (\hat{\Delta}_{SO(4,1)} + 1)^p \psi\right) \leqq k \left(\psi, (\Delta_{\mathcal{P}} + 1)^q \psi\right) \tag{*}$$

where $k < \infty$ is some constants, $q \geqq 2p$.

To show that $\hat{\Phi} \subset \tilde{\Phi}$, i.e., that $\tau_{\hat{\Phi}}$ is stronger than $\tilde{\tau}$, we have to show that every sequence converging to zero with respect to $\tau_{\hat{\Phi}}$ converges also to zero with respect to $\tilde{\tau}$:

Let $\psi_\nu \xrightarrow{\tau_{\hat{\Phi}}} 0 \iff ||\psi||_q \to 0$ for every q

$\iff (\psi_\nu, (\Delta \quad +1)^q \psi_\nu) \to 0$ for every q

$\implies$ with (*): $(\psi_\nu, (\Delta_{SO(4,1)} + 1)^p \psi_\nu) \to 0$ for every p

$\iff \langle \psi_\nu, \psi_\nu \rangle_p \to 0$ for every p $\iff \psi_\nu \overset{\tilde{\tau}}{\to} 0.$

q.e.d.

Remark: Δ_p is the direct sum of the Nelson operators in each $\mathcal{H}(s,m,+)$ of (16). As these are e.s.a. the deficiency indices of Δ_p are (0,0) and its closure is already maximal.

The above proof shows, in fact, that already for the completion $\overline{\Phi}$ of $\hat{\Phi}$ with respect to the direct sum topology of $\tau_{\hat{\Phi}(1)}$ and $\tau_{\hat{\Phi}(-1)}$, one has $\overline{\Phi} \subset \tilde{\Phi}$. We remark further that the possibility of the integrability of the SO(4,1)-Lie-algebra on $\overline{\Phi}$ has not been excluded.

Appendix B. Proof of (42).

According to the definition of closure, we have to show that for every $x_n \in D(S_\mu)$ there exists a sequence $x_n \in D = D(S_{\mu/D})$ such that $||x_n - x|| \to 0$ and $||S_{\mu/D}x - S_{\mu/D}x_m|| \to 0$ for $n, m \to \infty$. Let

$$S_\mu = \int_{m(s)>0}^{\infty} \lambda dE(\lambda) \tag{B1}$$

be the spectral resolution of S_μ.

Then we define for a given $x \in D(S_\mu)$

$$x_n = \int_{>0}^{n} dE(\lambda)x. \tag{B2}$$

Then clearly $x_n \xrightarrow{\tau_{\mathcal{H}}} x$ and $x_n \in D$ because

$$(x_n, S_\mu^p x_n) = \int_0^n \lambda^p (x, dE(\lambda)x) < \infty .$$

Further we have

$$||S_{\mu/D}x_{n+m} - S_{\mu/D}x_n|| = \int_n^{m+n} \lambda\, ||dE(\lambda)x|| \to 0.$$

So by (B2) we have constructed for every $x \in D(S_\mu)$ a sequence x_n which fulfills the above conditions, which proves (42).

References and Footnotes

1. Bengt Nagel, lectures in these proceedings.
2. K. Maurin, General Eigenfunction Expansions and Unitary Representations of Topological Groups, Warszawa (1968).
3. A. Böhm, Boulder Lectures in Theoretical Physics, Vol. 9A, 255 (1966). The notation given in this reference will be employed here.
4. See also D. Sternheimer, lectures in these proceedings.
5. E. P. Wigner, Ann. Math. 40, 199 (1939), Istanbul Lectures (1964), F. Gürsey, Editor.
6. M. A. Naimark, Linear Representations of the Lorentz Group, New York (1964). In distinction to the notation in Naimark's book, we choose

$$\nu = |c^{\text{Naimark}}|, \; k_o = \text{sign}\,(ic^{\text{Naimark}}) \cdot k_o^{\text{Naimark}}.$$

7. I. M. Gelfand, M. L. Zetlin, DAN SSSR 71, 1017 (1950).
8. We shall denote the representative of $L_{\alpha\beta}$ (and the elements of the enveloping algebra $\mathcal{E}(SO(4,1))$) in all the representation spaces again by $L_{\alpha\beta}$.
9. Where it does not lead to misunderstandings, we use the same symbol A for an operator in $\mathcal{H}$, $\bar{A}$, for its restriction to the subspace Φ, $\bar{A}/\Phi = A$, and for its extension to the conjugate space Φ^x, A^x, as well as for the restriction to any other subspace of $\mathcal{H}$ or any other extension. Where necessary, we indicate the domain D of an operator A by A/D.
10. E. Nelson, Ann. Math. 70, 572 (1959).
11. J. Dixmier, C. R. Acad. Sci., Paris 250, 3263 (1960), where it is shown for the reduction of u.i.r. of SO(n,1) with respect to SO(n).
12. H. D. Doebner, proceedings of the 1966 Istanbul Summer Institute.
13. L. H. Thomas, Ann. Math. 42, 113 (1941).
    T. D. Newton, Ann. Math. 51, 730 (1950).
    J. Dixmier, Bull. Soc. Math. France 89, 9 (1961).

14. I. M. Gelfand, M. I. Graev, Tsv. Akad. Nauk SSSR ser. Mat. 29, No. 6, 1329 (1965).
A. V. Nikolov, Funkt. Anal. Appl. 2, 94 (1969), Dubna preprint P5-3140 (1967).
F. Schwarz, Jour. Math. Phys. 12, 13 (1971).
15. Cf. also S. Ström, Arkiv f. Fysik 30, 451 (1965), Sec. III, where a different phase convention has been used.
16. A. Böhm, Jour. Math. Phys. 8, 1551 (1967), Appendix B.
17. E. Nelson, W. F. Stinespring, Ann. J. Math. 81, 547 (1959).
18. S. Ström, Arkiv for Fysik 40, 1 (1969). For the decomposition with respect to non-compact subgroup representation, see also N. Mukunda, J. Math. Phys. 9, 50 and 9, 417 (1968).
19. In analogy to the choice for SO(2,1) in G. Lindblad, B. Nagel, Ann. Inst. Poincaré, 13, 27 (1970).
20. H. Joos, Fortschr. Phys. 10, 65 (1962).
Chou Kuang-Chao, L. G. Zastavenko, Sov. Phys. JEPT 8, 990 (1956) and 35, 1417 (1958).
21. E. Inönü, E. P. Wigner, Proc. N.A.S. 39, 510(1953).
22. It was first pointed out by Harish-Chandra that there are subspaces of Φ (the set of differentiable vectors of a representation of a non-compact Lie group) which are invariant under the Lie-algebra without its closure being invariant under the group. This was the starting point for the introduction of the space of analytic vectors, which does not have this kind of pathologies. Harish-Chandra, Trans. Amer. Math. Soc. 75, 185 (1953).
23. Connections like (4) between the Poincaré group and SO(4,1) have been used extensively during the last years in the physics literature.
a) C. Fronsdal, Rev. Mod. Phys. 37, 221 (1965).
b) A. Sankaranarayanan, Nuovo Cim. 38, 1441 (1965).
c) A. Böhm, Phys. Rev. 145, 1212 (1966). Similar relations hold between all pseudo-orthogonal groups SO(p,q) and inhomogeneous pseudo-orthogonal groups of one lower dimension ISO(p-1,q), e.g., the "Gell-Mann formula" in R. Hermann Commun. Math. Phys. 2, 155 (1966) for the connection between ISO(3) and SO(3,1). See, e.g., John G. Nagel, "Expansions of Inhomogenisations of all the Classical Lie Algebras to Classical Lie Algebras" and references thereof, Ann. Inst. Henri Poincaré, 13, 1 (1970).
The here described mathematical problems arise when SO(p-1,q) is non-compact. The description given here is, of course, immediately generalized to the connection between ISO(n-1,1) and SO(n,1), and we have chosen n=4 only because in that case the representations are very familiar.

The application of relation (4) for the problem of the mass spectrum of elementary particles is described in A. Böhm, Phys. Rev. 175, 1767 (1968) and Phys. Rev. D3, 377 (1971).
24. More details of the calculation can be found in A. Böhm, Phys. Rev. 145, 1212 (1966), Chakrabarti, Levy-Nahas, Seneor Journ. Math. Phys. 9, 1274 (1968).
25. a) M. A. Naimark, Tzv. Akad. Nauk SSSR ser. Mat. 4, 53 (1940); 4, 277 (1940).
b) N. I. Akhiezer, I. M. Glasman, Theory of linear operators in Hilbert space, New York (1961), Appendix I.
26. As the deficiency indices of an in general non-closed operator A, we mean the deficiency indices of its closure. For the definition of deficiency indices and a brief introduction into the theory of extension (of the first kind) of symmetric operators, we refer to M. A. Naimark, Linear Differential Operators, Part II, Sec. 14, New York (1968) or Reference 25b, Ch. VII.
27. The choice of the scalar products (34^1) over the scalar products (34) is inessential and gives the same result.
28. a) M. Krein, Mat. Sbornik 20, 431 (1947).
b) F. Riesz, B Sz-Nagy Functional Analysis, New York (1955). The notation S_μ, etc. has been chosen in accordance with Reference 28a.

Acknowledgement

Several of my colleagues have helped to reduce the number of mistakes. I am particularly indebted to Professor B. Nagel and most of all to Professor S. Ström, who has checked Section I and has given me his advice in many problems connected with the material presented here.

THE INVERSE DECAY PROBLEM†

J. P. Marchand
Department of Mathematics
University of Denver
Denver, Colorado

I. Minimal Phenomenology for Reduced Motions

Let $\{\mathcal{H}, U(t)\}$ be a quantum mechanical system with $\mathcal{H}$ a separable Hilbert space and $U(t)$ a unitary representation of the one-parameter group $\{t\} = \mathbb{R}$ of time translations. Let P be a projector in $\mathcal{H}$ with n-dimensional range, and $U_p(t)$ the restriction of $PU(t)P$ to the subspace $P\mathcal{H}$. We call $U_p(t)$ the reduced motion of the subsystem $P\mathcal{H}$. It is an operator family of the positive type, i.e.,

$$\sum_{i,k} (\varphi_i, U_p(t_k - t_i)\varphi_k)_{P\mathcal{H}} \geqq 0$$

for all finite sequences $\varphi_i \in P\mathcal{H}$ and $t_i \in \mathbb{R}$.

Conversely we have the following:

Reconstruction Theorem.[1)] If $U_p(t)$ is a continuous operator family of positive type in a Hilbert space $\mathcal{H}'$, then there exists a triplet $\{\mathcal{H}, U(t), P\}$ where $U(t)$ is a continuous group of unitary operators in $\mathcal{H}$ and P a projector such that

$$\mathcal{H}' \approx P\mathcal{H}, \quad U_p(t) \approx PU(t)P\big|_{P\mathcal{H}}.$$

If $P\mathcal{H}$ is required to be cyclic in $\mathcal{H}$ under $U(t)$, i.e., if the closed linear space of the vectors $U(t)\varphi[\forall t \in \mathbb{R}, \varphi \in P\mathcal{H}]$ is $\mathcal{H}$, then the extension $\{\mathcal{H}, U(t); P\}$ is minimal and essentially unique.

The system $\{\mathcal{H}, U(t)\}$ may thus be considered as the minimal quantum-mechanical phenomenology consistent with the reduced motions $U_p(t)$ in the subspace $P\mathcal{H}$.

We first consider two extreme situations for the case $n=1$.

---

†Presented at the NATO Summer School in Mathematical Physics, Istanbul, 1970. This work has since been considerably extended and a detailed version published with L. A. Horwitz and J. A. LaVita (cf. Reference 3).

Example 1: Exponential decay.

Let $U_P(t) = e^{-\alpha t}$ $[t \geqq 0,\ \alpha > 0]$. With $U_P(-t) \equiv U_P(t)^{\dagger}$, we obtain the following complex-valued one-dimensional operator family of positive type:

$$U_P(t) = \begin{cases} e^{-\alpha t} & t \geqq 0 \\ e^{\alpha t} & t < 0 \end{cases} = \int_{\infty}^{\infty} e^{-i\lambda t} \frac{\alpha/\pi}{\lambda^2+\alpha^2}\, d\lambda .$$

On the other hand we have, according to Stone's theorem,

$$U(t) = e^{-iHt} = \int e^{-i\lambda t}\, dE_H(\lambda)$$

where H is the self-adjoint generator of $U(t)$ and $E_H(\lambda)$ its spectral family. Comparison shows that $PE_H(\lambda)P$ has the Radon-Nikodym derivative $(\alpha/\pi)/(\lambda^2+\alpha^2)$ with support $\mathbb{R}$. Hence the spectrum $\sigma(H)$ of H is the entire line $\mathbb{R}$ and absolutely continuous (and, of course, simple). The system $\{\mathcal{H}, U(t)\}$ is realized as

$$\mathcal{H} = L^2(\mathbb{R}), \qquad (U(t)f)(\lambda) = e^{-i\lambda t} f(\lambda).$$

Furthermore $P\mathcal{H}$ is cyclic.

Proof. Let φ be a unit vector which spans $P\mathcal{H}$. Then we have, with $\varphi(\lambda)$ the spectral representation of φ with respect to H,

$$U_P(t) = (\varphi, U_P(t)\varphi) = \int e^{-i\lambda t} |\varphi(\lambda)|^2 d\lambda; \quad |\varphi(\lambda)|^2 = \frac{\alpha/\pi}{\lambda^2+\alpha^2}.$$

We must show that $E_H(\Delta)\varphi \neq 0$ for all Borel sets $\Delta \subset \mathbb{R}$. But clearly

$$(\varphi, E_H(\Delta)\varphi) = \int_{\Delta} \frac{\alpha/\pi}{\lambda^2+\alpha^2}\, d\lambda \neq 0 \qquad \forall \Delta .$$

We finally remark that H is not defined on $P\mathcal{H}$. In fact, while $\varphi(\lambda) \in L^2_{\lambda}(\mathbb{R})$, $(H\varphi)(\lambda) = \lambda\varphi(\lambda) \notin L^2_{\lambda}(\mathbb{R})$.

Example 2: Periodic Motion.

Let $U_P(t) = \cos t$. This again is a continuous function of positive type, satisfying $U_P(-t) = U_P(t)^{\dagger}$. From

$$U_P(t) = \tfrac{1}{2}(e^{it} + e^{-it}) = \tfrac{1}{2}\int e^{-i\lambda t}(\delta(\lambda+1) + \delta(\lambda-1))\,d\lambda$$

$$= \int e^{-i\lambda t}\, dE_H(\lambda)$$

it follows that $\sigma(H)$ has two discrete simple eigenvalues ± 1. The triplet $\{\mathcal{H}, U(t), P\}$ can then be represented by

$$\mathcal{H} = V_2; \quad U(t) = \begin{pmatrix} \cos t & \sin t \\ -\sin t & \cos t \end{pmatrix}; \quad P = \begin{pmatrix} 1 & 0 \\ 0 & 0 \end{pmatrix}.$$

II. General Properties of Cyclic Reduced Motions

We discuss now reduced motions under the general assumption that $P\mathcal{H}$ is cyclic in $\mathcal{H}$ under $U(t)$.

A first basic distinction must be made as to whether H is defined on $P\mathcal{H}$ or not. If it is, the matrix element $(\varphi, H\varphi) = m_\varphi$ is finite for any φ $P\mathcal{H}$, i.e., all φ have finite masses.

The main result for finite masses is as follows:

Theorem 1. In the case of finite masses, $U_p(t)$ is not a semi-group.

The proof relies on two lemmas which are interesting in themselves and have been proved in Reference [2].

Lemma 1. $U_P(t)$ is not a group if $P \neq I$.

Lemma 2. The initial "decay rate"

$$\frac{d}{dt}\left|(\varphi, U_P(t)\varphi)\right|^2_{t=0}$$

exists for any $\varphi \in P\mathcal{H}$ and vanishes. The conclusion of the theorem then follows, because a semi-group which is not a group has never had vanishing initial decay rates for all $\varphi \in P\mathcal{H}$.

Another useful result is as follows: If the mass is finite, i.e., if H is defined on $P\mathcal{H}$, then the reduced motion $U_P(t)$ satisfies the Master Equation

$$i\frac{d}{dt}U_P(t) = PH\,U_P(t) - i\int_0^t d\tau\; PH\bar{P}\, e^{-i\bar{P}H\bar{P}\tau}\,\bar{P}H\,U_P(t-\tau);$$

$$\bar{P} = I - P.$$

For a proof, see Reference 2.

We now drop the requirement of finite masses and investigate the impact of the following conditions:

a) $\text{s-lim}_{t \to \infty} U_p(t) = 0$

b) $U_p(t)$ is a semi-group.

The interest of these conditions lies in the fact that they seem natural for a description of particle decay.

Theorem 2.[3] Condition a) implies $\sigma(H)$ continuous. Condition b) implies $\sigma(H) = \mathbb{R}$. Conditions a) and b) imply $\sigma(H) = \mathbb{R}$, $\sigma(H)$ absolutely continuous and of uniform multiplicity n.

* * *

We draw the main conclusions in the form of the following:

Corollary. The semi-group property for reduced motions is incompatible with finite mass for all $P\mathcal{H}$ (Theorem 1) and with positive energy (Theorem 2).

Note, however, that Example 1 provides the case of a reduced motion (with m not finite and $\sigma(H)$ not positive) which <u>is</u> a semi-group. Another example is the theory of Lax and Phillips.[4] Here the reduced motion $Z(t)$ of the subspace $P_K \mathcal{H} = K$ is a semi-group. Hence the spectrum $\sigma(H)$ of the generator of $U(t)$ is not positive. Furthermore, H is not defined on all of K.

III. Unstable Particles

It is clear that an arbitrary state φ does not describe an unstable particle; its decay should be approximately an exponential (compare the above examples 1 and 2!). More generally we may postulate, for the <u>reduced motion $U_p(t)$ of a subspace $P\mathcal{H}$ of unstable particles</u>, that it is approximately a semi-group for $t \geq 0$.

In order to obtain an appropriate criterion, one writes $U_p(t)$ in the form

$$U_p(t) = \frac{1}{2\pi i} \oint R_p(z) e^{-izt} dz, \quad (1)$$

where $R_p(z)$ is the reduction of the resolvent $R(z) = (z-H)^{-1}$ of H to the subspace $P\mathcal{H}$, and the integration is taken along a path enclosing $\sigma(H)$. The problem is now to get some insight into the analytical structure of $R_p(z)$.

In the case of <u>finite mass</u>, the point of departure is the following <u>decomposition of H</u>:

$$H = H_0 + V, \quad H_0 = PHP + \overline{P}H\overline{P}; \quad V = PH\overline{P} + \overline{P}HP. \quad (2)$$

This decomposition is canonical in the sense that H_0 represents the maximal self-adjoint part of H under which the subspace $P\mathcal{H}$ is stable. If dim P is finite, PH_0P can furthermore be diagonalized by

$$PH_0P = \sum_{i=1}^{n} m_i P_i ,$$

where the orthogonal projectors P_i project on the discrete eigenspaces of H_0 with eigenvalues m_i. Hence, the P_i represent stable states of H_0 with "free" masses m_i which, however, decay under the action of V.

Using the second resolvent formula, we obtain for the reduced resolvent[2)]

$$R_p(z) = \left(\sum_i (z-m_i)P_i - PV\overline{P}\, R_0(z)\overline{P}VP \right)^{-1}_{P\mathcal{H}}$$

where $R_0(z) = (z - H_0)^{-1}$ is the resolvent of H_0.

The analytic properties of $R_p(z)$ can be studied in this form. They have been discussed in the literature[2)] and may be summarized as follows: If the spectrum of $\overline{P}H_0\overline{P}$ is continuous on the interval $[0, \infty]$, say, and if $m_i > 0$, then, for sufficiently weak perturbation V, $\sigma(H) = [0, \infty]$ is continuous without discrete eigenvalues provided all $\overline{P}VP_i \neq 0$. $R_p(z)$ has a branch cut $\sigma(H)$ through which it may be analytically continued if dim $P\mathcal{H} = n$ is finite. It is regular in the first sheet and has n poles in the second sheet from above (converging to the n poles m_i of $R_0(z)$ whe the perturbation V goes to zero). By deformation of the path in (1) into the lower second sheet, one picks up these poles and obtains a decomposition of $U_p(t)$ into a sum of exponential residue terms plus corrections which are small with V.[5)]

Conclusion

$U_p(t)$ describes the decay of a subspace $P\mathcal{H}$ of unstable particles roughly if

a) V as defined by (2) is weak (in operator norm)

b) $\sigma(H_0) = [0, \infty] \cup \{m_i > 0\}$.

* * *

The system $\{H_0, H\}$ as obtained from the decomposition (2) forms a decay-scattering system, if n and $m_i > 0$ are finite. One obtains, then, a canonical scattering theory as in Reference 2. This can be termed as pure resonance scattering due to the fact that in (2) the perturbation V contains only that part of H which leads out of the subspace $P\mathcal{H}$. In the case of unstable particles, the solution of the inverse decay problem therefore provides the reconstruction of a canonical resonance scattering theory from the knowledge of the decay matrices $(f, U_P(t)g)$.

References

1. Generalization of the well-known theorem of Gel'fand-Raikov, cf. Appendix to Riesz and Sz.-Nagy, "Functional Analysis," Frederick Unger Publishing Co., New York.
2. L. P. Horwitz and J. P. Marchand, "The Decay-Scattering System," Rocky Mountain J. Math. 1, 225 (1971).
3. L. P. Horwitz, J. A. LaVita and J. P. Marchand, J. Math. Phys. 12 (1971).
4. P. Lax and R. D. Phillips, "Scattering Theory," Academic Press, New York.
5. Private communication from N. Bleistein and L. P. Horwitz.

ON OPERATOR EQUATIONS IN PARTICLE PHYSICS†

R. Rączka
Institute for Nuclear Research
Warsaw, Poland

1. Introduction

Many problems in Particle Physics and Quantum Field Theory reduces to the problem of finding solutions of certain operator equations. The most popular are operator equations for currents, e.g.,

$$[\, J_{\ell}(x), \;\; J_k(y)] \;=\; iC^{s}_{\ell k}\; \delta_3(\vec{x}-\vec{y})\; J_s(\vec{x}) \tag{1}$$

or dynamical equations for scalar quantum fields like

$$(\Box + m^2)\, \phi(x) \;=\; \lambda \phi^n(x)\,. \tag{2}$$

The commonly used method of solutions of operator equations like (1) is based on Weyl trick which he used for a classification of representations of Heisenberg commutation relations. It consists of the association with the commutation relations (1), the well-defined infinite-dimensional topological group G. Using then the well elaborated global representation theory of G and passing to the infinitesimal representations, one obtains a class of solutions of current commutation relations (1) (cf. Reference 1).

This method has, however, a severe limitation—namely, the operators entering into a given operator equation satisfy additionally certain supplementary conditions like continuity equation, Dashen-Gell-Mann angular condition, etc. Thus the problem of finding a solution extends to the problem of finding a representation of infinite-dimensional topological group, whose generators satisfy certain supplementary conditions. Such a problem is not solved in the Representation Theory even for simplest finite-dimensional groups.

Thus we propose to use another method which seems more effective in practice. It consists of introducing in the linear space

---

†Presented at the NATO Summer School in Mathematical Physics, Istanbul, 1970.

of operators a certain orthogonal operator basis; then expanding each operator in this basis and comparing the resulting coefficients at basic operators, we reduce a problem of a solution of operator equation to a more manageable problem of solution of differential (or functional) equation for c-number functions. This equation, as we show, can often be explicitly solved.

2. Expansion Theory of Operators

To illustrate the idea of the expansions of operators and a concept of orthogonal operator basis, we start first with the expansion theory of operators from the enveloping algebra.

Let G be a semi-simple Lie group and $g \to T_g$ a unitary representation of G in a Hilbert space H. Then we have

Theorem 1. There exists a nuclear space $\phi \subset H$ such that for every X in the enveloping algebra E of G holds

$$X\varphi = \int_G \mathrm{Tr}(X T_g^{-1})\, T_g \varphi\, dg. \tag{3}$$

(For the proof, cf. Reference 2.) The nuclear space ϕ in Theorem 1 can be taken as the inductive limit of spaces constructed out of Gårding space of differentiable vectors

$$\varphi(h) = \int_G h(g)\, T_g\, \varphi\, dg \qquad \varphi \in H, \quad h \in C_o^{\infty}(G) \tag{4}$$

or as a nuclear space defined by the system of semi-norms

$$P_n(\varphi) = \| (\bar{\Delta} + I)^n \varphi \| \tag{5}$$

where

$$\Delta = \sum_{i=1}^{\dim G} X_i^2$$

is the Nelson operator (cf. Reference 3 for the examples of nuclear spaces obtained by the second method).

The Formula (3) can also be written in basis-dependent form, which is convenient for introduction of operator basis. Let

$$H \to \hat{H} = \int_\Lambda \hat{H}(\lambda) d\rho(\lambda) \tag{6}$$

be an isometric isomorphism of H onto a direct integral $\hat{H}$ implied by the center of E. Then introducing an orthonormal basis $\{\hat{e}_p(\lambda)\}$ in each Hilbert space $\hat{H}(\lambda)$, we can rewrite Formula (3) in the form

$$X = \sum_{p,q} \int x_{pq}(\lambda) P^{\lambda}_{pq} d\lambda \tag{7}$$

where $x_{pq}(\lambda)$ is the matrix element of X in $\hat{H}(\lambda)$ and

$$P^{\lambda}_{pq} = \rho(\lambda) \int_G \overline{D}^{\lambda}_{pq}(g) T_g \, dg. \tag{8}$$

Here $\rho(\lambda)$ is a density of the spectral measure $d\rho(\lambda)$ and

$$D^{\lambda}_{pq}(g) = (T_g e_q(\lambda), e_p(\lambda)).$$

One readily verifies that operator distribution P^{λ}_{pq} have the following properties (cf. Reference 2, §4).

i) $$P^{\lambda}_{pq} P^{\lambda}_{p'q'} = \delta(\lambda-\lambda')\delta_{qp'} P^{\lambda}_{pq}, \tag{9}$$

ii) $$(P^{\lambda}_{pq})^* = P^{\lambda}_{qp} \tag{10}$$

iii) $$Tg\, P^{\lambda}_{pq} = D^{\lambda}_{rp}(g)\, P^{\lambda}_{rq}. \tag{11}$$

<u>Example 1</u>. Let $G = T^n$ be a translation group. Then

$$D^{\lambda}_{pq}(g) \sim e^{i\lambda g}$$

and

$$P^{\lambda} = \frac{1}{\sqrt{(2\pi)^n}} \int_{T^n} e^{-i\lambda g} T_g \, dg$$

The action of P^{λ} onto a function $\varphi(x) \in L^2(G)$ gives

$$(P^{\lambda}\varphi)(x) = \frac{1}{\sqrt{(2\pi)^n}} \int_{T^n} e^{-i\lambda g} \varphi(x+g)dg = \hat{\varphi}(\lambda)e^{i\lambda x}$$

where $\hat{\varphi}(\lambda)$ is the Fourier transform of the function φ. This shows that the operator P^{λ} is a generalized projection operator which satisfies the relations

$$P^{\lambda}P^{\lambda'} = \delta(\lambda-\lambda')P^{\lambda}, \qquad (P^{\lambda})^* = P^{\lambda}. \tag{12}$$

Hence the quantity P^{λ} should be treated as operator valued distribution which transform an element $\varphi \in \phi \subset H$ into an element of ϕ'. The smeared out operator

$$P(f) = \int_{\Lambda} f(\lambda)P^{\lambda}d\lambda$$

satisfies

$$||P(f)\varphi|| \leqq \max|f(\lambda)| \; ||\varphi||$$

Hence if $f \in C_0(\Lambda)$, then each $P(f)$ is bounded.

Using Eq. (7), we obtain the following decomposition of Casimir operators

$$C_i = \sum_p \int_{\Lambda} \hat{C}_1(\lambda)P^{\lambda}_{pp}d\lambda$$

where $\hat{C}_i(\lambda)$ is the spectrum of C_i. The non-invariant operators A_i from the maximal set of commuting operators in H have the following decomposition in terms of generalized projectors

$$A_i = \sum_p \int_{\Lambda} \hat{A}_i(P,\lambda)P^{\lambda}_{pp}d\lambda.$$

Comparing this with von Neumann spectral decomposition

$$X = \int_{\Lambda} \lambda \, dE(\lambda), \tag{13}$$

we see that the function

$$\sum_p P^{\lambda}_{pp} \, d\lambda$$

plays the role of the spectral function $dE(\lambda)$ for the maximal set of commuting operators in the carrier space H of the representation $g \to T_g$ of G.

The Formula (3) or (7) provides an expansion formula for every operator X from the enveloping algebra E of G. There arises the following interesting question: when the Formula (3) provides an expansion formula for an arbitrary self-adjoint operator in H. The satisfactory answer to this question is given by the von Neumann density theorem.

Theorem 2. Let $g \to T_g$ be a (*)-algebra generated by elements of the form

$$A = \sum_{i=1}^{n} c_i T_{gi}, \quad c_i \in C.$$

Let Q denote the commutant of $\mathcal{a}$ (i.e., the set of all bounded operators in H which commute with $\mathcal{a}$) and let Q' denote the bicommutant of $\mathcal{a}$. Then Q' is the strong closure of $\mathcal{a}$ (for the proof, cf. Reference 4).

The assertion of Theorem 2 states, in fact, that every element X in Q' has the expansion of the form

$$X = \int_G x(g)\, T_g \, dg. \tag{14}$$

Now if the representation T of G is irreducible in H, then Q reduces to the set $\{\lambda I\}$, $\lambda \in C$; thus Q'' coincides with the set of all bounded operators in H; therefore, a spectral function $E(\lambda)$ of an arbitrary self-adjoint operator in H may be expressed in the form (14). Consequently, by virtue of von Neumann spectral decomposition (13), every self-adjoint operator admits an expansion of the form (14).

As we shall see, Theorems 1 and 2 are sufficient for the applications in many physical problems. One could put, however, the following fundamental question: Let B(H) be a set of all bounded operators in a Hilbert space H; does there not exist a subalgebra $\mathcal{a}$ in H such that every element of B(H) can be expanded in terms of elements of $\mathcal{a}$? The positive answer to this question was given by Arveson who proved the following theorem.

Theorem 3. Let H be an infinite dimensional Hilbert space and let $\mathcal{a}$ be any subalgebra of B(H) with the following properties.

i) $\mathcal{A}$ has no proper invariant subspaces.

ii) $\mathcal{A}$ contains a countable decomposable self-adjoint maximal abelian subalgebra of B(H). Then $\mathcal{A}$ is strongly dense in B(H). (For the proof, cf. Reference 5.)

Example 2. Let $H = L^2(R)$ and let $\mathcal{A}$ be a subalgebra of B(H) spanned by the spectral resolutions of self-adjoint operators Q and P which satisfy the canonical commutation relations

$$[Q, P] = iI. \tag{15}$$

It is evident that the subalgebra $\mathcal{A}$ satisfies the condition i) and ii) of Theorem 3. Hence it provides an expansion for any operator $X \in B(H)$. It is interesting to see how in this case, looks like the expansion (3) for elements in E. In the present case we have

$$T_{g(\xi,\eta)} = e^{i(Q\xi+P\eta)}, \quad dg = d\xi\, d\eta \tag{16}$$

and

$$X = \int_{-\infty}^{\infty} \mathrm{Tr}\left(X T^{-1}_{g(\xi,\eta)}\right) e^{i(Q\xi+P\eta)}\, d\xi\, d\eta. \tag{17}$$

If $X = Q^n$, then

$$\mathrm{Tr}\left(Q^n e^{i(Q\xi+P\eta)}\right) = \left(i\frac{\partial}{\partial\xi}\right)^n \delta(\xi)\,\delta(\eta) \tag{18}$$

and if $X = Q^n P^m$, then

$$\mathrm{Tr}\left(Q^n P^m e^{i(Q\xi+P\eta)}\right) = e^{-i(\xi\eta/2)}\left(i\frac{\partial}{\partial\xi}\right)^n \delta(\xi)\,\frac{\partial}{\partial\eta^m}\,\delta(\eta). \tag{19}$$

One readily verifies that a Hilbert-Schmidt operator from B(H) has the square integrable transform $x(g) = \mathrm{Tr}(X T_g^{-1})$: On the other hand, Formulas (18) and (19) show that the transform x(g) of unbounded operators from the enveloping algebra represents distributions with point support.

3. Solutions of Dynamical Equations

Let $\phi(x)$ be a quantum scalar field satisfying the dynamical equation

$$(\Box + m^2)\,\phi(x) = \lambda\phi^n(x). \tag{20}$$

The right hand side of Eq. (20) is usually understood in the sense of normal product of operators. Suppose that a carrier Hilbert space H is a Fock space associated with an irreducible set of creation-annihilation operators

$$[a_i, a_k^*] = \delta_{ik} I.$$

We assumed for simplicity a box normalization; hence the set $\{a_i\}$ or $\{a_i^*\}$ represents a countable set of operators in H. Let α_i be a complex parameter and set

$$\alpha a^* - \bar{\alpha} a \equiv \sum_i \alpha_i a_i^* - \bar{\alpha}_i a_i \tag{21}$$

and

$$T_{g(\alpha)} = e^{\alpha a^* - \bar{\alpha} a}. \tag{22}$$

We have

$$T_{g(\alpha)} = T^*_{g(-\alpha)} = T^{-1}_{g(\alpha)} \tag{23}$$

and

$$\mathrm{Tr}(T_{g(\alpha)} T^*_{g(\alpha')}) = \delta(\alpha - \alpha'). \tag{24}$$

Since $T_{g(\alpha)}$ acts irreducibly in H, then by Theorem 2 every operator X in B(H) has expansion of the form (14). Thus in particular, a spectral resolution of smeared-out self-adjoint operator $\phi(f)$ admits an expansion of the form (14). The expansion (13) allows us, therefore, to write field operator $\phi(x)$ in the form*)

$$\phi(x) = \int \hat{\phi}(x, \alpha)\, T_{g(\alpha)}\, dg(\alpha). \tag{25}$$

Clearly this equality holds after smearing with a test function $f \in S(R^4)$.

Since the right hand side of Eq. (20) is understood in the sense of normal product and since

---

*The measure $dg(\alpha)$ is defined by means of Gaussian canonical measure. (See Appendix of Reference 8.)

$$T_{g(\alpha)} = e^{\alpha a^* - \bar{\alpha} a} = e^{-|\alpha|^2} e^{\alpha a^*} e^{-\bar{\alpha} a}, \tag{26}$$

it is more convenient to represent field $\phi(x)$ directly in the form of the normal product

$$\phi(x) = \int \hat{\phi}(x, \alpha) e^{\alpha a^*} e^{-\bar{\alpha} a} dg(\alpha). \tag{27}$$

Inserting this expression into Eq. (20), we obtain: (We illustrate the calculation for $n = 2$. The remaining cases $n = 3, 4, \ldots$ run similarly.)

$$(\Box + m^2) \int \hat{\phi}(x, \alpha) e^{\alpha a^*} e^{-\bar{\alpha} a} dg(\alpha) =$$

$$= \int \hat{\phi}(x, \alpha)\, \phi(\tilde{\alpha} - \alpha) e^{\tilde{\alpha} a^*} e^{-\tilde{\alpha} a} dg(\alpha) dg(\tilde{\alpha}). \tag{28}$$

Using now Eq. (26), multiplying both sides of Eq. (28) by $T^{-1}_{g(\alpha')}$, taking the trace and using orthogonality property (24), one obtains

$$(\Box + m^2)\, \hat{\phi}(x, \alpha') = \lambda \int \hat{\phi}(x, \alpha)\, \hat{\phi}(x, \alpha' - \alpha)\, dg(\alpha). \tag{29}$$

This is the integro-differential equation for c-number function $\hat{\phi}(x, \alpha')$. Taking the complex Fourier transform with respect to variable α', we convert the right hand side convolution into the product and obtain

$$(\Box + m^2)\, \hat{\hat{\phi}}(x, \beta) = \lambda \hat{\hat{\phi}}^2(x, \beta). \tag{30}$$

Thus the problem of a solution of dynamical equation for a quantum scalar field $\phi(x)$ is reduced to a problem of a solution of a classical non-linear relativistic equation for a c-number function $\hat{\hat{\phi}}(x, \beta)$.

Notice, however, that because of presence of parameters $\beta = (\beta_1, \beta_2, \ldots)$, the equation (30) is in fact a differential-functional equation. This shows a difference between ours and Segal's approach who tried to introduce a quantum field $\phi(x)$ as an operator acting in a tangent space of a manifold of solutions of classical non-linear equations (Reference 6).

$$(\Box + m^2)\, \varphi(x) = \lambda \varphi^n(x). \tag{31}$$

We now show that there exists a class of solutions of Eq. (30). Indeed, let

$$\hat{\hat{\phi}}(x,\beta) = \varphi(h(\beta)x) \qquad \text{with} \quad h(\beta) \neq 0. \tag{32}$$

Then setting $x'_\mu = h(\beta)x_\mu$, one obtains

$$\left(\Box_{x'} + \frac{m^2}{h^2}\right) \varphi(x') = \frac{\lambda}{h^2} \varphi^2(x'). \tag{33}$$

Putting $m'^2 = m^2/h^2$ and $\lambda' = \lambda/h^2$, we convert this equation into a non-linear classical relativistic equation with a "renormalized" mass m' and a coupling constant λ'. Thus we can now use the Segal theory (cf., e.g., Reference 7). He showed in particular that for any solution $\varphi_0(x')$ of a free equation at $t=-\infty$ there exists a solution $\varphi(x')$ of the non-linear equation (31) which at $t \to +\infty$ passes into another free solution $\varphi_1(x')$. This fundamental result shows the existence of an S-operator in a classical non-linear relativistic theory. By virtue of a representation (25) for $\phi(x)$, this also shows an existence of a certain $\hat{S}$-operator in a carrier Hilbert space H.

One can also consider a more general solution of the form

$$\hat{\hat{\phi}}(x,\beta) = h_1(\beta)\varphi(h_2(\beta)x).$$

The resulting equation for $\varphi(x')$ is in general a differential-functional equation, which may also be reduced to the equation (31). The solution in this case is given in terms of two arbitrary functions $h_1(\beta)$ and $h_2(\beta)$.

The detailed analysis of the properties of solutions of dynamical equation (20) are discussed in our work (Reference 8).

Discussion

We noticed that the first method of solutions of operator equations has a severe restriction due to difficulty of an accommodation of supplementary conditions into a framework of group Representation Theory.

On the other hand, in the second approach any supplementary condition imposes a new restriction on the c-number transform, which in general facilitates a finding of a solution of a resulting equation. Consequently it seems that a method of orthogonal expansions of operators is more effective in the solution of various operator equations, which we encounter in Particle Physics.

References

1. G. A. Goldin, Non-Relativistic Currents as Unitary Representations of Groups, Jour. Math. Phys. 12, 462 (1971).
2. R. Raczka, Operator Distributions in G.R.T. and Applications (preprint). Goteborg, Chalmers Technical University (1970).
3. B. Nagel, Lecture Notes on Group Representations (preprint). College de France (1970).
4. J. Dixmier, "C^*-Algebra and Their Representations" (1969).
5. W. B. Arveson, Duke Math. Jour. 34, 635 (1967).
6. I. Segal, Jour. Math. Phys. 5, 269 (1964).
7. I. Segal, International Congress of Mathematicians. Moscow (1966), p. 681.
8. R. Raczka, "Constructive Quantum Field Theory," preprint, University of Colorado (1972).

Acknowledgements

The author is grateful to Professor A. O. Barut for the kind hospitality extended to him during the Istanbul Summer School in Mathematical Physics. He expresses also his gratitude to Professors M. Flato and D. Sternheimer for valuable suggestions.

LECTURES IN NONLINEAR ALGEBRAIC TRANSFORMATIONS†

P. R. Stein‡
Los Alamos Scientific Laboratory
Los Alamos, New Mexico
and
S. Ulam
University of Colorado
Boulder, Colorado

Lectures 1 and 2

Our subject will be the study of certain special transformations, defined in the euclidean space, and the actual results will pertain mostly to the case of one, two and three dimensions. During the first few talks, I will present some elementary facts concerning transformations, in general, and Paul Stein will continue with more specific problems. In the main our studies concern properties of iterates of our transformations; that is to say, the behavior of sequences of points: $P, T(P)$, $T^2(P) = T(T(P))$, ..., $T^n(P)$, ...

Historically, the first mathematical studies of transformations arose in problems of geometry. Real valued functions of a real variable can, of course, be considered as transformations of the line into itself but most of analysis is not primarily concerned with this aspect of a function. It was in the problems of classical physics in the Nineteenth Century that the idea of transformations in the n-dimensional space proved very useful. If one considers a dynamical system consisting of n mass points, we may represent, for purposes of classical mechanics, the whole system as a single point in 6n dimensions, using for each point three coordinates to determine its "initial" position and three numbers for describing its initial momentum. Given the equations of motion (the Newtonian or Hamiltonian equations), the system will evolve in a definite way through time; the point of the 6n-dimensional space given initially will describe a trajectory, i.e., for every value of t, the time parameter, we will get new points. From the set of all possible initial conditions, we will obtain a class of trajectories. We will have a one-parameter

---

†Presented at the NATO Summer School in Mathematical Physics, Istanbul, 1970.

‡Work performed under the auspices of the U.S. Atomic Energy Commission.

family—a one-parameter group of transformations. If we denote the initial point by P we will have a family of transformations $T_t(P)$. $T_{r+s}(P) = T_r(T_s(P))$. The transformation T is well defined through the equations of dynamics. It is continuous and for conservative systems it is volume preserving (the theorem of Liouville). If we adopt a unit of time (a "second") we have a single transformation T and in integer multiples of our unit the transformations are simply iterates of the given transformation: $T_n(P) = T^n(P) = T(T^{n-1}(P))$. The positions of the given mechanical system, with all its possible initial conditions, forms a subset of the n-dimensional space—the phase space. There might be "integrals of motions"; for example, the constancy of total energy and other invariants which imply that the original set is divided into subsets, each of which is separately transformed into itself under the given transformation T. Some of the most fundamental problems of statistical mechanics concern the properties of the sequence of iterated points; for example, the question of whether such sequences are dense or even uniformly dense on the invariant manifolds on which they travel. We consider these discrete sequences rather than the continuous flow mainly for reasons of conciseness and simplicity of formulations—the properties of the continuous flow being derivable from the properties of the sequence of iterates. Apart from some number of singular points, the transformation is continuous, one to one, and volume preserving.

In the pages that follow we will describe, in an elementary way, some ideas and results in a general and systematic study of transformations. The "morphology" of a transformation—to begin with on a topological level—involves a few properties which are elementarily stated: Given a transformation T, one is interested in the existence of its fixed points; that is to say, points which do not move under the transformation: P such that $T(P) = P$. The next thing to ascertain is the existence, if any, of points which have a finite period. $T^k(P) = P$, where k is any integer. But one might suspect that if these exist, then, for the continuous flow, there exist closed orbits. The first curiosity then perhaps involves both the ascertaining of invariant points or invariant finite sets; then the existence of invariant subsets of more complicated nature (i.e., "curves" or "manifolds" of higher dimensions which transform into themselves under T). Such invariant sets can always be obtained, of course, by considering together with a point T all its iterates (for both positive and negative time), i.e., the points $T^i(P)$ for $i = \pm 2, \ldots \pm n \ldots$. This set is obviously invariant and, considering its closure, that is to say it, plus all accumulation points, one will obtain an invariant closed set. Given the transformation T, one might consider all transformations which are similar to it—certainly as far as the above properties are concerned—these transformations include all of the form $H(T(H^{-1}(P))$ where H is an arbitrary one to one continuous transformation and H^{-1} denotes its inverse (the conjugate transformations). Given a T, it is sometimes possible to find a "change of coordinates" H such that HTH^{-1} is analytically simpler or otherwise more amenable to analysis.

We all know the history of the ergodic problem: Boltzmann, one of the founders of statistical mechanics, stated the postulate that, in general, the trajectory of a single point, i.e., the set of all points $T_t(P)$ will go through <u>all</u> the available points of the phase space of the system. More precisely, he believed that if there are k integrals of the motion (including the energy integral), the 6nk-dimensional subspace will be <u>filled</u> by a trajectory of a single point. As is well known, this is topologically impossible. The trajectory forms a homeomorphic image of the real line and cannot fill a space of two or more dimensions. (The so-called Peano mappings are continuous but they are not one-to-one, not homeomorphisms and since the equations of dynamics insure a uniqueness of solutions, that is the one-to-one character of the mapping, this is impossible because of, for example, Brouser's Theorem on invariance of dimension. This was noticed by Rosenthal and others early in the century but I believe that Boltzmann, himself, expected that this covering of <u>every</u> point is impossible and stated that "quasi-ergodic" hypothesis: the trajectory of single points is, in general, <u>dense</u> in the 6 n-k dimensional phase space. This, of course, is entirely possible topologically. However, this property is, by itself, not sufficient to establish what one needs in the foundations of statistical mechanics, namely, the equality of the "time average" to the "space averages." The sequence of iterated points has to be not only dense but <u>uniformly</u> dense. This means that if we norm the volume, i.e., the measure of our whole phase space to 1 and if A is a region in phase space, then the sojourn time of a point in the region is equal to the volume or measure of the set A. In a concise notation, one wants to have:

$$\lim_{N=\infty} \frac{1}{N} \sum_{i=1}^{N} \chi_A(T^i(P)) = m(A),$$

where $\chi_A(P)$ is the characteristic function of the set A:

$$t_A(P) = 1 \quad \text{if} \quad P = A$$
$$= 0 \quad \text{if} \quad P \neq A.$$

This is to hold for almost every point P and all measurable sets A. Everything above is, of course, relative to the given "irreducible manifold" of the phase space, taking into account the integrals of the problem, i.e., in the 6 n-k dimensional subset on which we take the initial point. To continue with the rather well known story: The first theorems on measure-preserving flows were established in the early 1930's by von Neumann and G. D. Birkhoff. Birkhoff's strong ergodic theorem asserts that the time average limit exists for almost every point P (almost every in the sense of Lebesgue measure defined on our subspace of the phase space) and in order to prove that this limit is indeed equal to the space average, that is to say, to m(A), it is necessary and sufficient that the transformation be metrically transitive. This assumption states that there is no set of a positive measure, less than the measure of the whole space, which is invariant under the transformation T. Examples of such transformations were known: The simplest is on the circumference of a circle; the transformation being a rotation through an irrational angle. In higher dimensions one can consider a torus and a rotation through a set of angles, algebraically independent. The theorems of Kronecker and Weyl assert the uniform density of iterates of such rotations. Some other, sporadic, examples were known, e.g., a flow on geodesic lines on surfaces of constant negative curvature. A general theorem was proved in 1941 by Oxtoby and myself. This asserts that "most" continuous measure-preserving transformations on manifolds are metrically transitive. The expression "most" means that in the space of all such continuous transformations those which are not metrically transitive form a "thin" set—a set of first category, i.e., a union of countably many nowhere dense sets. In particular, the theorem establishes the existence of such metrically transitive transformations on every manifold and much more: arbitrarily near a given transformation there exist metrically transitive ones.

One can go further than ergodicity. As Professor Kruskal mentioned in one of his talks, one may be interested in stronger properties of the flow. One would like to know whether the "general" transformation has characteristics of mixing. The simplest example of an ergodic transformation, the rotation of a circle through an irrational angle, is ergodic but it is very regular: if one starts with two given points their images will always, for every iterate, be the same distance apart. The flow is exceedingly laminar. One would like to define mixing in a precise way. One way to do it is as follows: Suppose we start with any little sphere S_0 in our phase space and consider its image under some high iterate of the transformation T. This set will have the same volume as the volume of the original sphere but if the transformation is mixing, it should look like a very elongated and thin set which lies rather close to any point of the phase space. In fact, if we take any other sphere S then for a sufficiently high n, the image $T^n(S_O)$ should intersect this sphere S and the volume of this intersection should be approximately the same relative to the sphere S, as the volume of S_O was relative to the whole space. In the limit we should have

$$\lim_{n=\infty} m(T^n(S_O) \cap S) = m(S_O)\cdot m(S).$$

We expect, even in a steady but irregular flow of an incompressible liquid, to have this happen: A drop of "ink" put in the fluid which is in motion should, after a sufficiently long time, become a long and thin filament penetrating near to any other point of the fluid.

Here then is another property of the transformations which one should study, and we shall define and discuss briefly a few more.

All this very well known background is presented here only because, while the one-to-one transformations have been studied extensively and the literature is ample, much less is known about the transformations which, still continuous, are not necessarily one-to-one. This will be the case of the transformations which Paul Stein and I have considered. All these properties have to be studied for the many—one mappings. To illustrate the great difference in behavior between homeomorphisms and the not necessarily one-to-one continuous mappings, I shall start with an example in the simplest case, in one-space dimension, on the real line. Consider the function whose graph is a parabola $T(X) = 4X(1-X)$. It transforms the interval (0,1) into itself but not in a one-to-one manner. If we now take the iterates starting with an arbitrary point P on our interval, we obtain a sequence of points which, in general, will jump about irregularly throughout the interval. This is true of almost every point—obviously there are exceptions: e.g., the fixed point of the transformation, that is to say, the point where the diagonal cuts the graph, stays fixed under all the powers. The same is true of points which are fixed points of the powers of the transformation. There is a countably infinite number of such. But, some fifteen years ago or so, von Neumann and I observed that one can prove that for almost every point on the interval (in the sense of Lebesgue measure) the sequence of iterates will be dense in the interval. More than that, one can prove that the ergodic limit exists, i.e., given a sub-interval, the time of sojourn of almost every point in this sub-interval exists and is in fact easily computable. This can be easily seen if one observes that our transformation or function T is conjugate to a function represented by a broken line $S(X) = 2X$, for X between 0 and $\frac{1}{2}$, and $= 2 - 2X$ for X between $\frac{1}{2}$ and 1. In other words, there exists a one-to-one continuous transformation H (which, by the way, can be written down analytically ($H(X) = 2/\pi \sin^{-1}(\sqrt{X})$) so that $T = HSH^{-1}$. The study of iterates of T can thus be reduced to the study of iterates of S. Now, for S, the behavior of the iterates of S follows from the consideration of the "law of large numbers"—in the simple case of Bernouilli—the "shift" transformation on the binary development of a real number whose asymptotic behavior is therefore well known—their distribution is uniform on the interval and by conjugation with H we can now get the distribution of the iterates of T. Clearly, no one-to-one transformation of the real variable which is continuous can be ergodic in this sense—the sequence of iterates will be monotone—whether converging to a fixed point or, in the case of an infinite line, going perhaps to infinity.

The fact that we were able to prove something about the iterates of our special T is perhaps accidental. In general, given a function, even of a simple nature, say a cubic, it is not at all obvious what the iterates will do—in which cases one will have density of such sequences, perhaps on some sub-intervals, etc. One is tempted to conjecture that for functions which are sufficiently regular from the point of view of algebra or analysis, at least the ergodic limits will exist; what their values will be for given sub-intervals seems not at all easy to determine. This is, mind you, in one dimension; for two or more dimensions, even for a simple quadratic or cubic transformation, it appears very difficult to predict the properties of their iterations. As you will see on the examples which will be given by Paul Stein, very diverse phenomena can occur. There is a whole zoo of different possibilities for the same transformation, depending upon which points you start with—some will be periodic; some will converge; some others seem to generate peculiar-looking sets. I would like to mention here that the behavior of the iterates of certain quadratic transformations in one variable was found to schematize the qualitative behavior of some hydrodynamical problems. Professor Lorentz from M.I.T. has shown how some functionals of the long-range behavior of the circulation of the atmosphere of our globe—long-range meaning thousands or millions of years—can be mirrored, at least as the behavior of certain functionals of the motion is concerned, by the behavior of iterates of certain quadratic functions transforming the line into itself. This particular work stylizes severely, of course, the equations which are formulated for certain average properties of the motion.

Given a system of functional equations—differential or integral equations, for example, one can adopt a few which helps to geometrize some problems of analysis. It was Birkhoff and Kellogg who were first to consider functional operators as transformations of a space of functions into itself. If one considers an equation U(f) = 0 where U is some functional operator and f denotes a function defined on a set of points in space, we may take the operator V(f) = U(f) + f. The existence of the solution for U(f) = 0 is equivalent to the existence of a fixed point for the transformation V(f): V(f) = f. V(f) is a transformation of a function space into itself and, under proper conditions, one can use the topological theorems on the existence of fixed points in infinitely dimensional spaces to obtain the existence of solutions for a wide class of equations of analysis. The original paper of Birkhoff and Kellog initiated a whole class of investigations which produced new topological methods in mathematical analysis. The work of Schauder and Léray is very well known here and several of our speakers have already used this approach. The fact that, in some cases at least, one can use this geometrical approach is due to topological theorems on the existence of at least one fixed point for a transformation of the full n-dimensional cube into itself. This theorem of Brouwer has been generalized to infinitely many dimensional spaces, provided the transformation, in addition to being continuous, satisfies certain other properties (e.g., for the image of the sphere in a Banach space should be a compact subset of it—or if the transformation differs from the identity by such, etc., etc.). Brouwer's original theorem, by the way, can be proved very easily from a purely combinatorial lemma, established by Sperner, on a finite subdivision of a simplex into sub-simplices. This reasoning is "finitistic" which I am stressing here in order to underline the importance of "finite combinatorics" in investigations of analysis. Experimentation on computing machines can give further insight into the behavior of such finite arrangements! What we are interested in here in our sessions goes, of course, far beyond the mere existence of solutions; it concerns the behavior and properties of functional equations.

There is another, quite general, fixed point theorem which, in some cases, can be used to prove the existence of solutions of equations. In contrast to Brouwer's theorem, it is not restricted to a certain rather special class of topological spaces which have no "holes" (like the cube), but concerns a more special class of continuous transformations with a strong property of "contraction."

Suppose E is a general metric space with the property of compactness (for example, E is a closed bounded set in Euclidean space; the compact Hilbert cube is another example). Suppose T is a continuous transformation of this space into itself with the property of shrinking:

$$\rho(T(P_1), T(P_2)) < \lambda \cdot \rho(P_1, P_2) \qquad \lambda < \lambda_0 < 1$$

(ρ denotes the distance between points of the space E). It is easy to show that T must have a fixed point—in fact, this fixed point is unique and the iterated sequences converge to the fixed point. Indeed, consider for any P the sequence

$$P,\ T(P),\ T^2(P) \ldots T^N(P) \ldots$$

the distance between successive points must shrink to 0 because

$$\rho(T(P),\ T(Q)) = \rho(P,\ Q)$$

which is impossible if T is shrinking. From compactness, it follows that there is a limiting point. For this limiting point PT(P) = P, from continuity of T. Suppose now there would exist another fixed point Q. This leads to contradiction.

The above is the weakest form of the theorem for fixed points on shrinking transformations. One can weaken somewhat the assumptions defining shrinking and, in fact, with this assumption even compactness is not necessary. Completeness of space would guarantee the convergence. This theorem has wide applications. In many cases one can verify that the transformation satisfies such shrinking properties. To give just one example, the existence of a limiting distribution for finite Markoff chains follows easily from it; some functional operators happen to satisfy the shrinking condition, etc.

I would like now to introduce the motion of a quasi-fixed point for a transformation relative to a system of given functions. The general formulation is as follows: Suppose E is a space, T a transformation of E into itself and $f_1(P)$, $f_2(P) \ldots f_N(P)$ are given continuous functions defined on E. It may happen that there exists a point P^* such that for all i forms 1 to N we have $f_i(P^*)$. Obviously, if the space E has the fixed point property, then for any system of functions this fixed point satisfies the above condition.

There exists spaces which do not possess the fixed point property, i.e., the whole Euclidean space but given the functions $f_1 \ldots$ up to f_N there might exist a quasi-fixed point P^* (depending on T) for every continuous T. It might happen that for some functional operators we are unable to find a function which is its fixed point but our solution does not exist for the corresponding equation. Nevertheless, given any finite number of "properties" of a function and given the transformation, there might exist a function such that it has the same N properties as its transform under T. This would be, then, in a sense, a very weak solution—relative to the N given properties, i.e., N functionals of the solution. One can prove some special results on the existence of such quasi-fixed points. As an example, consider the following: Let T be a continuous transformation of the whole plane into itself. Let the two functions f_1, f_2 be the absolute value of the x-coordinate and the y-coordinate respectively of a point P in the plane. One can prove then that there must exist a quasi-fixed point; that is to say, a point P such that it and its transform under T have the same absolute values of both coordinates. (In other words there must be a point which goes into one of the four corners of the rectangle of which it is a corner.) In the space of higher dimension, one could establish an analogous quasi-fixed point for more than two functions of this sort. It seems to me worthwhile to investigate this generalization of a notion of a fixed point for functional transformations. We saw how the idea of Lax of weak solutions has had so many successful applications, many of these we have seen established and presented by Professor Germain and we heard in Professor Kruskal's and Zabuski's lectures of other interesting cases. I wanted to draw attention to this purely topological generalization of the notion of a fixed point because quite often, in problems of hydrodynamics, for example, one is not necessarily interested in obtaining knowledge of the position of every point of the fluid, i.e., the entire function which satisfies some functional equations but only perhaps in a few—a small number of properties of such a function. As an example of the such N given functionals, we may think of the first N moments of a distribution or of the first N coefficients in its development, in a Fourier series, etc. The quasi-solutions would perhaps exist for all N, but these need not converge to a strict solution in a classical sense!

We have discussed the problem of fixed points of a transformation or fixed points for a power of a transformation, i.e., the periodic points. Given the existence of a fixed point, one wants to know its nature. It might happen that, given a sufficiently small neighborhood of the fixed point, starting from any point in this neighborhood the sequence of iterated images of it will converge to this fixed point. In this case, we call the fixed point attractive. The opposite can take place. Starting from any point, in the fixed point itself, the iterate of the transformation may lead away from this fixed point. We would call it totally repellent. In general, neither may take place. There might be angles, i.e., regions of space such that in some of them there will be convergence to the fixed point which lies on the common boundary of these regions. Starting in other sectors, we may get points going away from it. From some still other regions there may be a "labile" behavior. The iterated points will form a bounded set without any convergence to the fixed point. In problems of numerical analysis—in attempts to find out the solution of a functional equation by a process of iteration, it is important to know which is the case. In the elementary procedures, solving a differential equation by a process of iteration is successful in the case of the solution being "attractive." The algebraic transformations which will be discussed by Paul Stein show that for a given transformation, for different fixed points, all alternatives can take place and in some cases it is not at all easy to establish either of them. In the next talk I will discuss again, on an elementary and general topological level, properties of stability of transformations and of their invariant points or manifolds. Briefly speaking, these questions are of the following sort: Given a transformation T which has a fixed point and given another transformation S, sufficiently near it, can we assert that S will have a fixed point near the one possessed by T? Suppose T is a continuous function $y = T(x)$ transforming the interval $(0,1)$ into itself such that its graph crosses the diagonal (of the square of the variables (x,y)). It is obvious then that any continuous S will have to have a fixed point nearby. Its graph has to also cross the diagonal. Should the graph of T merely touch, without crossing, the diagonal, then one can find, arbitrarily near T, transformations S which do not have in the neighborhood of the original fixed point of T any points in common with the diagonal.

Lecture 3

1. Introduction

The primary purpose of this and the following two lectures will be to illustrate some of the points discussed by Professor Ulam in the first two lectures of this course. We shall do this by exhibiting some properties of a few, very restricted, classes of transformations in one, two and three dimensions. Despite the formal simplicity of our special transformations, we have not yet succeeded in developing a theory which would enable us to treat their iterative behavior in a rigorous manner. For this reason most of our work has been of an experimental nature, our principal research tool being the electronic computer. In the course of this work we have found it necessary to use computers in a somewhat unorthodox way, relying heavily on visual display rather than on numerical output; much of our data in fact consists of photographs taken directly from an oscilloscope screen. A few of these will be reproduced here; for a more complete presentation we refer the reader to our article.[1)]

The plan of these lectures will be as follows. The first two will be devoted exclusively to polynomial transformations, i.e., transformations of the form:

$$x_i' = f_i(x_i, \dots x_n) \qquad i = 1, 2, \dots n, \qquad (1.1)$$

where the f_i are polynomials in the indicated variables; here x_i' denotes the value of the variable at the next iterative step (we shall never use a prime to mean the derivative). In other words, Eq. (1.1) is a system of first-order, non-linear difference equations. In the last lecture we shall discuss three peripherally related topics: (a) "broken-linear" transformations in two dimensions; (b) some aspects of one-dimensional iteration; and (c) a particular family of one-dimensional transformations on the unit circle for which an exact treatment is possible.

2. Polynomial Transformations in General

Consider the following quantity:

$$F = \left(\sum_{i=1}^{k} x_i\right)^m = \sum_{j=1}^{N_k^m} M_j \qquad (2.1)$$

where

$$N_k^m = \binom{m+k-1}{k-1} . \qquad (2.2)$$

Here the M_j are monomials of degree m in the variables $x_i, \dots x_k$, each multiplied by its appropriate multinomial coefficient. Let us now introduce a particular formal decomposition of F into disjoint sets:

$$\sum_{i=1}^{k} f_i = F, \tag{2.3}$$

with

$$f_i = \sum_{j=1}^{N_k^m} d_{ij} M_j, \qquad d_{ij} = 0 \text{ or } 1. \tag{2.4}$$

The particular decomposition is, of course, specified by assigning the values 1 or 0 to the d_{ij} in some given manner, subject to the requirement [Eq. (2.3)]; this means we must have

$$\sum_{i=1}^{k} d_{ij} = 1 \quad (\text{all } j). \tag{2.5}$$

Let us, in addition, introduce a restriction on the variables x_i, namely:

$$\sum_{i=1}^{k} x_i = 1, \qquad x_i \geqq 0 \ (\text{all } i). \tag{2.6}$$

We now construct the following transformation:

$$x_i' = f_i(x_1, x_2, \dots x_k), \quad i = 1, 2, \dots k. \tag{2.7}$$

This is evidently a transformation of the positive portion of the hyperplane $\Sigma x_i = 1$ into itself (note that by construction $\Sigma x_i' = \Sigma x_i$). Because the transformation is bounded, it is a particularly convenient object to study numerically.

It is clear that for given m and k there are only a finite number of possible decompositions [Eq. (2.3)]. Many of these, however, lead to transformations [Eq. (2.7)] which differ from each other only by permutations of the subscripts on the variables and must therefore be considered strictly identical. The number of _independent_ systems for given m and k is not known in general; we have calculated this number—call it T_k^m—by direct enumeration for just four cases: $m=2$, $k=2, 3, 4$ and $m=k=3$. It is easy to see, however, that T_k^m rapidly becomes astronomically large. A _lower limit_ T_k^{*m} to the number T_k^m is given by the number of ways of putting

$$N_k^m = \binom{m+k-1}{k-1}$$

objects (i.e., the monomials M_j) into k identical boxes (i.e., the k lines of the schema [Eq. (2.7)]), no box being empty:

$$T_k^{*m} = S_k^{N_k^m} \tag{2.8}$$

where the S_q^p are the well-known Stirling numbers of the second kind. (This counting underestimates T_k^m by assuming, in effect, that each transformation has k! non-identical copies, i.e., that no transformation is invariant under any permutation except the identity.) The trend is illustrated in the following table:

Table I

m	k	N_k^m	T_k^{*m}	T_k^m
2	3	6	90	97
3	3	10	9330	9370
2	4	10	34105	34337
4	3	15	2375101	---
2	5	15	210766920	---
3	4	20	45232115901	---

We can immediately generalize the above while still remaining within the class of positive, homogeneous, bounded polynomial transformations. We need only relax the condition that all the d_{ij} be 0 or 1; instead, we require merely that

$$\begin{aligned} d_{ij} &\geqq 0 \qquad (\text{all } i,j) \\ \sum_{i=1}^{k} d_{ij} &= 1 \qquad (\text{all } j). \end{aligned} \tag{2.9}$$

Then, for each m and k, we have a $(k-1)N_k^m$-parameter family of transformations. A few experiments on these more general systems will be reported during the course of these lectures.

A further interesting generalization is possible. Let us introduce a new parameter Δt, with $0 < \Delta t \leqq 1$. We then replace the system [Eq. (2.7)] by

$$x_i' = (1 - \Delta t)x_i + \Delta t\, f_i(x_1, x_2, \cdots x_k) \qquad i = 1, 2, \cdots k; \tag{2.10}$$

(here the f_i may be defined in general by Eq. (2.9), the restriction $d_{ij} = 0, 1$ not being necessary). In effect, we have added a linear term to the right hand side and have decreased the non-linear term correspondingly. Note that if $\Delta t = 1$, we recover the original system [Eq. (2.7)]; we exclude $\Delta t = 0$, since in that case the system reduces to the identity transformation $x_i' = x_i$ (all i). The "modified" system [Eq. (2.10)] will be observed to have the same fixed points as the original system. As we shall show in the next lecture, this circumstance allows us to find these fixed points in a particularly convenient manner.

3. The Iterative Properties of Polynomial Transformations

As Professor Ulam has remarked, the object of our investigation is to study and classify these transformations with regard to their iterative properties. More specifically, given some transformation $x_i' = f_i(x_1, x_2, \cdots x_k)$, $(i = 1, 2, \cdots k)$, we choose an initial point

$$\left(x_1^{(0)}, x_2^{(0)}, \cdots x_k^{(0)}\right)$$

on the hyperplane and generate a sequence of iterated images

$$\left(x_1^{(n)}, \cdots x_k^{(n)}\right), \quad n = 1, 2, \cdots.$$

What can we say about this sequence as $n \to \infty$? The mathematical literature contains very little on this problem except in the special case in which the transformation is 1 - 1 and has the "shrinking" property (cf. Lecture 1 for a definition of this term). The reason for this apparent neglect is, of course, that the problem is extremely difficult; techniques which work beautifully in the linear case are complete ineffective here. To be sure, in a suitable neighborhood of an attractive fixed point the analysis frequently reduces to a linear problem so that quite a lot may be said. We, however, are more interested in global behavior, particularly in those cases for which the "limit set" does not consist of a single point. Lacking sufficiently powerful techniques, we must at present be content with an experimental approach and with results which are mainly of a taxonomic nature.

Let us introduce the subject by considering the simplest non-trivial case, $m = k = 2$. As is easily seen, there are just three possible systems, viz.:

$$x_1' = x_1^2 \qquad\qquad (3.1)$$
$$x_2' = x_2^2 + 2x_1x_2$$

$$x_1' = x_2^2 \qquad\qquad (3.2)$$
$$x_2' = x_1^2 + 2x_1x_2$$

$$x_1' = 2x_1x_2 \qquad\qquad (3.3)$$
$$x_2' = x_1^2 + x_2^2.$$

Since $x_1 + x_2 = 1 = x_1' + x_2'$, these reduce to the equivalent one-dimensional difference equations:

$$x_1' = x_1^2 \quad \text{with fixed points} \quad x_1 = 1, x_1 = 0 \qquad (3.4)$$

$$x_1' = (1 - x_1)^2 \quad \text{with fixed point} \quad x_1 = \tfrac{1}{2}(3 - \sqrt{5}) \qquad (3.5)$$

$$x_1' = 2x_1(1 - x_1) \quad \text{with fixed points} \quad x_1 = 0, x_1 = \tfrac{1}{2}. \qquad (3.6)$$

What is the nature of these fixed points? To answer this question, it is perhaps worthwhile to draw the corresponding iteration diagrams:

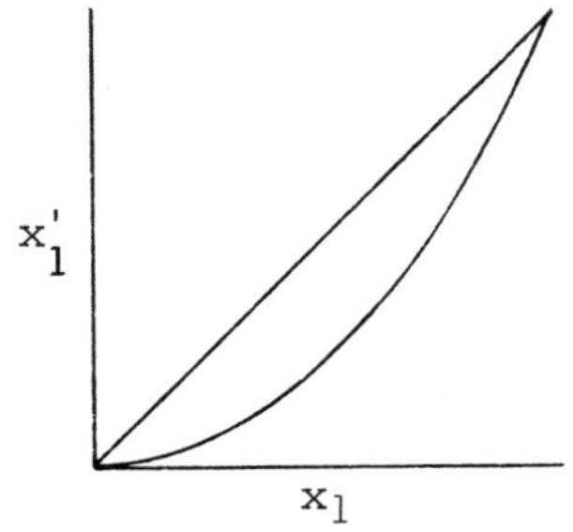

Figure 1-a

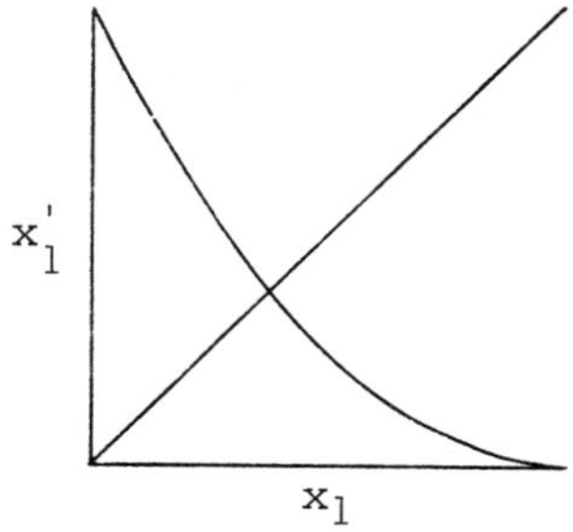

Figure 1-b

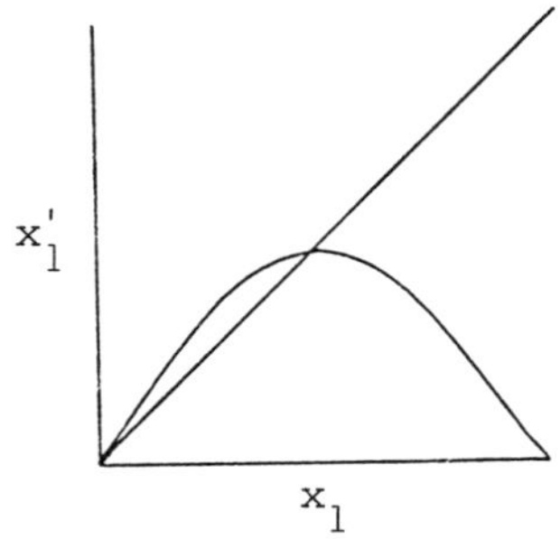

Figure 1-c

Iteration is equivalent to the following geometric construction: (1) choose a point on the x_1 axis and find the corresponding point x_1' on the curve; (2) draw a line through this point parallel to the x_1 axis and find the intersection with the diagonal $x_1' = x_1$; (3) draw a line parallel to the x_1' axis and find the new intersection with the curve, etc.

A fixed point x_f is called "attractive" if there exists a convex neighborhood $N(x_f)$ of x_f such that, for any $x \in N(x_f)$, the sequence of iterated images of x converges to x_f. If there is no such neighborhood, the fixed point is called "repellent" (this does not exclude the possibility that there exist certain "cones of convergence"—cf. Lecture 1). For one-dimensional transformations it is well-known (and, in fact, intuitively obvious) that the necessary and sufficient condition for a fixed point x_f to be attractive is that

$$\left| \frac{dx'}{dx} \right|_{x=x_f} < 1. \tag{3.7}$$

Applying this criterion to the simple cases above we see that the attractive fixed points are as follows:

for Eq. (3.4): $x_1 = 0$

for Eq. (3.5): none

for Eq. (3.6): $x_1 = \frac{1}{2}$.

A little more work shows that for the cases [Eq. (3.4)] and [Eq. (3.6)], these attractive fixed points are the unique limit sets for these transformations. On the other hand, for Eq. (3.5) the limit set consists of two points, $x_1 = 1$ and $x_1 = 0$. Such a limit set is frequently called a "period of order two" or a "2-period." To show that this is really a limit set, i.e., that the 2-period is attractive, we consider the "square" of the transformation:

$$x_1'' = [1 - x_1']^2 = [1 - (1 - x_1)^2]^2. \tag{3.8}$$

If the 2-period is to be attractive, then each of the points $x_1 = 1$ and $x_1 = 0$ must be an attractive fixed point of Eq. (3.8). But, by the chain rule,

$$\frac{dx_1''}{dx_1} = \frac{dx_1''}{dx_1'} \cdot \frac{dx_1'}{dx} = 4x_1(1 - x_1)(2 - x_1), \tag{3.9}$$

and this, in fact, is zero at both the points in question.

The simplicity of these examples is, of course, rather misleading. In the following section we shall discuss the next simplest case, $m = 2$, $k = 3$, which already presents certain difficulties. Before going on to this, however, we shall give the general criterion for an attractive fixed point, applicable to any finite-dimensional transformation which has continuous first partial derivatives in some convex neighborhood of the fixed point. The general form of this criterion is due to Ostrowski.[2)]

We consider a system of the form

$$x_i' = g_i(x_1, \cdots x_{k-1}) \qquad i = 1, 2, \cdots k-1. \tag{3.10}$$

(Here we have eliminated the redundant variable via the relation $\Sigma x_i = 1$.) We then form the Jacobian matrix J_f, evaluated at the fixed point $x_f \equiv (x_{1_f}, \cdots x_{k-1_f})$, and find its eigenvalues:

$$|J_f - \lambda I| = 0. \tag{3.11}$$

Let λ_A (Ostrowski's notation) be the eigenvalue which is largest in absolute value. According to whether $\lambda_A < 1$ or $\lambda_A > 1$, $x = x_f$ is the attractive or repellent. The criterion says nothing about the case $\lambda_A = 1$, nor does it specify the underlying neighborhood; obviously, a sequence will not converge to a given attractive fixed point if the initial point is chosen to lie outside the region of convergence. We shall see some examples of this in the next section.

4. Quadratic Transformations in Three Variables

As we remarked above, the number T_k^m of independent systems with $m=2$, $k=3$ ($d_{ij}=1$ or 0) is 97; all of these have been investigated on an electronic computer and the results published as a Los Alamos report.[3] The method of investigation needs little comment. In each case several initial points were chosen in a random manner and the transformation in question was numerically iterated until "convergence" in some sense was observed. As it turned out, two cases exhibited what appeared to be anomalous convergence behavior, while several cases had more than one limit set.

Before discussing individual cases, we must introduce a certain convention we have adopted in talking about 3-variable transformations. Because of the restriction $x_1+x_2+x_3=1$, these are really transformations in two dimensions. For much of the discussion it proves convenient to eliminate the redundant variable by a linear transformation:

$$\begin{aligned} x_1 &= s - \alpha \\ x_2 &= 2\alpha \qquad \text{or} \qquad s = \frac{1+x_1-x_3}{2} \\ x_3 &= 1 - s - \alpha \qquad \alpha = \frac{x_2}{2}\,. \end{aligned} \tag{4.1}$$

The basic region then becomes an isosceles triangle with base 1 ($0 \leqq s \leqq 1$) and height $\frac{1}{2}$ ($0 \leqq \alpha \leqq \frac{1}{2}$); we shall call this the "reference triangle." The general 3-variable transformation ($k=3$, m arbitrary) then takes the form:

$$\begin{aligned} s' &= F(s, \alpha) \\ \alpha' &= G(s, \alpha) \end{aligned} \tag{4.2}$$

where the F, G are still polynomials but no longer positive or homogeneous.

The fixed points of Eq. (4.2) are given by the solutions of the system

$$\begin{aligned} s &= F(s, \alpha) \equiv s_f \\ \alpha &= G(s, \alpha) \equiv \alpha_f . \end{aligned} \tag{4.3}$$

In this case we can solve explicitly for the eigenvalues of the Jacobian matrix:

$$\begin{pmatrix} \frac{\partial s'}{\partial s} & \frac{\partial s'}{\partial \alpha} \\ \frac{\partial \alpha'}{\partial s} & \frac{\partial \alpha'}{\partial \alpha} \end{pmatrix}_{\substack{s=s_f \\ \alpha=\alpha_f}} \tag{4.4}$$

In fact, in terms of the Trace of J_f:

$$T_f = \left(\frac{\partial s'}{\partial s} + \frac{\partial \alpha'}{\partial \alpha} \right)_{\substack{s=s_f \\ \alpha=\alpha_f}} , \tag{4.5}$$

we have

$$\lambda_f = \frac{T_f \pm \sqrt{T_f^2 - 4 \operatorname{Det}(J_f)}}{2} . \tag{4.6}$$

Let us now consider a few simple examples.

Example 1.

$$\begin{aligned} x_1' &= x_2^2 + x_3^2 + 2x_1x_3 \\ x_2' &= 2x_1x_2 + 2x_2x_3 \\ x_3' &= x_1^2 \end{aligned} \tag{4.7}$$

or, in the form [Eq. (4.2)]:

$$\begin{aligned} s' &= 1 + 3\alpha^2 - s^2 + 2\alpha s - 2\alpha \\ \alpha' &= 2\alpha(1 - 2\alpha). \end{aligned} \tag{4.8}$$

In this case, the first version is easier to deal with. We note first that if $x_2 = 0$ initially (corresponding to $\alpha = 0$), it remains so. Under these conditions Eq. (4.7) reduces to the 2-variable system [Eq. (3.2)] (by interchange of subscripts). This system has a fixed point which is repellent; thus no initial vector with $x_2 = 0$

can converge to this fixed point. We next note that Eq. (4.7) possesses the 2-period $x_1 = 1$, $x_3' = 1$, $x_1'' = 1$, etc. (corresponding to $s = 1$, $\alpha = 0$ and $s' = \alpha' = 0$). This is in fact the limit set for the 2-variable system [Eq. (3.2)], as we previously showed. Therefore we conclude that if $x_2^{(0)} = 0$ in Eq. (4.7), the limit set is just this 2-period. On the other hand, if $x_2^{(0)} \neq 0$, the second equation of Eq. (4.7) becomes

$$x_2' = 2x_2(1 - x_2) \tag{4.9}$$

which is equivalent to the 2-variable system [Eq. (3.3)] in the form [Eq. (3.6)]. The limit set for this system is $x_2 = \frac{1}{2}$ (independent of the other variable, provided that $x_1 + x_3 < 1$). Substituting back into Eq. (4.7), we find the (attractive) fixed point

$$\begin{aligned} x_1 &= \tfrac{1}{2}(\sqrt{3} - 1) & & \\ x_2 &= \tfrac{1}{2} & \text{or}\quad \alpha &= \tfrac{1}{4} \\ x_3 &= 1 - \tfrac{1}{2}\sqrt{3} & s &= \tfrac{1}{4}(2\sqrt{3} - 1). \end{aligned} \tag{4.10}$$

Thus we have found two limit sets for the transformation [Eq. (4.7)], associated respectively with initial vectors for which $x_2^{(0)} = 0$ or $x_2^{(0)} \neq 0$.

Example 2.

Let us now alter Eq. (4.7) in a seemingly trivial way, namely by interchanging the terms $2x_1x_2$ and $2x_1x_3$ on the right:

$$\begin{aligned} x_1' &= x_2^2 + x_3^2 + 2x_1x_2 & & \\ x_2' &= 2x_1x_3 + 2x_2x_3 & \text{or}\quad s' &\doteq 1 + 4\alpha - s - \alpha \\ x_3' &= x_1^2 & \alpha' &= s + \alpha - (s+\alpha)^2. \end{aligned} \tag{4.11}$$

By inspection we see that if any initial coordinate is equal to 1, we get the formal 2-period: $x_1 = 1$, $x_3' = 1$, $x_1'' = 1$, etc. This 2-period, however, can be shown to be repellent by our usual criterion applied to the square of the transformation. From the s, α form of Eq. (4.11), we find (for a general point):

$$J = \begin{pmatrix} 4\alpha - 1 & 4s - 1 \\ 1 - 2s - 2\alpha & 1 - 2s - 2\alpha \end{pmatrix}. \qquad (4.12)$$

Evaluating this at the two period points (which are fixed points of the square of the transformation), we then multiply the two matrices together to obtain J_f (an obvious generalization of the chain rule!). It turns out that $\lambda_A = 4$, so that the 2-period is highly repellent.

We next consider the ordinary fixed points. There is only one, given by the appropriate solution of the equation

$$2x_1^4 - 3x_1^2 - x_1 + 1 = 0. \qquad (4.13)$$

(This is obtained by elimination from Eq. (4.11).) We find

$$\begin{aligned} x_1 &= 0.45823825 \\ x_2 &= 0.33177946 \qquad (4.14) \\ x_3 &= 0.20998229. \end{aligned}$$

By the usual method we may show that this is also repellent. What, then, is the limit set associated with the transformation [Eq. (4.11)]? The answer given by the computing machine is that for "almost all" initial vectors the limit set is a 3-period (see Reference 3, page 93, for the actual coordinates). That the 3-period is indeed attractive can be verified by evaluating J at each of the three points and multiplying the matrices to get J_f. It seems to us that this 3-period would have been very hard to predict by working analytically with the transformation itself.

Example 3.

$$\begin{aligned} x_1' &= x_2^2 + x_3^2 + 2x_1x_2 \\ x_2' &= x_1^2 + 2x_2x_3 \qquad \text{or} \qquad s' = 1 + \tfrac{3}{2}s^2 - \tfrac{1}{2}\alpha^2 + 3\alpha s + 2\alpha - 2s \qquad (4.15) \\ x_3' &= 2x_1x_3 \qquad\qquad \alpha' = \tfrac{1}{2}s^2 - \tfrac{3}{2}\alpha^2 - 3\alpha s + 2\alpha. \end{aligned}$$

This has the 2-period $x_1 = 1$, $x_2' = 1$, $x_1'' = 1$, etc. It also has the fixed point

$$x_1 = \tfrac{1}{2},$$

$$x_2 = \frac{1}{\sqrt{8}},$$

$$x_3 = \frac{2-\sqrt{2}}{4}.$$

Both the fixed point and the 2-period are attractive, i.e., they are limit sets of Eq. (4.15). The question remains: What are the corresponding attractive neighborhoods? No complete theoretical treatment of this question has yet been given (although R. De Vogelaere has made some progress in this direction). The experimentally determined "separatrix" is shown in Figure 2:

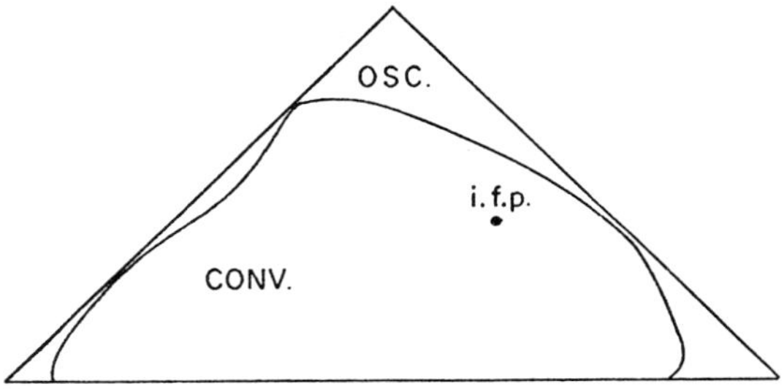

Figure 2

Example 4.

$$x_1' = x_1^2 + x_3^2 + 2x_1x_3$$

$$x_2' = x_2^2 + 2x_1x_2 \qquad \text{or} \qquad s' = 1 + 4\alpha^2 + 2\alpha s - 4\alpha \qquad (4.16)$$

$$x_3' = 2x_2x_3 \qquad\qquad \alpha' = 2\alpha s.$$

This system has a fixed point at

$$s_f = \tfrac{1}{2}$$

$$\alpha_f = \tfrac{1}{4}. \qquad (4.17)$$

In this case our criterion fails to tell us whether the fixed point is attractive or repellent since it turns out that $\lambda_A = 1$. If we introduce new variables centered around the fixed point:

$$x = s - \tfrac{1}{2}$$
$$y = \alpha - \tfrac{1}{4}, \tag{4.18}$$

we may write Eq. (4.16) in the form:

$$x' = -y + \tfrac{1}{2}x + 4y^2 + 2xy$$
$$y' = y + \tfrac{1}{2}x + 2xy, \tag{4.19}$$

with fixed point $x=y=0$. One might be tempted to think that the non-linear terms could be dropped if one were sufficiently close to the origin. In that case Eq. (4.19) becomes the linear system:

$$x' = -y + \tfrac{1}{2}x$$
$$y' = y + \tfrac{1}{2}x. \tag{4.20}$$

This transformation, however, merely generates the ellipse

$$x'^2 + x'y' + 2y'^2 = x^2 + xy + 2y^2. \tag{4.21}$$

Therefore, the linear system does <u>not</u> converge to the fixed point for <u>any</u> initial $(x,y) \neq 0$.

Now machine calculation indicates that the original system <u>does</u> converge to the fixed point (although the rate decreases as one gets closer). That the fixed point actually is attractive has been shown by E. Dahlberg (see Reference 8); he accomplished this by construction of an appropriate Lyapounov function.

<u>Example 5</u>.

$$x_1' = x_1^2 + 2x_1x_2$$
$$x_2' = x_2^2 + 2x_2x_3 \quad \text{or} \quad s' = s^2 - 3\alpha^2 + 2\alpha \tag{4.22}$$
$$x_3' = x_3^2 + 2x_1x_3 \qquad \alpha' = 2\alpha - 2\alpha s.$$

This system has four fixed points:

(a) $x_1 = 1$ $(s = 1, \alpha = 0)$

(b) $x_2 = 1$ $(s = \alpha = \frac{1}{2})$

(c) $x_3 = 1$ $(s = \alpha = 0)$

(d) $x_1 = x_2 = x_3 = \frac{1}{3}$ $(s = \frac{1}{2}, \alpha = \frac{1}{6})$.

These all prove to be repellent, but the first three have special directions along which they are attractive, namely:

(a) is attractive along the boundary $s + \alpha = 1$

(b) is attractive along the boundary $s = \alpha$

(c) is attractive along the boundary $\alpha = 0$.

What, then, happens to the sequence of iterates of a general interior point? Machine calculation indicates that the iterates will trace out a path which spirals out toward the boundaries of the reference triangle. Because of the limitations of numerical accuracy, some iterate will eventually lie sufficiently close to a boundary so that the computer will treat it as though it were a boundary point; numerical convergence to the corresponding repellent fixed point then follows rapidly. By the "principle of insufficient reason," however, the choice of boundary is inherently unpredictable (i.e., it will vary with the degree of accuracy of the computation). The only possible conclusion would seem to be that, in reality, the sequence does not converge at all! Blakely and Dixon[4)] have in fact shown that this is the case, using powerful methods which cannot be discussed here.

The examples given above indicate some of the complications that can arise even for a simple class of transformations $(m = 2, k = 3)$. There is, however, some evidence that in the framework of the more general case—in which the d_{ij} are no longer restricted to the values 0 or 1—our 97 basic quadratic transformations in three variables are exceptional rather than typical. Several hundred trials were made, choosing the coefficients at random subject only to the conditions [Eq. (2.9)]; in every case only simple convergence to a single fixed point was observed.

Lecture 4

Section 1.

In the previous lecture we studied some of the peculiarities of the class $m=2, k=3$, i.e., quadratics in three variables. Owing to the relative simplicity of the observed behavior in the majority of cases, it might in fact be possible to given an almost complete theory of this class. To our knowledge, this has not yet been attempted. Even if such a program were successfully carried through, it is doubtful that the techniques employed would be directly applicable to the classes of polynomial transformations which rank next in algebraic complication, namely cubics in three variables $(m=3, k=3)$ and quadratics in four variables $(m=2, k=4)$. As we shall see in this lecture, these two classes give rise to new and more complicated types of limit sets whose precise topological —let alone analytical—nature has not yet been established.

Let us consider a particular example of a 3-variable cubic:

$$\begin{aligned} x_1' &= x_2^3 + 3x_1x_3^2 + 3x_2x_3^2 + 3x_3x_2^2 + 6x_1x_2x_3 \\ x_2' &= x_1^3 + x_3^3 + 3x_2x_1^2 + 3x_3x_1^2 \\ x_3' &= 3x_1x_2^2. \end{aligned} \qquad (1.1)$$

At this point it is convenient to introduce a shorthand notation for these transformations. Writing out $(x_1+x_2+x_3)^3$, we get ten terms which we may denote by M_j $(j=1, 2, \cdots 10)$. We adopt the following conventional ordering:

$$\begin{aligned} &M_1 = x_1^3, \; M_2 = x_2^3, \; M_3 = x_3^3, \; M_4 = 3x_1x_2^2, \; M_5 = 3x_1x_3^2, \\ &M_6 = 3x_2x_1^2, \; M_7 = 3x_2x_3^2, \; M_8 = 3x_3x_1^2, \; M_9 = 3x_3x_2^2, \\ &M_{10} = 6x_1x_2x_3. \end{aligned} \qquad (1.2)$$

Any 3-variable cubic (with $d_{ij}=0$ or 1) can now be specified by giving the terms M_j—or, equivalently, the indices j—which occur in the first two lines of the three-line schema. Let us call these two index sets c_1 and c_2. Then, for example, the transformations [Eq. (1.1)] can be abbreviated as follows:

$$T_{c_1c_2}: \quad \begin{aligned} c_1 &= \{2,5,7,9,10\} \\ c_2 &= \{1,3,6,8\}. \end{aligned} \tag{1.3}$$

(The third line is, of course, determined by the first two.)

In the s, α notation this particular transformation reads:

$$\begin{aligned} s' &= \frac{3}{2}s^3 - \frac{3}{2}s^2\alpha - \frac{15}{2}s\alpha^2 + \frac{23}{2}\alpha^3 - 3s^2 - 3\alpha^2 + \frac{3}{2}s + \frac{3}{2}\alpha + \frac{1}{2} \\ \alpha' &= -\frac{3}{2}s^3 + \frac{3}{2}s^2\alpha - \frac{9}{2}s\alpha^2 + \frac{1}{2}\alpha^3 + 3s^2 + 3\alpha^2 - \frac{3}{2}s - \frac{3}{2}\alpha + \frac{1}{2}. \end{aligned} \tag{1.4}$$

There is a (repellent) fixed point with coordinates

$$s_f = 0.6149341, \qquad \alpha = 0.1943821. \tag{1.5}$$

(A simple numerical method of finding this fixed point will be explained later on in the lecture.) There is also an attractive 2-period with coordinates $s=1$, $\alpha=0$ and $s=\alpha=\frac{1}{2}$. For "most" initial points, however, the observed limit set is the "curve" shown in Figure 3:

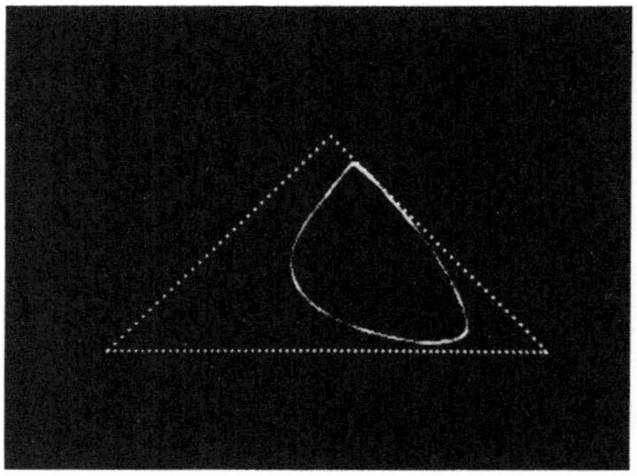

Figure 3

We put the word "curve" in quotes since it has not yet been shown that this actually is a curve; the photograph shows 900 consecutive iterates of a certain initial point, the 900 iterates being taken at a "late" time, when transient effects have disappeared. The results of various numerical tests[1)] make it probable that what we see is part of an infinite limit set which is, moreover, a closed curve. This infinite limit set (if such it be) is strongly attractive; if one starts with a set of initial points (rather than a single point) which lie on some other closed curve (e.g., a circle or a rectangle), the observed pattern assumes the form shown after relatively few iterations (the general shape is clearly recognizable after about 20 steps).

The curve shown is not traced out in "continuous" fashion by successive iterates, that is, the pth and $(p+1)^{th}$ iterates are not in general close to each other. If, however, we plot only every 71st iterate (which means we are plotting the 71st power of our transformation), then the successive iterates do trace out the curve in a continuous manner (and in the clockwise sense). If we take enough of these 71st iterates, the curve we get appears to be identical with the one shown, with precisely the same non-uniform density distribution. In this particular case it is likely that one would get the same picture by plotting any power of the transformation—which is equivalent to asserting that the limit set is actually infinite, and not merely a period of very high order. At the present time, however, the question remains open.

2. Other Types of Limit Sets

Of the 9370 basic cubic transformations in three variables, more than 96 percent were found to have only finite limit sets (attractive fixed points and k-periods). Some 334 transformations have what appear to be infinite limit sets, of which roughly three-fourths are "closed curves" (see Reference 1 for further examples of these). The remaining cases had infinite limit sets of the following two types:

(a) Sets of points confined to two sides of the reference triangle. The square of the transformation, restricted to points on these boundaries, then yields two one-dimensional transformations. There appear to be only a few of these one-dimensional types, so that many different cubics lead to the same pair. For example, consider transformations of the form:

$$x_1' = x_3^3 + 3a_1x_1x_3^2 + 3x_2x_3^2 + 3b_1x_3x_1^2 + 3x_3x_2^2 + 6c_1x_1x_2x_3$$

$$x_2' = x_1^3 + x_2^3 + 3a_2x_1x_3^2 + 3b_2x_3x_1^2 + 6c_2x_1x_2x_3 \tag{2.1}$$

$$x_3' = 3x_1x_2^2 + 3a_3x_1x_3^2 + 3x_2x_1^2 + 3b_3x_3x_1^2 + 6c_3x_1x_2x_3,$$

with non-negative a_i, b_i, c_i (not restricted to the values 0, 1) satisfying

$$\sum a_i = \sum b_i = \sum c_i = 1. \tag{2.2}$$

When restricted to initial points on the boundaries $x_1 = 0$, $x_3 = 0$, all such transformations lead to the pair of one-dimensional 6th order transformations:

$$y' = w[3 - 3w + w^2], \quad w \equiv 3y(1 - y)$$
$$u' = 3v(1 - v), \quad v \equiv u[3 - 3u + u^2]. \tag{2.3}$$

We reserve discussion of these cases to the following lecture and consider the next class:

(b) Complicated distributions of points with no recognizable orderly structure. We shall give three examples of this phenomenon; several other examples will be found in Reference 1.

Example 1.

$$T_{c_1c_2}: \quad \begin{aligned} c_1 &= \{3, 4, 6, 7, 9, 10\} \\ c_2 &= \{5, 8\}. \end{aligned} \tag{2.4}$$

The observed limit set is shown in Figure 4. As is evident from the photograph, it consists of seven separated pieces; each of these is invariant under the 7th power of the transformation. The segments themselves do not have a simple structure. This can be seen from the detailed plot of the upper left-hand segment (Fig. 5):

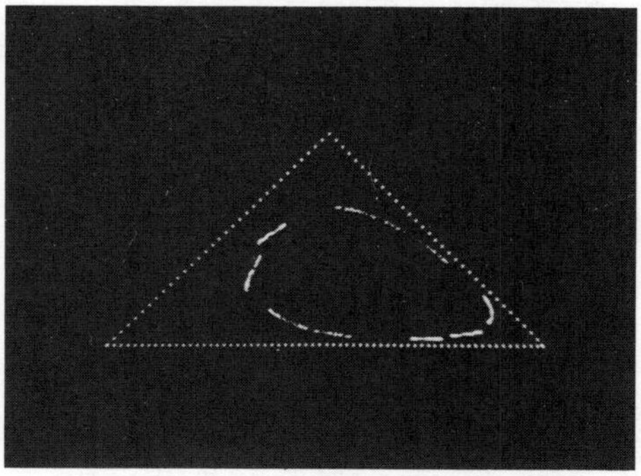

Figure 4

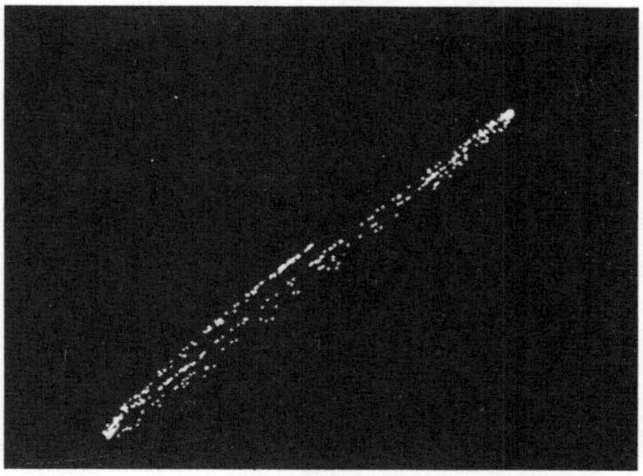

Figure 5

Example 2.

$$T_A = T_{c_1 c_2}: \quad \begin{array}{l} c_1 = \{3, 5, 7, 9, 10\} \\ c_2 = \{1, 2, 8\}. \end{array} \qquad (2.5)$$

We give this transformation the distinctive label "A," since we shall refer to it several times in the course of this lecture. T_A has two repellent fixed points. It is also of the form [Eq. (2.1)], so that it transforms two edges of the reference triangle into each other. For most initial points, the limit set—which we denote by

$L(T_A)$—is that shown in Figure 6 (the reference triangle has been suppressed to accommodate the larger format). We shall return to this case later.

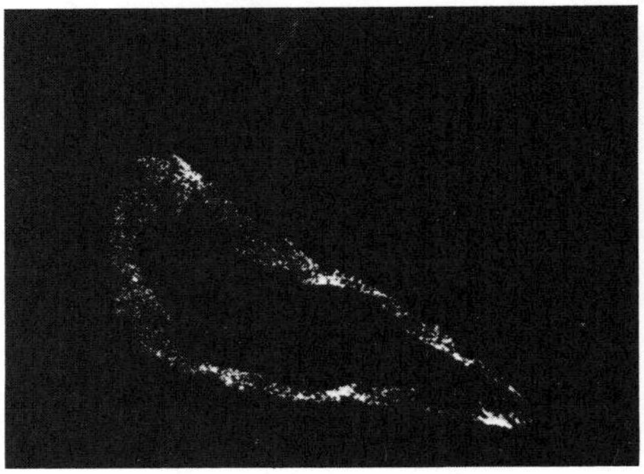

Figure 6

Example 3.

$$T_{c_1 c_2}: \begin{array}{l} c_1 = \{3, 5, 8, 10\} \\ c_2 = \{1, 7, 9\} \end{array} \qquad (2.6)$$

The infinite limit set is shown in Figure 7.

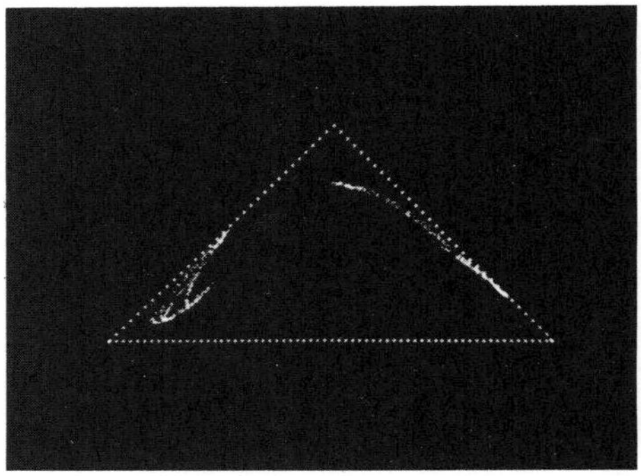

Figure 7

Now this transformation actually has an attractive fixed point ($s=0.6259977$, $\alpha=0.1107896$), so we have a situation analogous to that of Example 3 of the previous lecture. What is the "separatrix" or boundary between the attractive neighborhoods of these two limit sets? At present this can only be determined experimentally. The attractive neighborhood of the fixed point turns out to be rather complicated (Fig. 8); note that it is multiply connected.

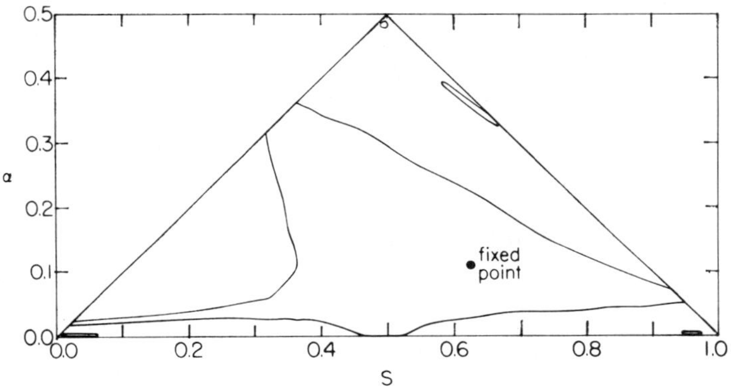

Figure 8

We have seen that there are at least three types of infinite limit sets that can arise in the 3-variable cubic case; there is actually a further type, but it apparently occurs only for transformations which are "generalized" by one of the methods mentioned in the previous lecture. We shall illustrate this new class in the next section. It is convenient to introduce a special nomenclature for our infinite limit sets as follows:

Class I: "closed curves"

Class II: limit sets restricted to two sides of the reference triangle

Class IV: complicated distributions of points like those illustrated in the three examples given above.

The designation "Class III" is reserved for what we have called "pseudo-periods"; these will be introduced in the next section.

3. <u>The "Δt-Modification"</u>

In Lecture 3 we mentioned a particular one-parameter generalization of our polynomial transformations which consists in adding a linear term to the right-hand side of the difference equation while simultaneously decreasing the size of the non-linear terms in an appropriate manner (Lecture 3, Eq. (2.10)). For a two-dimensional system this may be written

$$\begin{aligned} s' &= (1-\Delta t)s + \Delta t\, F(s,\alpha) \\ \alpha' &= (1-\Delta t)\alpha + \Delta t\, G(s,\alpha). \end{aligned} \qquad (0 < \Delta t \leqq 1) \tag{3.1}$$

For $\Delta t = 1$ we recover the original system:

$$\begin{aligned} s' &= F(s,\alpha) \\ \alpha' &= G(s,\alpha). \end{aligned} \tag{3.2}$$

This makes it natural to discuss the behavior of the family [Eq. (3.1)] in terms of the behavior of the "unmodified" system [Eq. (3.2)]. The relation between Eq. (3.1) and Eq. (3.2) is, in fact, a very special one: they have the same fixed points.

Let us write down the eigenvalues of the Jacobian matrix of the system [Eq. (3.1)] in terms of the trace T_f and determinant $\mathrm{Det}(J_f)$ of Eq. (3.2) (cf. Eq. (4.6) of Lecture 3):

$$\lambda_{\mathrm{mod}_f} = 1 - \Delta t + \frac{\Delta t}{2}\left[T_f \pm \sqrt{T_f^2 - 4\,\mathrm{Det}(J_f)}\right] = 1 - \Delta t + \Delta t\,\lambda_f. \tag{3.3}$$

If the eigenvalues λ_f are complex, that is, if $T_f^2 < 4\,\mathrm{Det}(J_f)$, then

$$|\lambda_A|^2 = |\lambda_{\mathrm{mod}_f}|^2 = 1 - \Delta t(2 - T_f) + \Delta t^2[1 - T_f + \mathrm{Det}(J_f)]. \tag{3.4}$$

Defining Δt_{lim} to be the value of Δt for which $|\lambda_{\mathrm{mod}_f}|^2 = 1$, we obtain

$$\Delta t_{\mathrm{lim}} = \frac{2 - T_f}{1 - T_f + \mathrm{Det}(J_f)}. \tag{3.5}$$

Thus, for the case of complex roots, we can make a repellent fixed point (s_f, α_f) <u>attractive</u> by choosing Δt such that $0 < \Delta t \leqq \Delta t_{\mathrm{lim}}$. This enables us to find these repellent fixed points by straightforward iteration—a much simpler procedure than solving a high-order algebraic system!

The method also works in the case of real roots provided that the root largest in absolute value is <u>negative</u>:

$$\Delta t_{lim} = \frac{2}{1+\lambda_A}; \tag{3.6}$$

we leave it to the reader to show that the method fails if the largest root is positive (cf. Reference 1, page 34). This case does not seem to arise for 3-variable cubics.

Suppose we choose some transformation [Eq. (3.2)] with a repellent fixed point and construct the modified system (Eq. (3.1)], using a value of Δt larger than Δt_{lim}, i.e., $\Delta t_{lim} < \Delta t < 1$. What will be the relationship of the modified limit sets to those of the original system? Unfortunately, this question can only be answered at present by resorting to experiment (since the effect of the parameter Δt cannot be studied analytically over many iterations). One would think that the limit sets should display some sort of continuity as a function of Δt, and, roughly speaking, this is so; for "sufficiently small" changes in Δt, two limit sets $L_{\Delta t}$ and $L_{\Delta t'}$ do look the same. At present, however, we have no precise measure of similarity between two infinite limit sets, nor do we know what "sufficiently small" means for changes in parameter space.

Let us illustrate the effect of the Δt-modification in a particular case—that of the transformation T_A (Eq. (2.5) above), which in the s, α variables takes the form:

$$\begin{aligned} s' &= s^3 - 6s^2\alpha - 3s\alpha^2 + 4\alpha^3 - \frac{3}{2}s^2 + 3s\alpha - \frac{3}{2}\alpha^2 + 1 \equiv F(s, \alpha) \\ \alpha' &= -s^3 + 3s^2\alpha + 2\alpha^3 + \frac{3}{2}s^2 - 3s\alpha + \frac{3}{2}\alpha^2 \equiv G(s, \alpha). \end{aligned} \tag{3.7}$$

The Class IV limit set was shown in Figure 6. If we construct the corresponding modified system [Eq. (3.1)] and set $\Delta t = 0.9930$, the observed limit set turns out to be a 7-period, the points of which correspond roughly to the bright areas in the original photograph. The limit set is actually a 7-period for a range of Δt values—$0.9772 \leqq \Delta t \leqq 0.9930$ approximately. Over this range the coordinates of the period shift in a continuous manner (cf. Figures 29 and 30 of Reference 1). On either side of this range, however, a new phenomenon is observed; for, say, $\Delta t = 0.99310$ and $\Delta t = 0.97713$, the underlying 7-period is repellent, and we observe instead a "pseudo-period" or "Class III" limit set. This has the form of seven dense clusters of points centered at the corresponding repellent period points (cf. Reference 1, Figure 28). In other words, as the 7-period becomes repellent, i.e., $\lambda_A^{(7)} > 1$, the limit set does not change wildly at first. This relative continuity of behavior holds out some hope for a successful theoretical treatment.

With a further decrease in Δt—e.g., $\Delta t = 0.9770$—we find a Class I limit set, i.e., a closed curve with the fixed point interior to it. As $\Delta t \to \Delta t_{lim}$ ($= 0.9180154$), the limit set remains of Class I type but the curve shrinks in diameter. For $\Delta t < \Delta t_{lim}$ the fixed point itself becomes the only limit set.

4. Modification of the Coefficients

The possibility of generalizing our polynomial transformations by relaxing the condition $d_{ij} = 1$ or 0 was mentioned in the previous lecture. In the present case this procedure generates a 20-parameter class of transformations; for each of these transformations we can also introduce the Δt-modification. It is obviously impossible to make any sort of "representative" study of this infinite class. Accordingly, in our experimental studies we reduced the 20 parameters to a single parameter, ϵ, which we introduced in the following manner:

Given some transformation $T_{C_1C_2}$, we multiplied one of the terms M_j by $1-\epsilon$ and added ϵM_j to one of the other two lines of the schema. For every transformation, this can be done in just 20 ways. Since we were interested in the behavior of transformations "near" some given transformation, we considered mainly "small" ϵ. The parameter range we chose was $0 < \epsilon \leqq 0.10$.

In the s, α coordinates, this one-parameter modification (the "ϵ-modification") takes the form

$$\begin{aligned} s' &= F(s, \alpha) + \epsilon f_{rs}(s, \alpha) \\ \alpha' &= G(s, \alpha) + \epsilon g_{rs}(s, \alpha) \end{aligned} \tag{4.1}$$

so that for $\epsilon = 0$ the original system is recovered. Here f, g are cubic polynomials in s, α like the F, G; the subscripts (r, s) specify which terms in the original system have been modified (see Reference 1, Sec. IV for details). The ϵ-modification differs from the Δt-modification in that the fixed point is a function of ϵ. Nevertheless, if ϵ is varied in sufficiently small steps, there appears to be a continuity in the limit sets which mirrors the corresponding continuity in the transformations themselves.

It is not possible within the scope of these lectures to describe the results of this part of our study; many examples will be found in Reference 1, especially in Appendix II. To give some idea of the complex behavior that can result from varying ϵ, we cite one particular example, a modification for T_A:

$$T_{A(2,1)\epsilon}: \quad \begin{aligned} s' &= F(s, \alpha) - 4\epsilon\alpha^3 \\ \alpha' &= G(s, \alpha) - 4\epsilon\alpha^3 \end{aligned} \tag{4.2}$$

where F, G are given by Eq. (3.7). A few results are given in the following table:

Table II

ϵ	limit set
0.03	class III
0.07	7-period
0.08	class I
0.09	class I
0.10	28-period

The case $\epsilon = 0.08$ looks like a typical pseudo-period when plotted to normal scale (Fig. 9); with a scale factor of ~77, however, each piece is seen to be a closed curve, one of which is shown in Figure 10. This is to be contrasted with a true pseudo-period ($\epsilon = 0.03$), a portion of which is shown in Figure 11 (scaled by ~5.6).

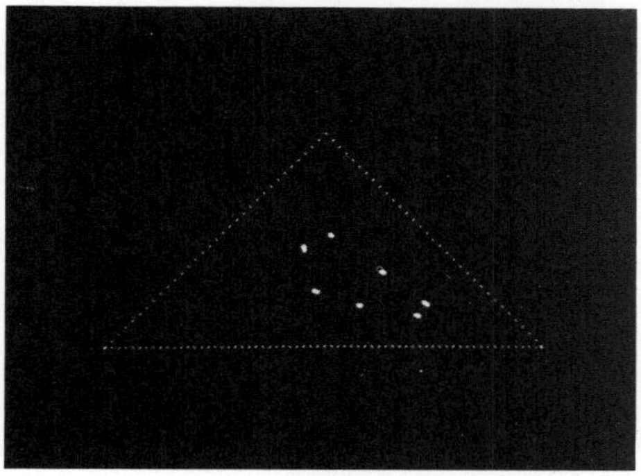

Figure 9

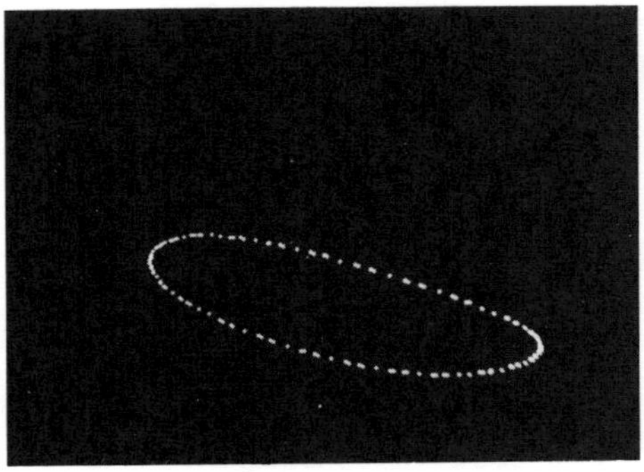

Figure 10

Figure 11

By modifying the single transformation T_A in various ways, k-periods of several orders were observed, viz:

$$k = 7, 9, 16, 23, 30, 37, 46, 62, 148.$$

The record period, $k=148$, is a limit set of the transformation:

$$T_A(5,0)\epsilon: \quad \begin{aligned} s' &= F(s,\alpha) - 3\epsilon(s-\alpha)(1-s-\alpha)^2 \\ \alpha' &= G(s,\alpha) \end{aligned} \tag{4.3}$$

with $\epsilon = 0.02$ and F, G again given by Eq. (3.7). It is not, however, the only limit set observed for this ϵ; there is also a Class IV limit set which is rather hard to distinguish visually from the period itself. The structure of these two limit sets would appear to be immensely complicated.

As can be imagined, introduction of the Δt parameter gives rise to further "pathological" (more properly, unexpected) behavior. In some cases the Δt and ϵ modifications, applied separately to the same basic cubic transformation, can lead to limit sets which are morphologically very similar. Furthermore, some transformations are very little affected (as regards their limit sets) by small parameter variations while others show a radical qualitative change; this is, no doubt, merely a reflection of our ignorance as to what "small" and "near" really mean in these cases.

It would be fruitless to attempt any generalizations from these particulars. The most that can be said is that the Class IV limit sets appear to be relatively unstable with regard to small parameter variations in the corresponding transformations. When the Δt or ϵ modifications are introduced, the Class IV limit sets seem to have a tendency to "condense" to finite periods; perhaps this is only a manifestation of the fact that small formal changes in the transformations may serve to make the underlying periods attractive. This part of the problem might be considered a purely algebraic question, but it is at present no less intractable on that account. Finite periods are, however, something we understand. Perhaps a perturbation theory approach could be used, taking the periods as the "unperturbed states." To our knowledge, this has not yet been attempted.

5. Quadratic Transformations in Four Variables

The new feature of this case—$m = 2$, $k = 4$—is of course that the limit sets are generally three-dimensional. Since there are 34,337 distinct basic transformations ($d_{ij} = 0, 1$), our experimental study was of necessity less detailed than it was in the cubic case. The finite limit sets need no special comment; they are of the same sort as those found for cubics, except that they are rarely plane sets. A few periods of rather high order ($k > 100$) were found, as well as many cases with 10 to 80 points. This probably should be expected in view of the greater variety of possible algebraic structures for this case.

About 2 percent of these transformations seem to possess infinite limit sets. These are mainly three-dimensional analogues of the Class I limit sets discussed above, i.e., space curves. Two examples are shown in Figures 12 and 13; these correspond, respectively, to the transformations:

$$\begin{aligned} x_1' &= x_1^2 + x_2^2 + 2x_2x_4 \\ x_2' &= x_4^2 + 2x_1x_4 + 2x_3x_4 \\ x_3' &= x_3^2 + 2x_1x_2 + 2x_1x_3 \\ x_4' &= 2x_2x_3; \end{aligned} \tag{5.1}$$

$$\begin{aligned} x_1' &= x_1^2 + x_3^2 + 2x_3x_4 \\ x_2' &= x_4^2 + 2x_1x_4 + 2x_2x_4 \\ x_3' &= x_2^2 + 2x_1x_2 \\ x_4' &= 2x_1x_3 + 2x_2x_3. \end{aligned} \tag{5.2}$$

The limit set belonging to Eq. (5.1) consists of two plane curves; one curve lies in the (x_1, x_3) plane, while the plane of the second curve makes an angle of 45° with that of the first. The limit set for Eq. (5.2) is, presumably, a twisted space curve.

A few examples of what may be Class IV limit sets were found; it is rather hard to be sure of this, since what one sees on the oscilloscope screen is strongly dependent on the chosen viewing angle or perspective. Our visual method has been to look at these limit sets from various angles (we produced stereo photographs of several of them), eventually choosing the one that seemed most revealing; unfortunately, it is quite easy to miss significant morphological details by proceeding in this manner.

The Δt and ϵ modifications can, of course, be applied to these quadratic systems; owing to the large number of potentially interesting cases, however, it has not yet proved feasible to carry through a systematic study.

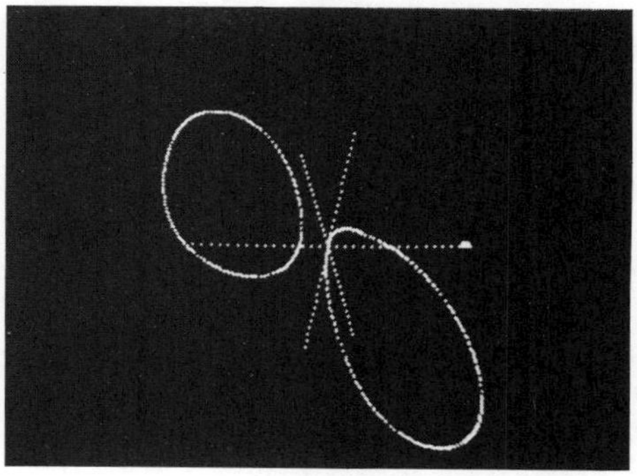

Figure 12

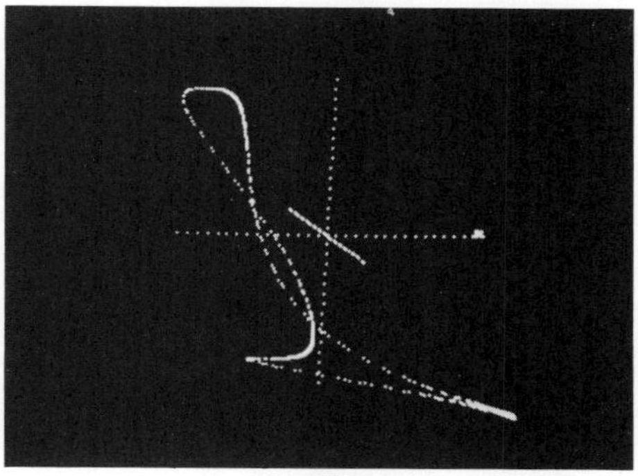

Figure 13

Lecture 5

1. Two-Dimensional Broken-Linear Transformations

For certain special quadratic transformations in one dimension, one can give an almost complete discussion of the iterative properties; this is possible because these transformations are conjugate to piece-wise linear ("broken-linear") mappings of the interval into itself. A well-known example is the Ulam-von Neumann transformation of Lecture 1: $x' = 4x(1-x)$, which is conjugate to the broken-linear transformation

$$x' = g(x) : g(x) = 2x, 0 \leqq x \leqq \tfrac{1}{2};\ g(x) = 2 - 2x, \tfrac{1}{2} \leqq x \leqq 1 .$$

The iterative properties of the latter can be obtained from a study of the law of large numbers for the elementary case of Bernonlli. Stated differently, the behavior of iterates of this simple quadratic transformation turns out to depend on the combinatorial rather than the analytic properties of the function. With this in mind, we tried to see whether an analogous situation would obtain in two dimensions. Our non-linear, polynomial transformations of the triangle into itself might, we thought, be similar to suitably chosen broken-linear mappings of a square into itself, at least as regards their asymptotic behavior.

One simple generalization to two dimensions of broken-linear transformations in one variable is a mapping:

$$\begin{aligned} x' &= f(x, y) \\ y' &= g(x, y) \end{aligned} \tag{1.1}$$

where each of the functions f and g is linear in regions of the plane. In other words, the graphs of these functions consist of planes fitted together to form pyramidal surfaces. The motivation for studying such transformations is the hope that their iterative properties will turn out to depend only on the folding of the plane along straight lines, or, more specifically, on the combinatorics of the overlap of the various linear regions which is generated by the mapping.

One simple, non-trivial construction is the following. We choose the unit square as our reference space and pick a point in the square with coordinates (x_1, y_1); at this point we erect a perpendicular of height d_1, $0 < d_1 \leqq 1$. This defines a surface consisting of four intersecting planes. We take this surface to be the function $f(x, y)$. Explicitly:

We define four regions I to IV by the bounding lines:

$$L_1:\ y = \frac{y_1}{x_1}x$$

$$L_2:\ y = \frac{y_1 - 1}{x_1}x + 1$$

$$L_3:\ y = \frac{1 - y_1}{1 - x_1}x + \frac{y_1 - x_1}{1 - x_1} \qquad (1.2)$$

$$L_4:\ y = \frac{y_1}{1 - x_1}(1 - x);$$

then:

Region I is bounded by L_1, L_2 and $x = 0$,
Region II is bounded by L_2, L_3 and $y = 1$,
Region III is bounded by L_3, L_4 and $x = 1$, (1.3)
Region IV is bounded by L_4, L_1 and $y = 0$;

$x' = f(x, y)$ in the various regions is then given by the following equations:

$$\text{Region I:}\quad x' = \frac{d_1}{x_1}x$$

$$\text{Region II:}\quad x' = \frac{d_1}{1 - y_1}(1 - y)$$

$$\text{Region III:}\quad x' = \frac{d_1}{1 - x_1}(1 - x) \qquad (1.4)$$

$$\text{Region IV:}\quad x' = \frac{d_1}{y_1}y.$$

Analogous equations hold for $y' = g(x, y)$ with parameters x_2, y_2, d_2. No systematic study of these transformations has yet been carried out, but we have investigated many special cases in detail. Unfortunately, the iterative properties do not seem to be noticeably less complex than those of the 3-variable cubics. We cite only one example. Consider the one-parameter family:

$$T_z: \begin{array}{l} x_1 = y_1 = z, \quad d_1 = 1 \\ x_2 = y_2 = 1 - z, \quad d_2 = 1 \end{array}, \quad 0 < z \leqq \tfrac{1}{2}. \tag{1.5}$$

For $z = \frac{1}{2}$, T_z reduces to the one-dimensional Ulam-von Neumann example; the limit set covers the diagonal of the reference square uniformly. As z is decreased, the limit set becomes a collection of segments of increasingly complicated structure; Figures 14 and 15 illustrate the cases $z = 0.36$ and $z = 0.27$ respectively. For $z = 0.25$ we see a densely covered rectangle with trailing edges (Fig. 16). With $z = 0.15$ the limit set splits into four rectangular regions (Fig. 17); this general configuration persists down to $z = 0.01$, the overall size of the pattern contracting with decreasing z. Below $z = 0.01$ the fixed point becomes attractive. (A more complete set of pictures will be found in Reference 1, Appendix II.)

It is possible that, for some range of the parameter z, T_z is ergodic in a two-dimensional subregion (cf. Fig. 16). If this turns out to be true, it will be the only simple, non-trivial example yet discovered.

2. Iteration in One-Dimension[9]

The process of iteration in one dimension is the subject of a considerable body of literature, most of it connected with the problem of solving equations in a single variable. This means that most of the effort has been directed toward the problem of speeding up convergence to a limit set which consists of a single point. As

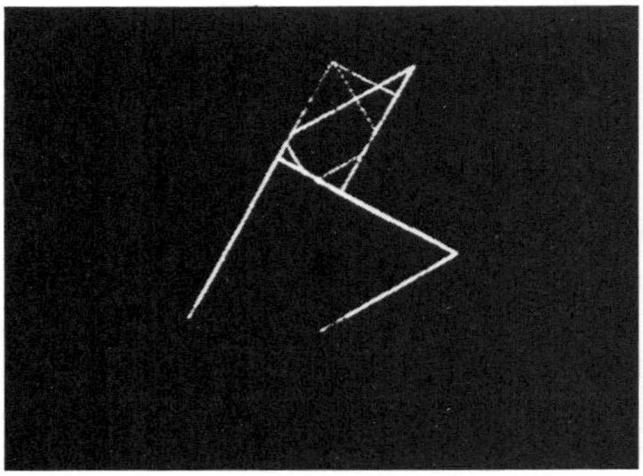

Figure 14

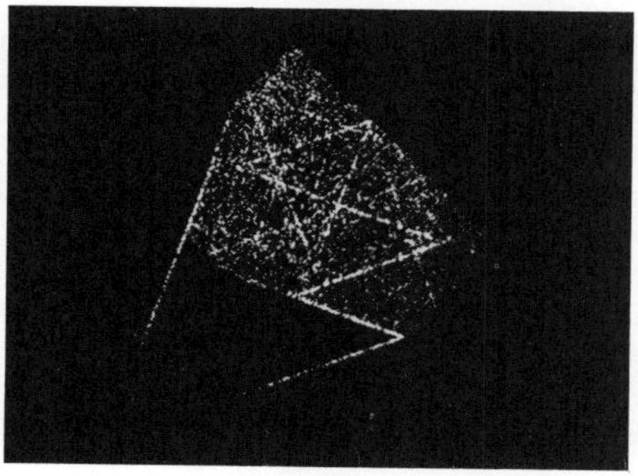

Figure 15

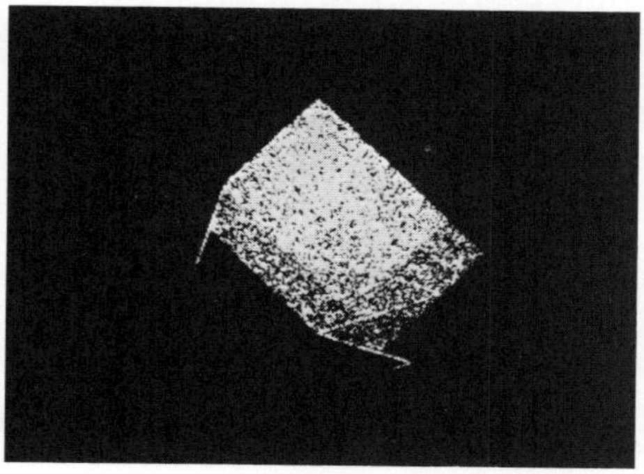

Figure 16

Figure 17

we have indicated, our interests lie in a different direction, since we want to study more complicated limiting behavior (infinite limit sets and k-periods). As we have seen, one-dimensional transformations are encountered as special cases of cubic polynomial systems in three variables (Lecture 4, Eq. (2.3)). For example, one such transformation would be

$$T_\sigma:\ y' = w(3 - 3w + \sigma w^2), \qquad w \equiv 3y(1 - y). \tag{2.1}$$

This transformation arises as a special case of the "generalized" system:

$$x_i' = \sum_{i=1}^{10} d_{ij} M_j \qquad i = 1, 2, 3 \tag{2.2}$$

where we set

$$d_{13} = \sigma, \quad d_{17} = d_{19} = d_{34} = d_{36} = 1, \qquad d_{31} = 0; \tag{2.3}$$

the other d_{ij} are arbitrary, subject to the usual condition:

$$\sum_{i=1}^{3} d_{ij} = 1 \quad (\text{all } j), \qquad 0 \leqq d_{ij} \leqq 1. \tag{2.4}$$

If we restrict ourselves to the subclass of initial points such that $x_3 = 0$ (i.e., the side $s + \alpha = 1$ of the reference triangle), then the square of Eq. (2.2) can be written in the form [Eq. (2.1)].

The family T_σ has many interesting iterative properties which will not be discussed here.[1] It appears to have very complicated limit sets for certain ranges of the parameter σ; these may be infinite, although the possibility that they are actually periods of high order cannot at present be excluded. If the limit sets are actually infinite they may be "pathological," i.e., Cantor sets; resolution of this question would be of the greatest interest (but see Ref. 9).

Iteration in one dimension is much easier to visualize than it is in two dimensions; it is also much easier to carry out numerically. Nevertheless, there is one pitfall which must be avoided when iterating one-dimensional transformations on computing machines, namely, the occurrence of "spurious" or "false" convergence. This manifests itself in the apparent existence of an attractive period of high order where, in fact, no such period exists. Such might be the case, for example, with the function $x' = 4x(1 - x)$, which is known not to possess any finite limit sets.

If we use a computing machine with fixed word length—equivalent, say, to eight decimal places—we are bound to find an exact period in not more than 10^8 steps. Actually, the probability that the chain will be cyclic long before the full theoretical run of 10^8 steps is very close to unity (e.g., it is $1 - 1/e$ for $\sim 10^4$ iterations). Two-dimensional iteration, on the other hand, is much less likely to lead one astray (in the first approximation one has two uncorrelated cycles to consider).

As a by-product of these considerations, an interesting problem arises. Consider the class of convex, symmetrical (about $x = \frac{1}{2}$), continuous functions $f(x)$ mapping the whole unit interval into itself. We have

$$f(0) = f(1) = 0, \quad f(\tfrac{1}{2}) = 1, \quad \left|\frac{df}{dx}\right|_{x=0,1} > 1, \quad \left|\frac{df}{dx}\right|_{x=\frac{1}{2}} = 0;$$

we also restrict $f(x)$ so that at the fixed point $x_f = f(x_f)$, we have

$$\left|\frac{df}{dx}\right|_{x=x_f} > 1.$$

Problem: Give a constructive characterization of the subclass which has no finite limit sets.

This problem has not been solved to our knowledge. It is clear that its solution will involve a precise formulation of the concept of "degree of flatness" in the neighborhood of the maximum; if the curve is "too flat" one can clearly find an attractive period of some order (but see Ref. 9).

3. A "Soluble" Example

In this section we shall discuss a particular one-parameter family of transformations for which explicit results can be established. We can only sketch the argument here; details will be found in the original paper.[5]

S. Ulam[6] has proposed the following problem: Let $a_i(\lambda_1, \cdots \lambda_n)$, $i = 1, 2, \cdots n$ denote the elementary symmetric functions of the n real numbers $\lambda_1, \lambda_2, \cdots \lambda_n$. Given an initial vector $\{\lambda_1^{(0)}, \cdots \lambda_n^{(0)}\}$, determine the limiting behavior of the nonlinear transformation:

$$\lambda_i' = (-1)^i a_i(\lambda_1, \cdots \lambda_n) \qquad i = 1, 2, \cdots n. \tag{3.1}$$

Not much is known about this system for general n, but all the fixed points have been determined.[7] A by-product of this determination is the following interesting fact: there are no polynomial equations of degree $n \geqq 5$, with real coefficients all different from zero, such that the coefficients are equal to the roots (the leading coefficient is taken to be 1).

The possible limit sets of Eq. (3.1) have not yet been investigated. The transformation is, of course, unbounded, which makes numerical investigation rather awkward. There is, however, a bounded modification of the case $n = 2$ which can easily be studied numerically, and even, as it turns out, theoretically.

Let

$$\begin{aligned} Z_1(x, y; \alpha) &= -(1-2\alpha)x - (1-\alpha)y \\ Z_2(x, y; \alpha) &= y[\alpha + (1-\alpha)x] \qquad (3.2) \\ D(x, y; \alpha) &= +\sqrt{Z_1^2 + Z_2^2}\,. \end{aligned}$$

We then consider the following transformation:

$$T_\alpha(x, y): \quad \begin{aligned} x' &= Z_1/D \\ y' &= Z_2/D. \end{aligned} \qquad (3.3)$$

Since T_α maps the plane on the circumference of the unit circle, it is no restriction to consider only points on the latter. Equation (3.3) is then effectively one-dimensional $(x^2 + y^2 = 1)$, or, if one wishes, "one-and-one-half-dimensional" (variables: x and the sign of y). The parameter α we shall allow to vary in the range

$$0 < \alpha < \alpha_L, \qquad (3.4)$$

where α_L is a number slightly less than 3/10; the reason for restricting α to this range will become apparent. Actually, we could, with little effort, include the range $-\infty < \alpha \leqq 0$, but this contributes nothing of interest.

For $0 < \alpha < \frac{1}{2}$, there is just one fixed point $x_F = T_\alpha(x_F)$, and this fixed point is independent of α (note that α is analogous to the Δt parameter we introduced into our polynomial transformation). The fixed point is the solution of the α-independent equation:

$$x^2 + x + y = 0 \qquad \text{or} \qquad x^3 + x^2 + x - 1 = 0, \tag{3.5}$$

namely,

$$\begin{aligned} x_F &= 0.5436890 \\ y_F &= -0.8392868. \end{aligned} \tag{3.6}$$

To discover whether this fixed point is attractive or repellent, we need the expression for the derivative. This turns out to be:

$$\frac{\partial x^{(1)}}{\partial x} = \frac{\alpha + (1-\alpha)x}{D^3}\left[-\alpha(1-2\alpha) - (1-\alpha)(1-2\alpha)x^3 + (1-\alpha)^2 y^3\right]. \tag{3.7}$$

We write $x^{(1)}$ instead of x' as being more consistent with the notation $x^{(k)}$ for the kth iterate; the partial derivation notation is used since we wish to treat the parameter α as a second variable. The α-derivative is

$$\frac{\partial x^{(1)}}{\partial \alpha} = \frac{y^2}{D^3}\,[\alpha + (1-\alpha)x][x + y + x^2]. \tag{3.8}$$

(Note that, by Eq. (3.5), this is zero at the fixed point.)

By setting $\partial x^{(1)}/\partial x = -1$ and using the fixed point equation, we can find the value of α for which the derivative at the fixed point is exactly -1. Calling this value α_c we find, after some laborious manipulation, the simple result

$$\alpha_c = x_F^2 = 0.2955977. \tag{3.9}$$

It is then easy to show that for $\alpha < \alpha_c$ the fixed point is repellent and therefore cannot constitute a limit set.

We now pass to a general description of the continuous, many-to-one mapping defined by T_α. As mentioned above, we limit ourselves to initial points on the circumference of the unit circle. Let

$$\begin{aligned} \beta &\equiv \frac{\alpha}{1-\alpha} \\ \gamma &\equiv \frac{1-\alpha}{\sqrt{(1-\alpha)^2 + (1-2\alpha)^2}} \end{aligned} \tag{3.10}$$

Further, let Q_i (i=1, 2, 3, 4) denote the open arc of the circle lying in the ith quadrant. By examining the sign of $(x^{(1)}, y^{(1)})$ relative to that of (x,y), we immediately derive the following mapping scheme:

$$\left.\begin{array}{lll} \text{(a) If } x \in Q_1: & x^{(1)} \in Q_2 & \\ \text{(b) If } x \in Q_2: & x^{(1)} \in Q_2 & \text{when } |x| < \beta; \\ & x^{(1)} \in Q_3 & \text{when } \beta < |x| < \gamma \\ & x^{(1)} \in Q_4 & \text{when } |x| > \gamma \\ \text{(c) If } x \in Q_3: & x^{(1)} \in Q_4 & \text{when } |x| > \beta; \\ & x^{(1)} \in Q_1 & \text{when } |x| > \beta; \\ \text{(d) If } x \in Q_4: & x^{(1)} \in Q_4 & \text{when } x < \gamma; \\ & x^{(1)} \in Q_3 & \text{when } x > \gamma. \end{array}\right\} \quad (3.11)$$

(For $|x| = \beta$ or $|x| = \gamma$, $(x^{(1)}, y^{(1)})$ is a boundary point of the quadrant.)

From Eq. (3.7) we see that $\partial x^{(1)}/\partial x = 0$ if and only if

$$\left.\begin{array}{l} x = -\beta \\ \text{or} \\ (1-\alpha)^2 y^3 = \alpha(1-2\alpha) + (1-\alpha)(1-2\alpha)x^3 . \end{array}\right\} \quad (3.12)$$

The first condition just yields inverses of the points $x = \pm 1$:

$$\begin{aligned} T_\alpha(-\beta, y > 0) &= (-1, 0) \\ T_\alpha(-\beta, y < 0) &= (1, 0). \end{aligned} \quad (3.13)$$

This, incidentally, shows that T_α is <u>not</u> one-to-one, since the points $x = \pm 1$ are mutual inverses; thus, e.g., (1,0) has the two inverses (-1,0) and $(-\beta, y < 0)$.

The second equation of Eq. (3.12) yields two additional points at which the derivative is zero, one in the third quadrant and one in the first; we call these x_0 and x_1 respectively. The first iterate of the third quadrant root x_0 will turn out to be of particular importance for the characterization of the limit sets; with this in mind, we introduce a special symbol, x_m, for this point:

$$x_m = T_\alpha(x_o). \tag{3.14}$$

It is easy to show that $|x_o| > \beta$, from which it follows (cf. Eq. (3.11)) that x_m is in the first quadrant. Since the mapping is continuous, we conclude that the region $x_m \leqq x^{(1)} \leqq 1$, $y \geqq 0$ is doubly covered as x varies from -1 to $-\beta$ in the third quadrant. It is also evident that points in the region $0 \leqq x < x_m$, $y > 0$ have no inverses on the unit circle. Thus x_m is the smallest value of $x \in Q_1$ which can be reached from any point on the circle. Similar remarks apply to the arc bounded by x_1 and $T_\alpha(x_1)$; since it can be shown, however, that $x_1 < x_m$ (and hence itself has no inverse), this region can be ignored if one is interested in limit sets. It is therefore appropriate to refer to the open arc bounded by x_m and its iterate $x_m^{(1)}$ as the "forbidden region." Correspondingly, the complement of the forbidden region will be referred to as the "allowed region" (closed arc). The boundary points are, of course, functions of the parameter γ. The situation is illustrated in Figure 18:

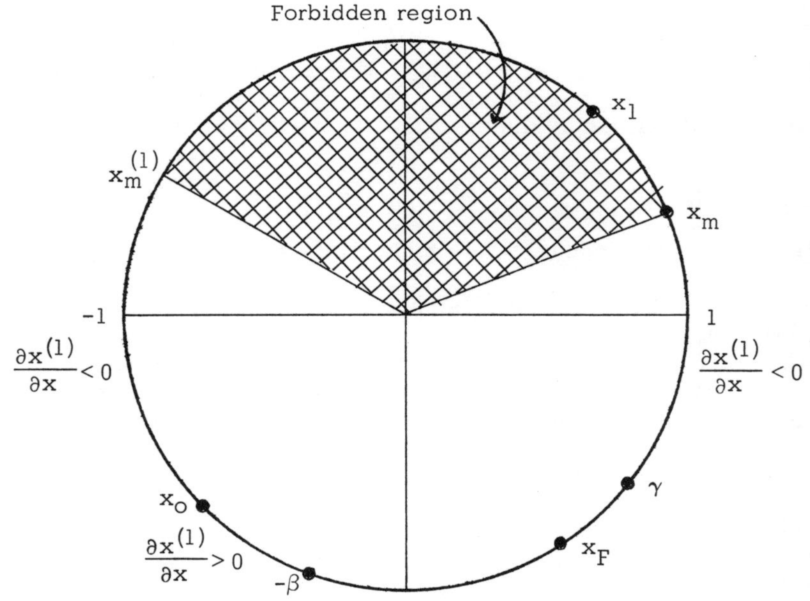

Figure 18

Let us now consider the behavior of an initial point in the allowed region of the first quadrant. Its first iterate lies in the second quadrant and its second iterate, $x^{(2)}$, lies in the fourth, with $x^{(2)} > x_F$ (the second inverse of x_F itself is in the forbidden region when it exists at all). Now the neighborhood of x_F is an "oscillating," repellent neighborhood. This means that the sequence $x^{(2)}, x^{(4)}, x^{(6)}, \cdots$ will diverge from the vicinity of x_F. Eventually, according to the mapping scheme [Eq. (3.11)], some iterate will again lie in the first quadrant. It is clear that the closer $x^{(2)}$ is to x_F, the more iterations will be required before this "return" occurs. The iteration step k for which $x^{(k)}$ is again in the first quadrant for the first time is called the first return index, k_μ. Clearly, k_μ is a function of both x and α. Without going into details (see Reference 5), let us say that continuity arguments and some technical lemmas allow us to assert the following:

(a) Every even integer $2n \geqq 4$ is a first-return index for some x and some α.

(b) For every such k_μ there exists an $\alpha = \alpha_\mu$ such that

$$x_m = T_{\alpha_\mu}^{(k_\mu)}(x_m). \tag{3.15}$$

This, however, is precisely the condition for the existence of an <u>attractive</u> period of order k_μ. The property of being attractive follows from the fact that x_m is the iterate of x_0; we recall that x_0 was defined by the condition $\partial x^{(1)}/\partial x = 0$, so

$$T_\alpha^{(k_\mu)}$$

has its derivative equal to zero for $x = x_m, \alpha = \alpha_\mu$ (since one of the factors is exactly zero).

This establishes the existence of k-periods of all even orders, but, in fact, much more can be shown. For a given order $k = 2n$ there are in general not one but many distinct attractive periods, each belonging to a different range of the parameter α. (There is only one value of α for which the derivative is zero, but the period is attractive over a finite range on either side of this value.) The number of these and the order of their occurrence with increasing α can be calculated in a <u>purely combinatorial manner</u>, using a certain linear model (the actual α values must, of course, be found numerically using T_α itself). Table III summarizes the combinatorial facts up to $k = 34$:

Table III

k	4	6	8	10	12	14	16	18	20	22	24	26	28	30	32	34
Number of distinct periods	1	1	2	3	5	9	16	28	51	93	170	315	585	1091	2048	3855

A detailed table giving the combinatorial "structure" and approximate α-range of all periods $k \leqq 20$ will be found in Reference 5.

By use of the linear model referred to above we can show the existence of an infinite sequence of "harmonics" for each distinct period. For example, consider one of the five periods with $k=12$. This is attractive over a certain α-range (the corresponding derivative actually goes continuously from 1 to -1 with increasing α). When $dx^{(12)}/dx$ becomes slightly less than -1, the first harmonic ($k = 24$) becomes attractive; when this becomes repellent, the second harmonic is observed ($k=48$), and so forth. As α increases, the sequence of harmonics converges to a countably infinite period of determinate combinatorial structure. This, however, is a one-sided limit; if we approach the corresponding limiting value of α from the right (i.e., as α decreases), we cannot say exactly what happens.

When $\alpha=\alpha_C$ the fixed point becomes attractive; nevertheless, we do not get convergence for initial points in the allowed region of the first quadrant. This is because $x_m^{(2)} > x_F$. As α increases beyond α_C, a "trapping" region (i.e., an attractive neighborhood of finite size) develops around x_F, and, eventually, $x_m^{(2)}$ will fall within it. The value of α for which $x_m^{(2)}$ just coincides with the right-hand boundary of the trapping region is $\alpha_L \cong 0.29927$. The behavior of limit sets for $\alpha > \alpha_L$ has not been investigated in detail.

The analogue of T_α can be constructed for any dimension. At present, nothing certain can be said about these higher dimensional cases. For $n=3$, however, computer experiments have uncovered a 3-period (on the surface of the unit sphere). This is a limit set for a certain α-range and a certain restricted set of initial points. For α's outside this range, the period becomes repellent and all sequences appear to converge to the equatorial plane ($Z=0$), thus falling under the analysis for the case $n=2$.

References

1. P. R. Stein and S. M. Ulam, "Non-Linear Transformation Studies on Electronic Computers," Rozprawy Matematyczne, No. 39 (1964).
2. A. M. Ostrowski, Solutions of Equations and Systems of Equations (Academic Press, 1960).
3. M. T. Menzel, P. R. Stein and S. M. Ulam, "Quadratic Transformations, Part I," Los Alamos Scientific Laboratory Report (unclassified), LA-2305 (1959).
4. G. R. Blakley and R. D. Dixon, "The Sequences of Iterates of a Non-Negative Non-Linear Transformation, II," University of Illinois preprint (1964).
5. N. Metropolis, M. L. Stein and P. R. Stein, "Stable States of a Non-Linear Transformation," Numerische Mathematik 10, 1-19 (1967).
6. S. M. Ulam, A Collection of Mathematical Problems (Interscience, 1960).
7. P. R. Stein, "On Polynomial Equations with Coefficients Equal to Their Roots," Am. Math. Monthly 73, 272-4 (1965).
8. E. Dahlberg, "On Lyapounov Functions for Linearly Indeterminate Fixed Points," Jour. Math. Anal. and Applications 22, 465-489 (1968).
9. Since these lectures were first given (July, 1966), the problem of one-dimensional iteration has been studied in greater detail. Much of the complicated behavior referred to under the discussion of equation (2.1) of Lecture 5 has turned out to be spurious. In addition, the problem posed in this section can now be at least partially answered. For a full discussion of the finite limit sets involved, see: N. Metropolis, M. L. Stein, P. R. Stein, "On Finite Limit Sets for Transformations on the Unit Interval," Los Alamos Scientific Laboratory preprint LA-DO 12449 (to appear in Jour. Comb. Theory).

Editor's Note: These lectures were first presented at the Munich Conference on Nonlinear Problems in July, 1966.

THE NEW MATHEMATICAL AND PHYSICAL PROBLEMS OF SUPERSTRONG COUPLING OF MAGNETIC MONOPOLES AND COMPOSITE HADRONS†

A. O. Barut
University of Colorado
Boulder, Colorado

A new physical problem can be attacked sometimes by the existing mathematical techniques; at other times it needs an entirely new mathematical framework. The relation between the concepts of theoretical physics and the mathematical entities is an intriguing one: the mathematician is often intrigued by how the physicist sets up a physical theory, i.e., a correspondence between physical notions and mathematical symbols obeying certain rules of combinations. The physicist, on the other hand, is intrigued by the abstract formalization of mathematics.

I should like to consider an entirely new physical situation and try to study the process of establishing a physical theory and the methods of developing physical notions and the question as to whether or not it will lead to new mathematical ideas as well.

The general problem is simply the study of the interactions of two magnetic monopoles and the resultant composite systems.

A. The Concept of "Magnetic Charge" Resulting from a Generalization of Maxwell's Equations

One basic source of new physical concepts is our ability to recognize generalizations of the existing and established physical notions. Other sources we shall discuss as we go along. A beautiful example of "generalization" is provided in the case of Maxwell's equations. The usual Maxwell equations describe the interactions of the electromagnetic tensor field $F_{\mu\nu}(x)$, $x \in$ Minkowski space-time, with the material electric current $j_\mu^{(e)}$, and have the form

†Presented at the NATO Summer School in Mathematical Physics, Istanbul, 1970.

$$F_{\mu\nu}{}^{,\nu} = j_\mu^{(e)}$$
$$\tilde{F}_{\mu\nu}{}^{,\nu} = 0 \tag{1}$$

where $\tilde{F}_{\mu\nu} = \epsilon_{\mu\nu\sigma\lambda} F^{\sigma\lambda}$ is the dual tensor. Equations (1) imply, in particular, that there are no point magnetic charges, i.e., all magnetic field-lines (with tangent vectors F_{ij} ($\rightarrow \underset{\sim}{B}$)) are closed, in contrast to the electric field lines (with tangent vectors F_{oi}($\rightarrow \underset{\sim}{E}$)) which spring or end in point sources or sinks of electric charges.

Clearly, Eqs. (1) can be naturally generalized to

$$F_{\mu\nu}{}^{,\nu} = j_\mu^{(e)}$$
$$\tilde{F}_{\mu\nu}{}^{,\nu} = j_\mu^{(m)} , \tag{2}$$

by introducing the new current $j_\mu^{(m)}$, the magnetic current, with the implication that now there are point magnetic charges as well. Equations (2) are more symmetrical than (1) (Dirac, 1931). We are certainly allowed mathematically to study the interesting system (2). Is the idea of "symmetry" of physical equations, which underly the generalization (2), a sufficient reason to make the system (2) a physical theory, i.e., a mathematical system for which we can find an isomorphism with the natural phenomena?

B. The Need for Generalization

At present the phenomena of nuclear forces and weak interactions are attributed to forces of new kinds outside the electromagnetic forces resulting from system (1) which apply to atomic phenomena. This separation stems from convenience and is in order to achieve simplicity. However, it turns out that, although convenient in practice approximately, the complete separation of electromagnetic, nuclear (or strong) and weak forces does not seem to be possible in principle. The same is true with respect to gravitational theories. Hence the idea of the so-called "unified theories" prevailed in theoretical physics for a long time. In classical theories, attempts to unify the gravitational and electromagnetic fields by Einstein, Weyl, Klein-Kaluza, and others are well-known. Thus, the generalization of the system (1) to the system (2) may correspond to some of the observed generalizations in the nature of the electromagnetic phenomena (Schwinger, 1968; Barut, 1969).

C. New Problems

Once a generalization has been recognized and appears to be applicable to a new class of phenomena, the next step is the mathematical study of the new set of equations (2). The new laws bring, quite unexpectedly, entirely new mathematical and physical problems as well as a number of striking and remarkable results.

1. Mathematical Problems.

a) The usual introduction of the electromagnetic vector potentials by the equation

$$F_{\mu\nu} = A_{\nu,\mu} - A_{\mu,\nu} \tag{3}$$

is now in conflict with the second equation of (2) if $A_\mu(x)$ is everywhere a differentiable function. Because the quantum mechanical wave equations are formulated in terms of potentials A_μ rather than $F_{\mu\nu}$, one way to introduce the potentials is again by Eq. (3), but to allow a singular function $A_\mu(x)$. It is sufficient to have a singularity in R^3 from origin to infinity, for example. With such singularities we can have (3) and, similarly,

$$\tilde{F}_{\mu\nu} = \tilde{A}_{\nu,\mu} - \tilde{A}_{\mu,\nu}. \tag{4}$$

b) Action principle: It seems to be still an unsolved problem whether and how a new Lagrangian field theory can be constructed for the generalized electrodynamics.

2. Physical Problems.

For the electromagnetic fields with sources where both electric and magnetic charges are present, Eqs. (2) imply that the magnetic charge g be pseudoscalar quantity (an axial charge) whereas the electric charge e is scalar quantity. Thus, although Eqs. (2) are more symmetric as compared to (1), the symmetry between e and g is a more subtle one (as between the electric field $\underset{\sim}{E}$ and magnetic field $\underset{\sim}{B}$). The experimental observation of the pseudoscalar charge g is a much more different and difficult problem than that of e; it seems that appropriate search for g has not yet been carried out.

More interesting are the problems associated with the interactions of two (or more) particles having both electric and magnetic charges. Without going into derivation, we state the nonrelativistic Hamiltonian of a particle with charges (e_1, g_1) in the field of another particle at the origin and with charges (e_2, g_2):

$$H = \frac{1}{2m}\,(\underset{\sim}{p} - \mu\,\underset{\sim}{D}(\underset{\sim}{r},\underset{\sim}{n}))^2 - \frac{\alpha}{r} \tag{5}$$

where

$$\mu = e_1 g_2 - e_2 g_1 \;, \; \alpha = e_1 e_2 + g_1 g_2$$

and

$$\underset{\sim}{D}(\underset{\sim}{r},\underset{\sim}{n}) = \frac{\underset{\sim}{r} \times \underset{\sim}{n}\; \underset{\sim}{r}\cdot\hat{\underset{\sim}{n}}}{r[r^2 - (\underset{\sim}{r}\cdot\underset{\sim}{n})^2]} \;. \tag{6}$$

Here $\underset{\sim}{p}$ is the relative momentum of the particle, m its reduced mass, $\underset{\sim}{r}$ the relative position, and, finally $\underset{\sim}{n}$ is a fixed unit vector, indicating, by the way, the singular line in the vector potential mentioned above.

It follows from the quantization of the angular momentum $\underset{\sim}{J} = \underset{\sim}{r} \times (\underset{\sim}{p} - \mu\underset{\sim}{D}) - \mu\hat{r}$ associated with the Hamiltonian (5) that

$$\underset{\sim}{J}\cdot\hat{\underset{\sim}{r}} = -\mu = 0,\ \pm\tfrac{1}{2},\ \pm 1,\ \ldots\ . \tag{7}$$

Consequently, if e has the value $1/\sqrt{137}$ (in units with $\hbar = c = 1$), then g must be large and of the order of $\sqrt{137}$. Hence the coupling constant α in (5) is of the order of 137, a superstrong coupling when compared to the usual electrostatic coupling in atoms.

The unsolved problems are related with this strong coupling. Although the nonrelativistic Hamiltonian (5) is exactly soluble, the real problem is not so simple. In the face of such a strong coupling and the resulting tight binding, we cannot treat the two-body problem nonrelativistically. Unfortunately, there is no complete relativistic theory of two interacting particles even for small coupling constants. However, for small coupling constants we can use the perturbation theory of renormalized quantum electrodynamics and obtain practical results although we do not know the final solution. In the strong coupling case this procedure is not possible, nor is any other modified perturbation theory known (expansion in power of $1/g^2$, for example).

Nonperturbative treatment of the two-body problem via the Dirac or Klein-Gordon equation with the potentials as in Eq. (5) is also not satisfactory for strong coupling case. In fact, the Dirac Hamiltonian with a potential α/r can be proved to be essentially self-adjoint only if

$$\alpha < \frac{\pi}{4}\ \text{(KATO)}.$$

Even if some self-adjoint extension can be found, the Dirac Hamiltonian does not contain recoil corrections and does not reflect the true physical problem.

It is for these reasons that a new approach is necessary. The physicist has to be open minded to develop new theories and concepts under such new circumstances. It may be that our age-old concept of "particles in interaction" is inadequate, for under strong interactions the particles may use their identity and become one unit. The second age-old concept of "fields interacting locally" [that is, coupling of field quantities of the form $\varphi(x)\psi(x)$] may equally be inadequate, for the system may not be describable as resulting from just point interactions.

Thus, a system containing superstrong interactions inside its structure is best treated as a whole as one unit. An approach of this kind is to write a so-called infinite-component wave equation for the composite system as a whole. This is a relativistic wave equation and the infinitely many components indicate the infinitely many internal degrees of freedom of the system.

Or, one can treat the superstrongly bound system as a quantum fluid, a hadronic matter and try to find the laws governing such a system.

The future will show us whether the new model of hadron can be described by the same quantum theory and we may need a new theory to go with the new model. For physics needs both types of advances: new models of matter and new dynamical equations; for example, the model of the H-atom and the corresponding quantum mechanics.

In any case, we have here a beautiful example where new physics can lead to new physical ideas and mathematical theories, and physics and mathematics can enrich each other.

References

1. P. A. M. Dirac, Proc. Roy. Soc. A133, 60 (1931); Phys. Rev. 74, 817 (1948).
2. J. Schwinger, Science, 165, 757 (1969).
3. A. O. Barut, Proc. Coral Gables Conference (Gordon & Breach, 1970); Phys. Rev. D3, 1747 (1971); Phys. Letters 38B, 97 (1972).